C0-AWT-911

Introduction to
Approximation Theory

E. W. CHENEY

Professor of Mathematics

University of Texas

Chelsea Publishing Company

New York, N. Y.

Second Edition

Library of Congress Catalog Card Number 81-67708

International Standard Book Number 0-8284-0317-1

Printed on 'Long-life' Acid-free Paper

Printed in the United States of America

Preface to the second edition

The publication of a new edition has made it possible to incorporate many suggestions and corrections which have been communicated to me by astute readers. I am indebted to all those persons who took the trouble of bringing errors to my attention, particularly Michael J. D. Powell, Herbert E. Salzer, Henry L. Loeb, Charles C. Miao, and G. Adomian. I am especially appreciative of the efforts of a study group in Peking composed of Ying-kuang Shih, Chia-kai Li, Kuei-ching Hsiung, and Dr. Hsien-yu Hsu.

E. W. Cheney

Austin, Texas
September 1, 1981

Preface to the first edition

In writing this book I have sought to bring a large segment of approximation theory within the comfortable grasp of undergraduate and beginning graduate mathematics students. This goal would not be a reasonable one were it not for the fact that traditionally much of the activity in this discipline has been directed toward uniform approximation on the real line, and that many of the most appealing results in the theory are based upon elementary real-variable arguments. Nevertheless, this pedagogical framework dictates a certain bias in the choice of topics. In particular, approximation with integral norms has been deemphasized in favor of uniform (or "Tchebycheff") approximation. This has made it unnecessary to presuppose the Lebesgue integral or measure theory. Further, in order not to require of the reader a knowledge of complex function theory, the discussion has been directed to problems involving only real-valued functions of real variables. We do not presuppose any "modern" algebra except elementary linear space theory.

What *is* prerequisite of the reader is a familiarity with such topics as sequences, vector spaces, series, uniform convergence, continuity, and the mean-value theorem—all of which are normally acquired in a good calculus course. If he is thus equipped, the reader should find the book self-contained. As a practical matter, a course in advanced calculus or real analysis would smooth the way. Since some material in the text (as, for example, the Bolzano-Weierstrass theorem) is included solely for completeness, the reader should be prepared to skip boldly over what is familiar to him.

The reader's attention should be directed to the large supply of problems (approximately 430 of them) and to the extensive notes and literature references at the end. Some of the problems are admittedly of a trivial nature, but, on the other hand, many others concern matters of considerable importance. Inclusion of such material in the text proper would have forced me to abandon the responsibility of distinguishing between what is of primary and what is of secondary interest in an *introduction* to the subject. Those problems which are known to be of greater-than-average difficulty or involve nonelementary concepts are distinguished by asterisks. The sections entitled Notes and References are especially long because I intended them to restore some of the historical perspective which is inevitably lost when the subject is presented in the dry manner of a *fait accompli*.

A preliminary version of the book was prepared during the 1961 and 1962 summer institutes in numerical analysis which were held at the University of California (Los Angeles) under National Science Foundation sponsorship. I am indebted to Professor Peter Henrici, then director of the institute, for an invitation to lecture on approximation. The book benefited from classroom exposure then as well as during subsequent courses which I taught at the Universities of California and Texas. During various periods of work on the manuscript I received support, which I gratefully acknowledge, from the Air Force Office of Scientific Research and from the Boeing Scientific Research Laboratories.

I am grateful for helpful comments and suggestions from a number of persons, notably Professors Philip Curtis, Ky Fan, Michael Golomb, and Robert Lynch. I especially thank two colleagues, Drs. Allen Goldstein and Henry Loeb, for enjoyable collaboration over the past several years on matters related to this book. I owe a great debt of gratitude to my wife Beth for typing the manuscript and for faithful assistance in many other ways.

E. W. Cheney

Austin, Texas
May 1, 1966

Contents

Chapter 3
Tchebycheff approximation by polynomials and other linear families

Chapter 4
Least-squares approximation and related topics

Chapter 5

Rational approximation

Chapter 6

Some additional topics

Introduction to Approximation Theory

Chapter 1

Introduction

1. Examples and Prospectus

In a subject like approximation theory which is very strongly influenced by our need to solve practical problems of computation it seems appropriate to begin by considering some concrete examples of such problems. The reader will observe in these examples a basic similarity in that each involves the selection from a prescribed *class* of functions of an element that is, in some sense, "close to" a certain *fixed* function.

(*i*) Determine a polynomial p of minimum degree such that on the interval $[0,\pi/2]$ we will have $|p(x) - \sin x| \leq 10^{-8}$.

(*ii*) More generally, given a function f and a positive number ϵ, determine a polynomial p such that $|p(x) - f(x)| \leq \epsilon$ on some given interval $[a,b]$.

(*iii*) Determine a polynomial p of degree not exceeding 7 such that on the interval $[0,\pi/2]$ we will have $|p(x) - \sin x| \leq \epsilon \sin x$, with ϵ a minimum.

(*iv*) Determine two polynomials p and q the sum of whose degrees is a minimum such that on the interval $[0,1]$ we will have $|p(x)/q(x) - \arctan x| \leq 10^{-16}$.

(*v*) Determine a vector $x = (x_1,\ldots,x_n)$ which is a best approximate solution in the sense of least squares to a system of linear equations $\sum_{j=1}^{n} a_{ij}x_j = b_i$ $(i = 1, \ldots, m)$. That is, determine x to minimize the expression $\sum_{i=1}^{m} (\sum_{j=1}^{n} a_{ij}x_j - b_i)^2$.

(*vi*) As a variant of (*v*), we require x to minimize the expression

$$\max_{1 \leq i \leq m} \left| \sum_{j=1}^{n} a_{ij}x_j - b_i \right|.$$

(*vii*) As a variant of (*ii*), we require that $\int_a^b | p(x) - f(x) |^2 \, dx < \epsilon.$

These are typical problems of computation in approximation theory, and one of our goals is to develop *algorithms* (numerical procedures) for their solution. A great many questions of deeper mathematical interest arise in a natural way from the problems just cited. For example we may ask: Can a polynomial always be found solving problem (*i*)? What exactly is the class of functions f for which problem (*ii*) can always (i.e., for any $\epsilon > 0$) be solved? For a fixed function f in problem (*ii*) how does the *degree* of the polynomial advance as ϵ decreases to zero? What relationship exists between the polynomials which solve the two problems (*ii*) and (*vii*)? Do other families of functions possess the advantages of polynomials for the purposes of approximation? If problem (*ii*) turns out to be difficult computationally, are there approximations that can be more easily obtained even if they are not quite optimal?

These theoretical questions will occupy our attention throughout most of this book.

Before embarking on our study, let us stop to recall the definitions of several terms and symbols which occur frequently. We usually describe *sets* by the notation

$$\{x: \cdots\}$$

which is read, "the set of all x such that . . . ." If A is a set, we write $x \in A$ when x is an element (member) of A, and we write $x \notin A$ in the contrary case. We also put

$$A \cup B = \{x: x \in A \quad \textit{or} \quad x \in B\} \qquad \text{(union)}$$

$$A \cap B = \{x: x \in A \quad \textit{and} \quad x \in B\} \qquad \text{(intersection)}$$

$$A \sim B = \{x: x \in A \quad \textit{and} \quad x \notin B\} \qquad \text{(difference)}$$

The notation $A \subset B$ means that $A \cap B = A$ (every element of A is an element of B). If A is a set of real numbers, we write $\lambda = \sup A$ or $\lambda = \max A$ if λ is the least number such that $x \leq \lambda$ for all $x \in A$. If $P(x)$ is a *proposition* involving x and if f is a real-valued function, we often replace the cumbersome $\sup \{ f(x): P(x) \}$ by $\sup_{P(x)} f(x)$. Similarly, we write $\mu = \inf A$ or $\mu = \min A$ if μ is the largest number such that $\mu \leq x$ for all $x \in A$. We admit the possibility that λ may be $+\infty$ and μ may be $-\infty$. (The words *inf* and *sup* are abbreviations for *infimum* and *supremum*, respectively.)

The symbol $\Rightarrow$ is read "implies." The symbol ∎ is read "this is the end of the proof."

2. Metric Spaces

The reader has now seen that a number of typical problems in approximation involve the selection from a set of elements of one that is in some sense close to a prescribed element not in the set. Thus, the given set may consist of polynomials while the element to be approximated is a function such as $\arcsin x$ which is assuredly *not* a polynomial. Such an approximation problem does not become precise until we have decided how the distance between two elements is to be measured. We usually require for this distance a non-negative real number which is zero when the elements are identical, and positive otherwise. It is usual to require further that the distance be the same from x to y as from y to x, and finally that the distance measured directly from x to y be no greater than the distance measured via another, intermediate, point z. The abstract embodiment of these ideas is the *metric space*. By this we understand a pair (X,d) where X is a set and d a real-valued function defined for pairs of points in X in such a way that the following postulates are satisfied for all x, y, z in X:

(*i*)	$d(x,x) = 0$	(reflexitivity)
(*ii*)	$d(x,y) > 0$ if $x \neq y$	(positivity)
(*iii*)	$d(x,y) = d(y,x)$	(symmetry)
(*iv*)	$d(x,y) \leq d(x,z) + d(z,y)$	(triangle inequality)

For example, if X is the set of real numbers the "usual" metric is $d(x,y) = |x - y|$. In this case, properties (*i*) to (*iv*) simply reflect familiar features of the absolute-value function. A less useful metric may be defined by setting $d(x,y) = 1$ if $x \neq y$ and $d(x,x) = 0$. This metric may be introduced into any set whatsoever; this proves that, given a set X, there exists a metric d such that (X,d) is a metric space. In an abuse of language we say often that X itself is a metric space.

Now that the concept of a metric space is available, we may formulate a basic problem of approximation theory: Given a point g and a set M in a metric space, determine a point of M of minimum distance from g. Such a closest point may or may not exist. We shall therefore be interested in *existence theorems*, which give conditions sufficient to guarantee existence of closest points. We will also consider the question of *unicity* of closest points and their *characterization*. Finally, we shall not neglect the important subject of *algorithms* for the practical computation of closest points.

In a metric space it is possible to define a number of concepts belonging to the study of topology. These concepts are somewhat simpler in a metric space than in general topology, and the reader should be warned that definitions and theorems given for a metric space are generally *not* to be carried over without alteration to arbitrary topological spaces. First of all, the notion of a convergent sequence of points may be defined as follows. A sequence of points $x_1, x_2, x_3, \ldots$ in a metric space is said to *converge* to a point x^* (and we write $x_n \to x^*$) if the distance from x_n to x^* approaches zero as $n \to \infty$. Formally, $\lim x_n = x^*$ iff $\lim d(x_n,x^*) = 0$, both limits being taken as $n \to \infty$. A subset K of X is said to be *compact* if every sequence of points in K has a subsequence which converges to a point of K. Intuitively, a sequence $y_1, y_2, y_3, \ldots$ is a *subsequence* of another sequence $x_1, x_2, x_3, \ldots$ if the y's are obtained by deleting some of the x's and preserving the order. Formally, there must exist an increasing function n such that $y_k = x_{n(k)}$. In this context it is common to write n_k for $n(k)$ so that a subsequence in general is of the form $x_{n_1}, x_{n_2}, x_{n_3}, \ldots$ where $n_1 < n_2 < n_3 < \cdots$. We are now ready to give a basic existence theorem.

Theorem on Existence of Best Approximations in a Metric Space. *Let K denote a compact set in a metric space. To each point p of the space there corresponds a point in K of minimum distance from p.*

Proof. Let $\delta = \inf \{d(p,x): x \in K\}$. From the definition of an infimum, we may find a sequence of points $x_1, x_2, x_3, \ldots$ in K with the property that $d(p,x_n) \to \delta$ as $n \to \infty$. By the compactness of K, we may assume that the sequence converges to a point x^* of K; for if necessary we may extract from the given sequence a subsequence with this property. We will show that x^* is a point of K of minimum distance from p. Indeed, by the triangle inequality —postulate (*iv*) for the metric—we have $d(p,x^*) \leq d(p,x_n) + d(x_n,x^*)$. The left member of this inequality is independent of n, and the right member approaches δ as $n \to \infty$. Therefore $d(p,x^*) \leq \delta$. Since $x^* \in K$, $d(p,x^*) \geq \delta$. Hence $d(p,x^*) = \delta$. ■

A few other topological concepts will be useful. A set F in a metric space is said to be *closed* if every convergent sequence in F has its limit in F too. Thus $x_n \in F$ and $x_n \to x$ together imply that $x \in F$. The complement of a closed set is said to be *open*. The set A is said to be *dense in the set B* if to each $x \in B$ and to each $\epsilon > 0$ there corresponds a $y \in A$ such that $d(x,y) < \epsilon$. (Thus the elements of B may be approximated by the elements of A.) A mapping ϕ from one metric space to another is said to be *continuous at a point* x if for every sequence $x_n \to x$ it follows that $\phi(x_n) \to \phi(x)$. If ϕ is continuous at every point of its domain, it is said simply to be *continuous*. For real-valued functions it is understood that the usual metric $| x - y |$ will be employed, so that in this instance we may state that ϕ is continuous if $d(x^*,x_n) \to 0$ implies $| \phi(x^*) - \phi(x_n) | \to 0$. A property of continuous real functions which enters into many proofs is established next.

Theorem on Attainment of Extreme Values. *A continuous real-valued function defined on a compact set in a metric space achieves its infimum and supremum on that set.*

Proof. Let ϕ be a continuous real-valued function defined on a compact set K in a metric space. Let $m = \inf \{\phi(x) : x \in K\}$. Then there exists a sequence of points $x_1, x_2, x_3, \ldots$ in K such that $\phi(x_n) \to m$. By the compactness of K we may assume that the sequence converges to a point x^* of K. By the continuity of ϕ, $\phi(x^*) = m$, as was to be proved. The proof for the supremum now follows by observing that $\sup \phi = -\inf (-\phi)$. ∎

Theorem. *A closed subset of a compact set in a metric space is compact.*

Proof. Let F be a closed subset of a compact set M. If $x_1, x_2, \ldots$ is any sequence in F, we may—by the compactness of M—find a subsequence x_{n_k} which converges to a point $x^* \in M$. Since F is closed, x^* in fact belongs to F. ∎

Theorem. *A continuous mapping from one metric space into another carries compact sets into compact sets.*

Proof. Let ϕ be such a map and M a compact set in the domain of ϕ. If $y_1, y_2, \ldots$ is any sequence in the image set $\phi(M)$, then we may find $x_n \in M$ such that $\phi(x_n) = y_n$. By the compactness of M, we may find a subsequence such that $x_{n_k} \to x \in M$. By the continuity of ϕ, it then follows that $y_{n_k} \to \phi(x)$. ∎

Bolzano-Weierstrass Theorem. *A closed interval $[a,b]$ is compact.*

Proof. We give the proof for $[0,1]$ first. Consider an arbitrary sequence $\Lambda = \{\lambda_1, \lambda_2, \ldots\}$ in $[0,1]$. If some real number ξ occurs an infinite number of times in Λ, then there exists a subsequence $\lambda_{k_1}, \lambda_{k_2}, \ldots$ in which each element is equal to ξ. Such a subsequence converges to ξ, and it is clear that $\xi \in [0,1]$. In the other case no point occurs more than a finite number of times in Λ. Let each member of Λ be written in decimal form, $\lambda = 0.a_1a_2a_3 \ldots$. If we group the elements of Λ into ten subsets according to the first digit in their decimal expansions, then one of these ten subsets—call it S_1—must contain an infinite number of elements. The set S_1 thus defined contains infinitely many points of Λ, and these have the same first digit. Now let the elements of S_1 be grouped into classes according to their second decimal digit. One of these classes—call it S_2—must contain an infinite number of elements. The set S_2 thus defined contains an infinite number of points from Λ, and these have the same first two decimal digits. If we continue in this way, we define a sequence of sets $S_1, S_2, \ldots$ such that for each k, S_k contains an infinite number of points of Λ, all with the same first k decimal digits. Now select an index k_1 such that $\lambda_{k_1} \in S_1$, an index $k_2 > k_1$ such that $\lambda_{k_2} \in S_2$, etc. The subsequence generated in this way converges to a number ξ which agrees in its first digit with the points of S_1, in its first two digits with the points of S_2, and so forth. Thus $| \xi - \lambda_{k_i} | \leq 10^{-i}$. It is obvious that $\xi \in [0,1]$. Thus $[0,1]$ is compact.

Now let $[a,b]$ be an arbitrary closed interval. It is the image of $[0,1]$ under the continuous map $\phi(x) = bx + (1 - x)a$, and thus the preceding theorem implies that $[a,b]$ is compact. ∎

If a function ϕ from one metric space M to another is continuous, then it is continuous at each point $x \in M$, and consequently to each $\epsilon > 0$ there corresponds a $\delta > 0$ such that for all $y \in M$,

$$d(x,y) < \delta \Rightarrow d_1[\phi(x),\phi(y)] < \epsilon \tag{1}$$

The numbers δ generally depend on x as well as on ϵ. If, however, it is possible to determine a $\delta > 0$ in correspondence with ϵ in such a way that (1) holds for all x and y, then ϕ is said to be *uniformly* continuous on M.

Theorem. *A continuous function from one metric space to another is uniformly continuous on every compact set.*

Proof. If the theorem were false, we could find a continuous function ϕ on a compact metric space M and a number $\epsilon > 0$ to which no $\delta > 0$ corresponded as above. Then there would exist a sequence of pairs of points (x_n,y_n) such that $d(x_n,y_n) \to 0$ while $d_1[\phi(x_n),\phi(y_n)] \geq \epsilon$. By passing to a subsequence if necessary, we could assume that $x_n \to x \in M$. Then $y_n \to x$ also. But now the continuity of ϕ at x is contradicted by the inequality

$$\epsilon \leq d_1[\phi(x_n),\phi(y_n)] \leq d_1[\phi(x_n),\phi(x)] + d_1[\phi(x),\phi(y_n)]$$ ∎

Problems

1. Consider the metric space (X,d) where $d(x,y) = 1$ when $x \neq y$ and $d(x,x) = 0$. What are the convergent sequences, the closed sets, the compact sets, the continuous real-valued functions?
2. Which of the following equations define metrics on the real line?
   (*a*) $d(x,y) = \log(1 + | x - y |)$
   (*b*) $d(x,y) = e^{|x-y|} - 1$
   (*c*) $d(x,y) = | x - y |^2$
3. Give some convenient sufficient conditions on a function ϕ in order that the equation $d^*(x,y) = \phi[d(x,y)]$ shall define a metric d^* whenever d is a metric.
4. Discuss the existence, unicity, and characterization of closest points for compact sets in the metric space of Prob. 1.
5. Prove that $d(p,x)$ for fixed p is a continuous function of x in a metric space (X,d). Use this fact to derive the first theorem of this section as a corollary of the second theorem.

6. A convergent sequence of points in a metric space has a unique limit.
*7. A real-valued function ϕ defined on a metric space is said to be *lower semicontinuous* if for each real λ the set $\{x: \phi(x) \leq \lambda\}$ is closed. Prove that a lower semicontinuous function attains its infimum on any compact set.
8. Prove that a positive continuous real-valued function defined on a compact set has a positive infimum.
9. A real-valued function ϕ is *upper semicontinuous* if each set $\{x: \phi(x) \geq \lambda\}$ is closed. Prove that ϕ is continuous if and only if it is both upper and lower semicontinuous.
10. (On cluster points) A point x^* is said to be a *cluster point* of a sequence $x_1, x_2, \ldots$ if some subsequence $x_{n_1}, x_{n_2}, \ldots$ converges to x^*. Prove that a subset K of a metric space is compact if and only if every sequence in K has a cluster point in K. Prove that if a sequence in a compact set has only one cluster point, then the sequence converges to this cluster point.
11. If $x_n \to x^*$, then the set $\{x^*,x_1,x_2,\ldots\}$ is compact.
12. In the definition of a subsequence it is sometimes *not* required that $n_1 < n_2 < \cdots$ but only that $\lim_{k\to\infty} n_k = \infty$. Is the concept of compactness changed if we adopt this definition of a subsequence?
13. A *Cauchy* sequence is a sequence $\{x_1,x_2,\ldots\}$ with the property $\lim_{n\to\infty} \max_{i,j\geq n} d(x_i,x_j) = 0$. If every Cauchy sequence converges, the metric space is said to be *complete*. Show that every compact metric space is complete.
14. Is the union of two compact sets compact?
15. From the theorem on attainment of extreme values it follows that a continuous real-valued function is bounded on a compact set.
16. If x is a point lying in an open set $\mathcal{O}$, then for some $\epsilon > 0$, $\{y: d(x,y) \leq \epsilon\} \subset \mathcal{O}$.

3. Normed Linear Spaces

The setting for many problems of approximation is a metric space of a particular kind called a *normed linear space*. The reader will recall from his study of algebra that a *linear space* has the following three ingredients:

(*i*) a set E of elements called *vectors* or *points*
(*ii*) an operation in E called *addition* which obeys the usual rules of arithmetic
(*iii*) an operation of multiplying vectors by real numbers, obeying the usual rules of arithmetic

What we have chosen to call the "usual rules of arithmetic" may be summarized as follows. For all f, g, and h in E and for all real numbers λ and μ: $f+g = g+f$; $f+(g+h) = (f+g)+h$; $\lambda(f+g) = \lambda f + \lambda g$; $(\lambda+\mu)f = \lambda f + \mu f$; $(\lambda\mu)f = \lambda(\mu f)$; $1\cdot f = f$; there exists a vector 0 such that $0 + f = f$; there exists in correspondence with f a vector $-f$ such that $-f + f = 0$.

A *normed* linear space is a linear space in which there is defined a real-valued function on vectors called a *norm*, denoted by $\|\cdot\|$, and having the properties

$$(i) \quad \|f\| > 0 \text{ unless } f = 0 \qquad \text{(positivity)}$$
$$(ii) \quad \|\lambda f\| = |\lambda|\,\|f\| \qquad \text{(homogeneity)}$$
$$(iii) \quad \|f+g\| \leq \|f\| + \|g\| \qquad \text{(triangle inequality)}$$

Although the concepts of a Banach space and a Hilbert space play only minor roles in this book, it is appropriate to mention here that a *Banach* space is a normed linear space in which the axiom of completeness is satisfied:

(iv) If a sequence of vectors $f_1, f_2, \ldots$ has the Cauchy property, $\lim_{n,m\to\infty} \|f_n - f_m\| = 0$, then there exists a vector g such that $\lim_{n\to\infty} \|f_n - g\| = 0$.

Then we can state that a *Hilbert* space is a Banach space in which the parallelogram law is valid:

$$(v) \quad \|f+g\|^2 + \|f-g\|^2 = 2\|f\|^2 + 2\|g\|^2$$

We observe that properties *(i)*, *(ii)*, *(iii)* of the norm are again familiar facts for the absolute-value function, and the norm may therefore be interpreted as a generalized or abstract *magnitude*. Just as $|x - y|$ was seen to be a metric for the real numbers, so $\|x - y\|$ will now be proved to be a metric in any normed linear space.

Theorem. *In a normed linear space, the formula $d(f,g) = \|f - g\|$ defines a metric. Addition, scalar multiplication, and the norm are all continuous with this metric.*

Proof. Properties *(i)*, *(ii)*, *(iii)* for a metric are quickly verified: $d(f,f) = \|f - f\| = \|0\cdot f\| = |0|\cdot\|f\| = 0$; if $f \neq g$, then $f - g \neq 0$ so that $d(f,g) = \|f-g\| > 0$; $d(f,g) = \|f-g\| = |-1|\,\|f-g\| = \|g-f\| = d(g,f)$; $d(f,g) = \|f-g\| = \|f-h+h-g\| \leq \|f-h\| + \|h-g\| = d(f,h) + d(h,g)$. For the continuity of addition, we have $d(f^*+g^*, f_n+g_n) = \|f^*+g^* - (f_n+g_n)\| \leq \|f^*-f_n\| + \|g^*-g_n\| = d(f^*,f_n) + d(g^*,g_n)$. Thus if $g_n \to g^*$ and $f_n \to f^*$, then $f_n + g_n \to f^* + g^*$. For the continuity of scalar multiplication we have $d(\lambda^* f^*, \lambda_n f_n) = \|\lambda^* f^* - \lambda_n f_n\| = \|\lambda^* f^* - \lambda^* f_n + \lambda^* f_n - \lambda_n f_n\| \leq |\lambda^*|\,\|f^* - f_n\| + |\lambda^* - \lambda_n|\,\|f_n\|$. Thus if $\lambda_n \to \lambda^*$ and $f_n \to f^*$, then $\lambda_n f_n \to \lambda^* f^*$. Finally, $\|f\| \leq \|f-g\| + \|g\|$ so that $\|f\| - \|g\| \leq \|f-g\|$. By interchanging f and g we obtain $|\,\|f\| - \|g\|\,| \leq \|f-g\|$, which proves the continuity of $\|f\|$. ∎

As the reader is well aware, a linear space of exceptional importance is n-dimensional space R_n, the vectors of which are n-tuples of real numbers $f = [\xi_1, \xi_2, \ldots, \xi_n]$, $g = [\eta_1, \eta_2, \ldots, \eta_n]$, etc. We define $f + g = [\xi_1 + \eta_1, \ldots, \xi_n + \eta_n]$

and $\lambda f = [\lambda\xi_1, \ldots, \lambda\xi_n]$. It is possible to make R_n into a normed linear space in a variety of ways by introducing different norms. For example,

$$\|f\| = \max\{|\xi_1|, \ldots, |\xi_n|\}$$

$$\|f\| = |\xi_1| + \cdots + |\xi_n|$$

These two norms are the cases $p = \infty$ and $p = 1$ of the family of norms

$$\|f\| = \sqrt[p]{|\xi_1|^p + \cdots + |\xi_n|^p} \qquad (p \geq 1)$$

The case $p = 2$ is the so-called "Euclidean" norm.

Another especially important normed linear space is the space $C[a,b]$ of continuous real-valued functions defined on the interval $[a,b]$. In this space we define

Addition	$(f + g)(x) = f(x) + g(x)$
Scalar multiplication	$(\lambda f)(x) = \lambda f(x)$
Norm	$\|f\| = \max\limits_{a \leq x \leq b} \|f(x)\|$

We recommend as an exercise the verification that $C[a,b]$ is a normed linear space. Actually, $C[a,b]$ is a Banach space, a fact which we now prove.

Theorem. *The space $C[a,b]$ is complete.*

Proof. Let the sequence of vectors $f_1, f_2, \ldots$ have the Cauchy property. Then to each $\epsilon > 0$ there corresponds an N such that $\|f_n - f_m\| < \epsilon$ whenever $n, m \geq N$. From the definition of this norm it follows that for each fixed x, $|f_n(x) - f_m(x)| < \epsilon$ whenever $n, m \geq N$. Thus the sequence of real numbers $f_1(x), f_2(x), \ldots$ converges by the Cauchy test to a number which (in view of its dependence on x) we may denote by $f(x)$. The proof will be completed by showing that $f \in C[a,b]$. Observe first that by letting $m \to \infty$ in the preceding inequality we obtain the result that to each ϵ there corresponds an n such that $|f_n(x) - f(x)| \leq \epsilon$ *for all* x. By the continuity of f_n at x_0 there exists a $\delta > 0$ such that $|f_n(x_0) - f_n(y)| < \epsilon$ whenever $|x_0 - y| < \delta$. Thus for all such y,

$$\begin{aligned}|f(x_0) - f(y)| &\leq |f(x_0) - f_n(x_0)| + |f_n(x_0) - f_n(y)| + |f_n(y) - f(y)| \\ &< 3\epsilon\end{aligned}$$

Since ϵ was arbitrary, f is continuous at x_0. Since x_0 was arbitrary in $[a,b]$, f is continuous on $[a,b]$. ■

Several elementary criteria for compactness in a metric space were cited in the preceding section. In a normed linear space an additional criterion is available, and we turn to this next. A set in a normed linear space is said to be *bounded* if it is contained in some set of the form $\{f: \|f\| \leq c\}$.

Lemma. *In the space of n-tuples $\lambda = [\lambda_1,\ldots,\lambda_n]$ with norm $\|\lambda\| = \max|\lambda_i|$, every closed and bounded set is compact.*

Proof. We show first by induction on the dimension n that the sets $M_n = \{\lambda: \|\lambda\| \le 1\}$ are compact. Since $M_1 = [-1,1]$, the Bolzano-Weierstrass theorem (Sec. 2) implies that M_1 is compact. Now if M_n is compact, it will follow that M_{n+1} is compact. Indeed, let $\lambda^{(1)}, \lambda^{(2)}, \ldots$ be any sequence of vectors in M_{n+1}. Each of these vectors is of the form $\lambda^{(k)} = [\lambda_1^{(k)},\ldots,\lambda_{n+1}^{(k)}]$. Since the components of these vectors lie in $[-1,1]$ we may assume, passing to a subsequence if necessary, that $\lambda_{n+1}^{(k)}$ converges as $k \to \infty$. By the hypothesis that M_n is compact, we may pass to a further subsequence such that the first n components of $\lambda^{(k)}$ converge. If all the components converge, then the vectors themselves converge. Hence M_{n+1} is compact. Now let F denote an arbitrary closed and bounded set in n space. Then for some c, cM_n contains F. Since the mapping $\lambda \to c\lambda$ is continuous, cM_n is compact (theorem on page 5). Since F is a closed subset of a compact set, it too is compact (theorem on page 5). ■

Theorem. *Every closed, bounded, finite-dimensional set in a normed linear space is compact.*

Proof. Let F be such a set. Then there exists an independent set $\{g_1,\ldots,g_n\}$ with the property that each element $f \in F$ is expressible uniquely in the form $f = \sum \lambda_i g_i$. Let T denote the mapping $\lambda \to f$. If n space is normed by $\|\lambda\| = \max|\lambda_i|$, then T is continuous, because $\|T\lambda - T\mu\| = \|\sum \lambda_i g_i - \sum \mu_i g_i\| = \|\sum(\lambda_i - \mu_i)g_i\| \le \sum|\lambda_i - \mu_i|\,\|g_i\| \le \|\lambda - \mu\| \sum \|g_i\|$. The set F is the image, under the map T, of the set $M = \{\lambda: T\lambda \in F\}$. By a theorem on page 5, the compactness of F would follow from that of M. By the preceding lemma it will suffice to prove that M is closed and bounded. If $\lambda^{(k)} \in M$ and $\lambda^{(k)} \to \lambda$, then $T\lambda = T(\lim \lambda^{(k)}) = \lim T(\lambda^{(k)})$. Since F is closed, $T\lambda \in F$, whence $\lambda \in M$. This shows that M is closed. Now let us prove that M is bounded. Since the set $\{\lambda: \|\lambda\| = 1\}$ is compact and T is continuous, the infimum, α, of $\|T\lambda\|$ is attained on that set. Since $\{g_1,\ldots,g_n\}$ is independent, $\alpha > 0$. Thus for any $\lambda \ne 0$, $\|T\lambda\| = \|T(\lambda/\|\lambda\|)\| \cdot \|\lambda\| \ge \alpha\|\lambda\|$. Since $\|T\lambda\|$ is bounded on M, $\|\lambda\|$ is bounded on M. ■

Corollary. *Every finite-dimensional normed linear space is complete.*

Proof. Let $\{f_n\}$ be any Cauchy sequence in such a space. It must be bounded because if we take an index N such that

$$n \ge N \Rightarrow \|f_n - f_N\| \le 1$$

then the inequality $\|f_n\| \le \|f_n - f_N\| + \|f_N\|$ shows that $\|f_n\| \le 1 +$

$\max_{k \leq N} \| f_k \| \equiv M$. By the compactness of the set $\{ f: \| f \| \leq M \}$, the sequence $\{ f_n \}$ contains a converging subsequence f_{n_k}, with limit, say, f^*. From the inequality $\| f_n - f^* \| \leq \| f_n - f_{n_k} \| + \| f_{n_k} - f^* \|$ and the Cauchy property, it follows that $f_n \to f^*$. ∎

Problems

1. Prove the following properties of norms: (*a*) $\| 0 \| = 0$; (*b*) $\| f + g \| \geq | \| f \| - \| g \| |$; (*c*) if $\| f_n - g \| \to 0$ and $\| f_n - h \| \to 0$, then $g = h$; (*d*) $\| \sum_{n=1}^{m} f_n \| \leq \sum_{n=1}^{m} \| f_n \|$; (*e*) if f is of norm $\lambda \neq 0$, then f/λ is of norm 1.
2. Prove that the following are norms on R_n: (*a*) $\| f \|_T = \max_k | \xi_k |$, (*b*) $\| f \|_1 = \sum | \xi_k |$
3. Look up the proof of Minkowski's inequality in a book such as [Hardy, Littlewood, Pólya, 1934]:
$$\| f + g \|_p \leq \| f \|_p + \| g \|_p$$
Here $\| f \|_p = (\sum | \xi_k |^p)^{1/p}$.
4. Prove that the equation $\| f \| = \max_x | f(x) |$ defines a norm on $C[a,b]$.
5. In $C[a,b]$, $\| f_n - g \| \to 0$ if and only if f_n converges uniformly to g; i.e., to each $\epsilon > 0$ there corresponds an N such that *for all* x, $n \geq N \Rightarrow | f_n(x) - g(x) | < \epsilon$.
6. Is there a norm for *pointwise* convergence of functions? *Hint*: Consider functions whose graphs are of the form

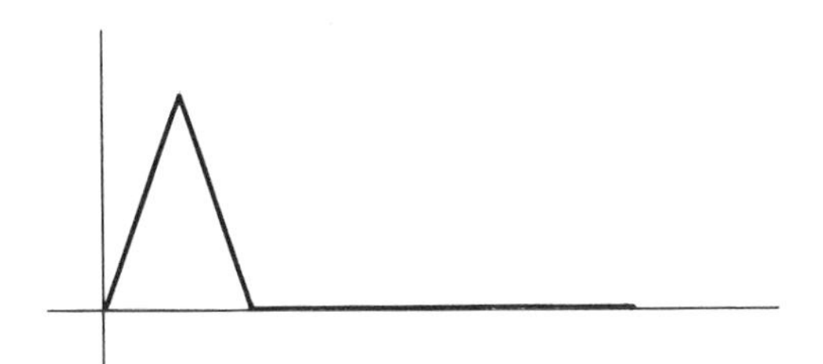

7. Find a necessary and sufficient condition on a set $G \subset R_n$ such that the equation
$N(f) = \sup_{g \in G} | \sum_{k=1}^{n} \xi_k \eta_k |$ shall define a norm. Here $f = [\xi_1, \ldots, \xi_n]$ and $g = [\eta_1, \ldots, \eta_n]$.
Are the norms of Prob. 2 of this type? *Hint*: N is always a seminorm. (See Prob. 13 for the definition of *seminorm*.)
8. Show that if the roots be omitted from Minkowski's inequality (Prob. 3), the direction of the inequality is reversed. That is, $\sum (| u_i | + | v_i |)^p \geq \sum | u_i |^p + \sum | v_i |^p$. Give a proof which is valid whether p is an integer or not.
9. In R_n, given any two norms N_1 and N_2, there exists a positive constant α such that $N_1(f) \leq \alpha N_2(f)$. Compute the least α in some special cases. *Hint*: If the desired inequality is true when $N_2(f) = 1$, then it is true in general. Try
$$\alpha = \sup \{ N_1(f): N_2(f) = 1 \}$$

10. Why is $(\sum |\xi_k|^p)^{1/p}$ not a norm when $p < 1$?

11. Can a norm always be defined in a linear space? You may assume the existence of a linearly independent set B whose finite linear combinations exhaust the space. Such a set is termed a *Hamel base*; the proof of its existence requires the axiom of choice.

12. In $C[a,b]$, $\|fg\| \leq \|f\| \, \|g\|$.

13. A *seminorm* is like a norm except that we do not insist on the property $\|f\| = 0 \Rightarrow f = 0$. Prove that $\|f\| = |f(0)| + |f'(0)| + |f''(0)|$ defines a seminorm on the linear space of polynomials. When is $\|f\| = 0$?

14. If $\|\cdot\|$ is a seminorm on a linear space E, then $N = \{f: \|f\| = 0\}$ is a subspace. In the quotient space E/N, the formula $\|f + N\| = \inf \|f - n\|$ defines a norm.

15. All norms on a finite-dimensional linear space are *equivalent* in the sense that corresponding to norms N_i there are constants $\alpha_i > 0$ such that $N_1 \leq \alpha_1 N_2$ and $N_2 \leq \alpha_2 N_1$.

16. Show that the equation $\|f\| = \max_{1 \leq i \leq m} |f(x_i)|$ defines a seminorm in $C[a,b]$. What about $\|f\| = \sup_i |f(r_i)|$ where $r_1, r_2 \ldots$ are all the rationals in $[a,b]$?

17. If d is a metric in a linear space, does it follow that $\|x\| = d(x,0)$ is a norm?

18. Let the sequence of functions $f_1, f_2, \ldots \in C[a,b]$ converge "simply uniformly" to a function f. This means that to each pair (ϵ, N) there corresponds an $n > N$ such that $\|f - f_n\| < \epsilon$. Show that f is continuous. Give an example of such a sequence which is not uniformly convergent. [Dini]

4. Inner-product Spaces

It is frequently desirable to introduce into a linear space E another operation, that of the *inner product*. The inner product of two vectors f and g is a real number denoted by $\langle f,g \rangle$ which obeys the following postulates:

(*i*)	$\langle f,f \rangle > 0$ unless $f = 0$	(positivity)
(*ii*)	$\langle f,g \rangle = \langle g,f \rangle$	(symmetry)
(*iii*)	$\langle f, \lambda g + \mu h \rangle = \lambda \langle f,g \rangle + \mu \langle f,h \rangle$	(linearity)

For example, in n space, if $f = [\xi_1, \ldots, \xi_n]$ and $g = [\eta_1, \ldots, \eta_n]$, we may define

$$\langle f,g \rangle = \xi_1 \eta_1 + \cdots + \xi_n \eta_n$$

In the space $C[a,b]$ of continuous functions on $[a,b]$ we may define

$$\langle f,g \rangle = \int_a^b f(x) g(x) \, dx$$

These are not the only inner products that may be defined in these spaces but are representative.

In any linear space furnished with an inner product the equation

$$\|f\| = +\sqrt{\langle f,f \rangle}$$

defines a norm. The nontrivial part of this assertion is included in the following theorem.

Theorem. *In any inner-product space, define* $\|f\| = +\sqrt{\langle f,f\rangle}$. *Then the following are true:*

(*i*) $|\langle f,g\rangle| \le \|f\|\,\|g\|$ (*Cauchy-Schwarz inequality*)

(*ii*) $\|f+g\| \le \|f\| + \|g\|$ (*triangle inequality*)

(*iii*) $\|f+g\|^2 + \|f-g\|^2 = 2\|f\|^2 + 2\|g\|^2$ (*parallelogram law*)

(*iv*) $\langle f,g\rangle = 0 \Rightarrow \|f+g\|^2 = \|f\|^2 + \|g\|^2$ (*Pythagorean law*)

Proof. If there exists a pair (f,g) such that $|\langle f,g\rangle| > \|f\|\,\|g\|$, then $g \neq 0$ and we may assume that $\|g\| = 1$. Postulate (*i*) for the inner product is now violated by putting $\lambda = \langle f,g\rangle$ and then observing that

$$\begin{aligned}\langle f-\lambda g, f-\lambda g\rangle &= \langle f,f\rangle - 2\lambda\langle f,g\rangle + \lambda^2\langle g,g\rangle \\ &= \langle f,f\rangle - 2\langle f,g\rangle^2 + \langle f,g\rangle^2 \\ &= \|f\|^2 - \langle f,g\rangle^2 < 0\end{aligned}$$

The triangle inequality now follows by writing

$$\begin{aligned}\|f+g\|^2 = \langle f+g, f+g\rangle &= \langle f,f\rangle + 2\langle f,g\rangle + \langle g,g\rangle \\ &\le \|f\|^2 + 2\|f\|\,\|g\| + \|g\|^2 \\ &= \{\|f\| + \|g\|\}^2\end{aligned}$$

The Pythagorean law is a special case of what we have just written, and the parallelogram law is a direct calculation which we omit. ∎

In analytic geometry, it is proved that two vectors $f = [\xi_1,\xi_2,\xi_3]$ and $g = [\eta_1,\eta_2,\eta_3]$ are perpendicular to each other if and only if

$$\xi_1\eta_1 + \xi_2\eta_2 + \xi_3\eta_3 = 0$$

With the inner product defined above, this might be written briefly as

$$\langle f,g\rangle = 0$$

We are thus led to define orthogonality in any inner-product space as follows: A set of vectors in an inner-product space is said to be *orthogonal* if $\langle f,g\rangle = 0$ for any two distinct elements f and g from the set. For example, the three functions $f(x) = 1$, $g(x) = x$, $h(x) = x^2 - \frac{1}{3}$ form an orthogonal set in $C[-1,1]$

with the inner product introduced above, as may be verified by computing

$$\langle f,g \rangle = \int_{-1}^{1} x\,dx = 0$$

$$\langle f,h \rangle = \int_{-1}^{1} (x^2 - \tfrac{1}{3})\,dx = 0$$

and

$$\langle g,h \rangle = \int_{-1}^{1} \left(x^3 - \frac{x}{3}\right) dx = 0$$

We sometimes write $f \perp g$ when $\langle f,g \rangle = 0$, and say "f is orthogonal to g."

An orthogonal set of vectors may have the additional property that each element is of unit norm: $\| f \| = 1$. In this case the set is said to be *orthonormal.* An orthogonal set of nonzero vectors may be changed into an orthonormal set by dividing each element by its norm. If we define the *Kronecker delta* by the equation

$$\delta_{ij} = \begin{cases} 1 & (i = j) \\ 0 & (i \neq j) \end{cases}$$

then the orthonormality property of a set of vectors $\{ f_1, f_2, \ldots \}$ is expressed by the simple equation

$$\langle f_n, f_m \rangle = \delta_{nm} \qquad (n, m = 1, 2, \ldots)$$

It is easy to prove that every orthonormal set is linearly independent. For if $\{ f_1, f_2, \ldots \}$ is orthonormal and $\sum \lambda_n f_n = 0$, then for fixed but arbitrary m,

$$0 = \langle 0, f_m \rangle = \langle \sum \lambda_n f_n, f_m \rangle = \sum \lambda_n \langle f_n, f_m \rangle = \sum \lambda_n \delta_{nm} = \lambda_m$$

so that each $\lambda_m = 0$.

From the standpoint of approximation theory, orthonormal sets of vectors in an inner-product space have a very pleasant property: They enable us to give an *explicit* solution to the problem of locating the point on a subspace of minimum distance from an external point, in accordance with the following theorem.

Theorem. *Let $\{g_1, \ldots, g_n\}$ denote an orthonormal set in an inner-product space with norm defined by $\| h \| = \sqrt{\langle h,h \rangle}$. The expression $\| \sum c_i g_i - f \|$ will be a minimum if and only if $c_i = \langle f, g_i \rangle$.*

Proof. With this choice for c_i, let $g = \sum c_i g_i$. Let h be any other linear combination of the g_i. Then $g - f$ is orthogonal to each g_i because $\langle g - f, g_i \rangle = \langle g, g_i \rangle - \langle f, g_i \rangle = \sum_j c_j \langle g_j, g_i \rangle - c_i = \sum_j c_j \delta_{ij} - c_i = 0$. Thus $g - f$ is orthogonal to $h - g$, because $h - g$ is a linear combination of the g_i. Hence $\| h - f \|^2 = \| (h - g) + (g - f) \|^2 = \| h - g \|^2 + 2\langle h - g, g - f\rangle + \| g - f \|^2 \geq \| g - f \|^2$. This last inequality will be *strict* unless $h = g$, which proves the unicity of the coefficients c_i. ■

Corollary. *If f is in the linear span of an orthonormal set $\{g_1, \ldots, g_n\}$,* then $f = \sum \langle f, g_i \rangle g_i$.

To conclude this section, we give a famous theorem, the proof of which contains a method for constructing orthonormal sets.

Gram-Schmidt Theorem. *Let $\{f_1, f_2, \ldots\}$ be a linearly independent set of vectors in an inner-product space. For each n it is possible to define a vector g_n as a linear combination of $f_1, \ldots, f_n$ in such a way that $\{g_1, g_2, \ldots\}$ is orthonormal.*

Proof. (By induction) To get the process started, we define $g_1 = f_1/\| f_1 \|$. Then $\{g_1\}$ is an orthonormal set. Of course, $f_1 \neq 0$ because of the independence assumption. Suppose now, as an induction hypothesis, that $g_1, \ldots, g_{n-1}$ have been successfully defined. That is, $\{g_1, \ldots, g_{n-1}\}$ is orthonormal, and for each $k < n$, g_k is a linear combination of $\{f_1, \ldots, f_k\}$. Define $u = f_n - \sum_{k<n} \langle f_n, g_k \rangle g_k$. It is easy to see that u is orthogonal to $g_1, \ldots, g_{n-1}$. Indeed, for $i < n$, we have

$$\langle u, g_i \rangle = \langle f_n, g_i \rangle - \sum_{k<n} \langle f_n, g_k \rangle \langle g_k, g_i \rangle = \langle f_n, g_i \rangle - \sum_{k<n} \langle f_n, g_k \rangle \delta_{ki} = 0$$

From the definition of u and from the induction hypothesis, u is a linear combination of $f_1, \ldots, f_n$. Finally, we note that $u \neq 0$; for if $u = 0$, then f_n must be a linear combination of $g_1, \ldots, g_{n-1}$ and thus indirectly a linear combination of $f_1, \ldots, f_{n-1}$, contradicting the independence of the f_i. Hence we may set $g_n = u/\| u \|$ to make $\{g_1, \ldots, g_n\}$ an orthonormal set. ■

Problems

1. Analyze carefully the cases of *equality* in the Cauchy-Schwarz and triangle inequalities.

2. $\langle \sum_i \alpha_i f_i, \sum_j \beta_j g_j \rangle = \sum_{ij} \alpha_i \beta_j \langle f_i, g_j \rangle$.

3. A slightly different proof of the Cauchy-Schwarz inequality proceeds as follows. Write $0 \leq \langle f + \lambda g, f + \lambda g \rangle = \lambda^2 \| g \|^2 + 2\lambda \langle f, g \rangle + \| f \|^2$. The discriminant of this quadratic function of λ is therefore nonpositive.

4. Consider the normed linear space of continuous functions on $[0,2]$ with norm $\|f\|^2 = \int_0^2 f^2(x)\,dx$. Do the functions f_n graphed here form a Cauchy sequence? Is this space complete?

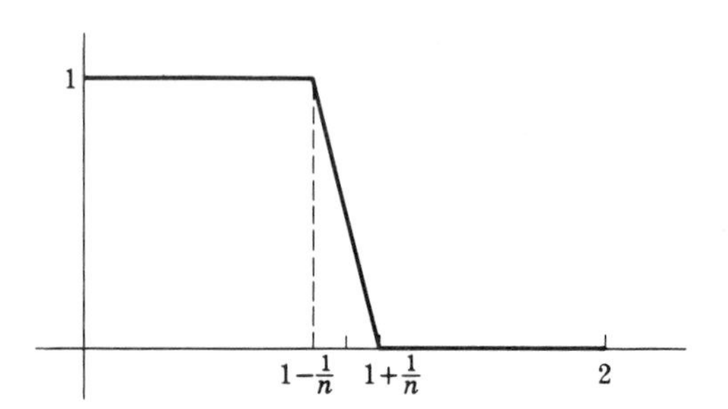

5. Prove that $\int_a^b f(x)g(x)\,dx$ is an inner product in $C[a,b]$.

6. Prove that the set of functions $f_n(x) = \cos nx$ $(n = 0, 1, \ldots)$ is orthogonal with respect to the inner product of Prob. 5, on $[0,\pi]$.

7. What conditions must be placed upon a function w if the equation $\int_a^b f(x)g(x)w(x)\,dx = \langle f,g\rangle$ is to define an inner product in $C[a,b]$?

*8. Using the theorem that every linear space has a basis, prove that an inner product can always be defined. (Cf. Prob. 11, Sec. 3.)

9. If $\{g_1,\ldots,g_n\}$ is an orthogonal set of vectors, how are the c's to be chosen to minimize $\| c_1g_1 + \cdots + c_ng_n - f \|$?

*10. A converse is known for the theorem on page 10: If every closed bounded set in the normed linear space E is compact, then E is finite-dimensional. A proof in the special case that E is an inner-product space goes as follows. If E is infinite-dimensional, then it contains an independent sequence $\{x_n\}$. By the Gram-Schmidt process we obtain an orthonormal sequence $\{u_n\}$. The set $\{u_n\}$ is closed and bounded but contains no converging subsequence. [Banach, 1932, p. 84]

*11. Prove that if a norm satisfies the parallelogram law, then the equation

$$2\langle f,g\rangle = \|f+g\|^2 - \|f\|^2 - \|g\|^2$$

defines an inner product, and $\langle f,f\rangle = \|f\|^2$.

*12. Given a_i, b_i, c_i satisfying $\sum a_ib_i^2 = \sum a_ic_i^2 = 0$ and $\sum a_ib_ic_i = 1$, we define orthogonality in any normed linear space by saying $x \perp y$ when $\sum a_i \| b_ix + c_iy \|^2 = 0$. Show that this generalizes ordinary orthogonality. [Carlsson, 1962]

5. Convexity

In the Cartesian plane, a set is said to be convex if with each two of its points it contains also the line segment joining them. Thus in the sketch, sets A and

B are convex, while C and D are not:

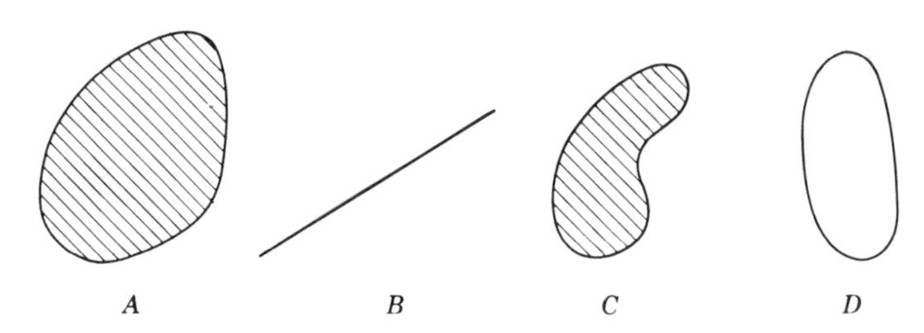

In order to extend this concept to an arbitrary linear space, we first note that in R_2 the line segment joining points f and g consists of all points of the form $\theta f + (1 - \theta)g$ for $\theta \in [0,1]$. Indeed, when $\theta = 0$, this expression yields the point g, and when $\theta = 1$, it yields the point f, while for intermediate points the coordinates depend *linearly* upon θ. In an arbitrary linear space, we adopt precisely the same definition for a line segment. Thus, generally, we shall say that a set K is *convex* if whenever f and g belong to K and when $\theta \in [0,1]$, $\theta f + (1 - \theta)g$ belongs also to K. By induction on m the reader may prove that $\sum_{i=1}^{m} \theta_i f_i \in K$ whenever $f_i \in K$, $\theta_i \geq 0$, and $\sum \theta_i = 1$.

Corresponding to *any* set A in a linear space there is a convex set $\mathcal{H}(A)$ called its *convex hull*, which is defined as the set of points g which are expressible as finite sums of the form $g = \sum \theta_i f_i$ with $f_i \in A$, $\theta_i \geq 0$, and $\sum \theta_i = 1$. Such linear combinations are conveniently termed *convex* linear combinations. Clearly $\mathcal{H}(A)$ contains A. It is also easy to see that any convex set containing A must contain $\mathcal{H}(A)$ as well. Thus $\mathcal{H}(A)$ is a minimal convex set containing A. The number of points f_i which enter into the expression of a point $g \in \mathcal{H}(A)$ must be finite, but may be arbitrarily great. If the linear space is of dimension n, however, each g is expressible as a convex linear combination of some set of $n + 1$ points from A, in accordance with the following theorem.

Theorem of Carathéodory. *Let A be a subset of an n-dimensional linear space. Every point of the convex hull of A is expressible as a convex linear combination of $n + 1$ (or fewer) elements of A.*

Proof. Let g be a point of $\mathcal{H}(A)$. Then g is expressible in the form $g = \sum_{i=0}^{k} \theta_i f_i$ with $\theta_i \geq 0$, $\sum_{i=0}^{k} \theta_i = 1$, $f_i \in A$. If k is minimal, $\theta_i > 0$. The set of vectors $g_i = f_i - g$ $(i = 0, \ldots, k)$ is linearly dependent because $\sum_{i=0}^{k} \theta_i g_i = 0$. If

$k > n$, then $\{g_1, \ldots, g_k\}$ is a set of more than n vectors in a space of dimension n and must therefore be dependent also. Suppose that $\sum_{i=1}^{k} \alpha_i g_i = 0$, where $\sum_{i=1}^{k} |\alpha_i| \neq 0$. Defining $\alpha_0 = 0$, we have for all λ, $\sum_{i=0}^{k} (\theta_i + \lambda\alpha_i) g_i = 0$. Let λ be chosen so that $|\lambda|$ is as small as possible under the condition that one of the coefficients $\theta_i + \lambda\alpha_i$ vanish. The remaining coefficients are non-negative, and do not all vanish because $\theta_0 + \lambda\alpha_0 = \theta_0 > 0$. But with this λ, one term (at least) drops out of the sum $\sum_{i=0}^{k} (\theta_i + \lambda\alpha_i) g_i$. Replacing g_i by $f_i - g$, we have $g \sum_{i=0}^{k} (\theta_i + \lambda\alpha_i) = \sum_{i=0}^{k} (\theta_i + \lambda\alpha_i) f_i$. Dividing by $\sum_{i=0}^{k} (\theta_i + \lambda\alpha_i)$, we obtain an expression for g that contradicts the assumed minimality of k. ∎

Corollary. *The convex hull of a compact set in n space is compact.*

Proof. Let X be a compact set in n space. Let $v_1, v_2, \ldots$ be a sequence of points in $\mathcal{H}(X)$. By Carathéodory's theorem, each v_k can be expressed in the form $v_k = \sum_{i=0}^{n} \theta_{ki} x_{ki}$ where $\theta_{ki} \geq 0$, $\sum_{i=0}^{n} \theta_{ki} = 1$, and $x_{ki} \in X$. By the compactness of X and of the set $\{(\theta_0, \ldots, \theta_n): \theta_i \geq 0, \sum_{i=0}^{n} \theta_i = 1\}$ there exists a sequence $k_1, k_2, \ldots$ such that the limits $\lim_{j\to\infty} \theta_{k_j i} = \theta_i$ and $\lim_{j\to\infty} x_{k_j i} = x_i$ exist for $i = 0, \ldots, n$. Clearly $\theta_i \geq 0$, $\sum \theta_i = 1$, and $x_i \in X$. Thus the sequence $\{v_k\}$ has a subsequence, $\{v_{k_j}\}$ which converges to a point of $\mathcal{H}(X)$, showing that $\mathcal{H}(X)$ is compact. ∎

Theorem. *Every closed convex subset of Euclidean n space possesses a unique point of minimum norm.*

Proof. Let K be such a set. Take $x_1, x_2, \ldots \in K$ such that $\lim_{i\to\infty} \|x_i\| = d = \inf_{x\in K} \|x\|$. Then by the parallelogram law (page 13) $\|x_i - x_j\|^2 = 2\|x_i\|^2 + 2\|x_j\|^2 - 4\|\tfrac{1}{2}(x_i + x_j)\|^2$. Since K is convex, $\tfrac{1}{2}(x_i + x_j) \in K$, so that $\|\tfrac{1}{2}(x_i + x_j)\| \geq d$. Hence $\|x_i - x_j\|^2 \leq 2\|x_i\|^2 + 2\|x_j\|^2 - 4d^2$. As i and $j \to \infty$, the right-hand side of this inequality tends to zero. Thus $\{x_i\}$ is a Cauchy sequence and has a limit point x. Since K is closed, $x \in K$. Since the norm is continuous, $\|x\| = d$. For the uniqueness, observe by the parallelogram law that if $\|x\| = \|y\| = d$ and $x \neq y$, then $\|\tfrac{1}{2}(x + y)\| < d$. ∎

A generalization of this theorem will be given later (page 22). The present proof is valid in Hilbert space.

Theorem on Linear Inequalities. *Let U be a compact subset of R_n. A necessary and sufficient condition that the system of linear inequalities $\langle u,z\rangle > 0$ $(u \in U)$ be inconsistent is that $0 \in \mathcal{H}(U)$.*

Proof. For the sufficiency, suppose $0 \in \mathcal{H}(U)$ so that $0 = \sum_{i=1}^{m} \lambda_i u_i$ with $\lambda_i \geq 0$, $\sum \lambda_i = 1$, and $u_i \in U$. Thus for all z, $0 = \sum \lambda_i \langle u_i,z\rangle$. This equation would be violated if $\langle u_i,z\rangle > 0$ for all $i = 1, \ldots, m$. For the necessity assume $0 \notin \mathcal{H}(U)$. By the preceding theorems there exists a point $z \in \mathcal{H}(U)$ for which $\| z \|$ is a minimum. Let u be arbitrary in U. By convexity, $\theta u + (1 - \theta)z \in \mathcal{H}(U)$ when $0 \leq \theta \leq 1$, and consequently $0 \leq \| \theta u + (1 - \theta)z \|^2 - \| z \|^2 = \theta^2 \| u - z \|^2 + 2\theta\langle u - z, z\rangle$. But this inequality cannot be valid for small positive θ unless $\langle u - z, z\rangle \geq 0$. Thus $\langle u,z\rangle \geq \langle z,z\rangle > 0$, showing that z is a solution of the inequalities. ∎

Theorem of Helly. *Let $\{K_i\}_0^m$ be a collection of compact convex sets in R_n. In order that $\bigcap_{i=0}^{m} K_i$ be nonempty, it is necessary and sufficient that every family of $n + 1$ sets from $\{K_i\}$ have a point in common.*

Proof. The necessity being trivial, we proceed to the sufficiency. Defining the distance from a point x to a set K by the formula dist $(x,K) = \inf_{y \in K} \| x - y \|$, let us set $f(x) = \max_i \text{dist}\,(x,K_i)$. Assume that $\bigcap_{i=0}^{m} K_i$ is empty. It is easy to see that f is continuous. Since $f(x) \to \infty$ if $\| x \| \to \infty$, the infimum of f is achieved by some point, y; since $\bigcap K_i = \phi$, $\rho \equiv f(y) > 0$. By a translation, we may assume $y = 0$. For each i select $x_i \in K_i$ such that $\| x_i \| = \text{dist}\,(y,K_i)$. By a renumbering we may assume $\| x_i \| = \rho$ for $i = 0, \ldots, q$ and $\| x_i \| < \rho$ for $i = q + 1, \ldots, m$. If $0 \notin \mathcal{H}\{x_0, \ldots, x_q\}$, then $f(y)$ could be decreased by moving y slightly toward $\mathcal{H}\{x_0, \ldots, x_q\}$. Thus $0 \in \mathcal{H}\{x_0, \ldots, x_q\}$. By Carathéodory's theorem we may assume (with another renumbering) that $q \leq n$, and hence the system $\langle z,x_i\rangle \geq \rho^2$ $(i = 0, \ldots, q)$ is inconsistent. But each point of K_i $(i = 0, \ldots, q)$ satisfies $\langle z,x_i\rangle \geq \rho^2$. Thus $\bigcap_{i=0}^{q} K_i = \phi$. ∎

Problems

1. In a normed linear space every closed *sphere* $\{f\colon \| f - g \| \leq \lambda\}$ is convex, g and λ being prescribed. In particular, the *unit sphere* $\{f\colon \| f \| \leq 1\}$ is convex.

2. In an inner-product space every closed *half space* $\{f: \langle f, g\rangle \geq \lambda\}$ is convex.
3. The intersection of a family of convex sets is itself convex.
4. Prove that the convex hull of a set is convex.
5. Any set of the form $\{\sum_{i=1}^{n}\alpha_i g_i : \alpha \in K\}$ is convex, the g's being prescribed vectors, and K being a convex subset of R_n.

*6. In applying Carathéodory's theorem in the normed linear space R_n, a point g of $\mathcal{H}(A)$ which lies on the boundary is expressible as a convex linear combination of n points of A.

*7. If a closed set S is "midpoint convex" ($x, y \in S \Rightarrow \frac{1}{2}x + \frac{1}{2}y \in S$), then it is convex.

8. A closed convex set K in n space possesses a unique point $\mathfrak{J}x$ closest to a given point x, if we use the Euclidean metric.

*9. The mapping $\mathfrak{J}$ defined in Prob. 8 satisfies the Lipschitz condition $\|\mathfrak{J}x - \mathfrak{J}y\| \leq \|x - y\|$.

10. The intersection of all convex sets containing A is identical with the convex hull of A.
11. Let U be compact in R_n. If the system of inequalities $\langle u,x\rangle > 0$ $(u \in U)$ is inconsistent, then it possesses an inconsistent subsystem of at most $n + 1$ inequalities.
12. Can a bounded set have a convex complement?
13. Prove the corollary to Carathéodory's theorem by establishing that the map $(\theta_0, \ldots, \theta_n, x_0, \ldots, x_n) \to \sum \theta_i x_i$ is a continuous map on a compact set.
14. Helly's theorem can be generalized easily to the following. If $\{K_\alpha\}$ is a collection of closed convex sets in n space such that every finite subfamily has a *bounded* intersection and every subfamily of $n + 1$ sets has a *nonvoid* intersection, then $\bigcap K_\alpha$ is nonvoid.

6. Existence and Unicity of Best Approximations

In Sec. 2 a general problem of approximation was posed, viz., to determine a point or points in a given subset of a metric space which lie at minimum distance from a certain fixed point. In general, such closest points do not exist. It is only by adding hypotheses about the subset or the space that the existence theorem can be established. In Sec. 2 one theorem of this type was given; compactness of the subset was the additional hypothesis employed there. In this section we turn our attention to normed linear spaces for some further existence theorems.

Existence Theorem. *A finite-dimensional linear subspace of a normed linear space contains at least one point of minimum distance from a fixed point.*

Proof. Let M be such a subspace and g the prescribed point. Let f_0 be an arbitrary point of M. Then the point we seek lies in the set

$$\{f: f \in M, \|f - g\| \leq \|f_0 - g\|\}$$

By the theorem on page 10, this set is compact. Hence, by the theorem on page 4, it contains a point of minimum distance from g. ∎

A natural question is suggested by this theorem: Is the finite-dimensionality hypothesis really necessary? We will answer this question with an example and a theorem. The example will show that finite dimensionality cannot be omitted from the above theorem, and the theorem will show that other hypotheses may take its place. In the example we consider the space (c_0) of infinite sequences $f = (\xi_1, \xi_2, \ldots)$ such that $\xi_n \to 0$. With the norm $\| f \| = \max | \xi_n |$, this becomes a Banach space.

Example. *In the space (c_0) the subspace M of points f for which $\sum_{k=1}^{\infty} 2^{-k}\xi_k = 0$ does not contain a closest point to any external point.*

Proof. Let $g = (\eta_1, \eta_2, \ldots)$ be any point of (c_0) not in M. Then the number $\lambda = \sum_{k=1}^{\infty} 2^{-k}\eta_k$ is different from zero. We prove first that the distance from g to M is no greater than $| \lambda |$. In fact, the following points belong to M,

$$f_1 = -\tfrac{2}{1}(\lambda, 0, 0, \ldots) + g$$

$$f_2 = -\tfrac{4}{3}(\lambda, \lambda, 0, 0, \ldots) + g$$

$$f_3 = -\tfrac{8}{7}(\lambda, \lambda, \lambda, 0, 0, \ldots) + g, \text{ etc.}$$

as may be easily verified. Further $\| f_n - g \| = [2^n/(2^n - 1)] | \lambda | \to | \lambda |$. It now remains to prove that no point of M is of distance $| \lambda |$ or less from g. If $f = (\xi_1, \xi_2, \ldots)$ is arbitrary in M, select n such that $| \xi_k - \eta_k | < \frac{1}{2} | \lambda |$ whenever $k \geq n$. This is possible since the elements of (c_0) are sequences converging to zero. Suppose $\| g - f \| \leq | \lambda |$. Then $| \sum 2^{-k}\eta_k | = | \sum 2^{-k}(\eta_k - \xi_k) | \leq \sum 2^{-k} | \eta_k - \xi_k | \leq | \lambda | \sum_{k<n} 2^{-k} + \frac{1}{2} | \lambda | \sum_{k \geq n} 2^{-k} < | \lambda |$, a contradiction. ■

In the next theorem we consider normed linear spaces which are *uniformly convex*. This means that to each $\epsilon > 0$ there corresponds a $\delta > 0$ such that $\| f - g \| < \epsilon$ whenever $\| f \| = \| g \| = 1$ and $\| \frac{1}{2}(f + g) \| > 1 - \delta$. This is a geometric property of the unit sphere of the space: if the *midpoint* of a line segment with end points on the surface of the sphere approaches the surface, then the end points must come closer together.

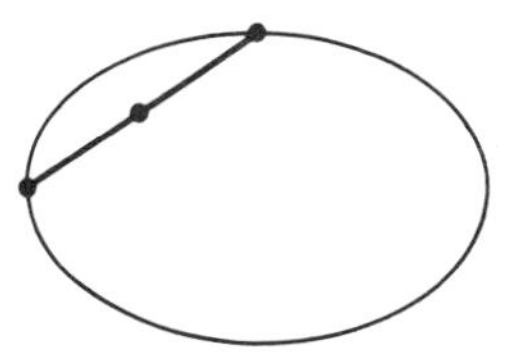

Uniformly convex

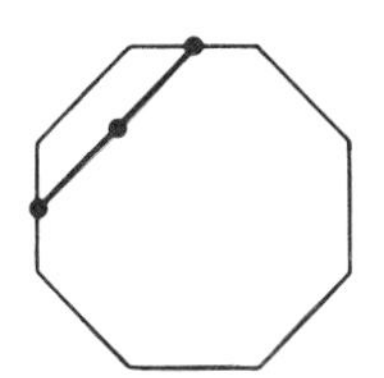

Not uniformly convex

Theorem. *A closed convex set in a uniformly convex Banach space possesses a unique point closest to a given point.*

Proof. Let K be such a set and g an arbitrary point. By making the change of variable $f \to f - g$ we may assume $g = 0$. Put $D = \inf_{f \in K} \| f - g \|$. If $D = 0$, then because K is closed, $g \in K$, and the conclusion is trivial. Otherwise, by making the change of variable $f \to D^{-1}f$ we may assume $D = 1$. Now select a sequence $f_n \in K$ such that $\lim_{n \to \infty} \| f_n \| = 1$. Given $\epsilon > 0$, let δ be taken as in the definition of uniform convexity. Select N so that $\| f_n \| - 1 < \delta$ whenever $n \geq N$. Let $n, m \geq N$, and for convenience of notation, let $\lambda_n = \| f_n \|^{-1}$. Then, using the triangle inequality and the convexity of K, we have

$$\begin{aligned} \tfrac{1}{2} \| \lambda_n f_n + \lambda_m f_m \| &= \tfrac{1}{2} \| f_n + f_m - (1 - \lambda_n) f_n - (1 - \lambda_m) f_m \| \\ &\geq \tfrac{1}{2} \| f_n + f_m \| - \tfrac{1}{2}(1 - \lambda_n) \| f_n \| - \tfrac{1}{2}(1 - \lambda_m) \| f_m \| \\ &> 1 - \delta \end{aligned}$$

By uniform convexity, $\| \lambda_n f_n - \lambda_m f_m \| < \epsilon$, and the sequence $\lambda_n f_n$ is a Cauchy sequence. By the completeness of the space, $\lambda_n f_n \to f$ for some f. Since $\| f_n - f \| \leq \| f_n - \lambda_n f_n \| + \| \lambda_n f_n - f \| \leq \| f_n \| (1 - \lambda_n) + \| \lambda_n f_n - f \|$, $f_n \to f$. Since K is closed, $f \in K$. Since $\| \lambda_n f_n \| = 1$, $\| f \| = 1$. The unicity is left to the reader. ∎

In general, the point whose existence is guaranteed by the first theorem of this section is not unique. It is easy to give an example in R_2. In fact, let us use the norm $\| f \| = \max \{ | \xi_1 |, | \xi_2 | \}$, where $f = [\xi_1, \xi_2]$. Let M denote the subspace of points of the form $[\xi_1, 0]$. Let g be the point $[0,1]$. Then the

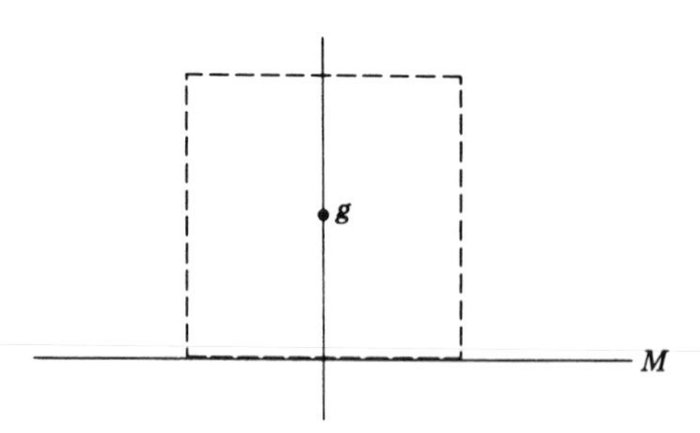

distance from g to M is 1 and is achieved by taking $[\xi, 0]$ with $| \xi | \leq 1$. This example suggests that the curvature of the unit sphere has a bearing on unicity. In fact, unicity may be proved for the closest point if the space is *strictly convex*. This means geometrically that the unit sphere contains no line segments on its surface. Algebraically, we require that

$$\| x \| = \| y \| = \| \tfrac{1}{2}(x + y) \| = 1 \Rightarrow x = y$$

Unicity Theorem. *In a strictly convex normed linear space a finite-dimensional subspace contains a unique point closest to any given point.*

Proof. The existence has already been proved. Suppose then that f and f' are two points of the subspace M of minimum distance λ from g. Then $\|\frac{1}{2}(f+f')-g\| \leq \frac{1}{2}\|f-g\| + \frac{1}{2}\|f'-g\| = \lambda$. Since M is a linear subspace, $\frac{1}{2}(f+f') \in M$, and consequently the left member of this inequality is $\geq \lambda$. Now if $\lambda = 0$, it is clear that $f = f' = g$. If $\lambda \neq 0$, then the vectors $(f-g)/\lambda$, $(f'-g)/\lambda$ and their midpoints are all of norm 1, and by the strict convexity, $f = f'$. ∎

To complete this topic we investigate the relationship between the two special types of spaces considered.

Theorem. *Every uniformly convex normed linear space is strictly convex. Every finite-dimensional strictly convex space is uniformly convex.*

Proof. Let E be a uniformly convex normed linear space. Suppose $\|f\| = \|g\| = \|\frac{1}{2}(f+g)\| = 1$. The definition of uniform convexity immediately yields $\|f-g\| < \epsilon$ for every ϵ, and consequently $f = g$.

Now let E be a finite-dimensional strictly convex normed linear space. Let $\epsilon > 0$ be given, and consider the set S of pairs (f,g) such that $f \in E$, $g \in E$, $\|f\| = \|g\| = 1$, and $\|f-g\| \geq \epsilon$. This is a closed and bounded set and hence compact by the theorem on page 10. On this set the expression $1 - \frac{1}{2}\|f+g\|$ is continuous and positive by the strict convexity of E. Hence it has a positive infimum, which we denote by δ. Thus for pairs (f,g) with $\|f\| = \|g\| = 1$ we have

$$\|f-g\| \geq \epsilon \Rightarrow 1 - \tfrac{1}{2}\|f+g\| \geq \delta$$

or in another form,

$$\tfrac{1}{2}\|f+g\| > 1 - \delta \Rightarrow \|f-g\| < \epsilon$$ ∎

Continuity Theorem. *Let M be a compact set in a metric space X. If each point x in X has a unique closest point $\mathfrak{J}x$ in M, then $\mathfrak{J}$ is a continuous operator.*

Proof. If $\mathfrak{J}$ is discontinuous at x_0, say, then there exists a sequence $x_n \to x_0$ such that $d(\mathfrak{J}x_n,\mathfrak{J}x_0) > \epsilon$. By passing to a subsequence if necessary, we may assume that $\mathfrak{J}x_n$ converges to a point $z \neq \mathfrak{J}x_0$. Now $d(x,\mathfrak{J}x)$ is a continuous function of x, as may be seen from the inequality

$$d(x,\mathfrak{J}x) \leq d(x,\mathfrak{J}y) \leq d(x,y) + d(y,\mathfrak{J}y)$$

Consequently, if we pass to the limit in the inequality

$$d(x_0,z) \leq d(x_0,x_n) + d(x_n,\mathfrak{I}x_n) + d(\mathfrak{I}x_n,z)$$

we obtain $d(x_0,z) \leq d(x_0,\mathfrak{I}x_0)$, which contradicts the unicity of $\mathfrak{I}x_0$. ∎

We shall see in Chap. 5 that if M is a certain class of rational functions in $C[a,b]$, then generally $\mathfrak{I}$ is discontinuous, although M always contains a unique element of best approximation to a given continuous function. This M, of course, cannot be compact.

Problems

1. If $1 < p < \infty$, the norm $\|f\| = (\sum_{k=1}^{n} |\xi_k|^p)^{1/p}$ makes R_n into a strictly convex space. If $p = 1$ or $p = \infty$, this is not true. Look in a book such as [Hardy, Littlewood, Pólya, 1934].
2. Every Hilbert space is uniformly convex.
3. In the example on page 21, the coefficients 2^{-k} may be replaced by any others λ_k as long as $\sum_{k=1}^{\infty} |\lambda_k| < \infty$ and an infinite number of the λ_k are nonzero.
4. A finite-dimensional closed set in a normed linear space possesses a point closest to a given point.

*5. In R_n, with norm $\|f\| = \max_i |\xi_i|$, what properties of a closed convex set will ensure that it possess a *unique* closest point to each given point?

6. Prove the second theorem of this section without the simplifying changes of variable.

*7. Does the unicity theorem have a converse? That is, can we infer strict convexity from a knowledge that every finite-dimensional subspace contains a unique point closest to a given point?

*8. Try to prove theorems similar to those in this section for metric linear spaces.

9. A *seminorm* would be sufficient for the first theorem of this section.
10. Let G be a linear subspace of a normed linear space. Write $f \perp G$ if $\|f\| \leq \|f + g\|$ for all $g \in G$. Prove that in an inner-product space this notion of orthogonality agrees with the original one. Prove that a point $g^* \in G$ is a best approximation of f if and only if $f - g^* \perp G$. See [Pták, 1958].
11. From the example of this section it must follow that the space (c_0) is not uniformly convex. Prove this directly.
12. Let $\{g_1,\ldots,g_n\}$ be independent in a normed linear space E, and let S be any closed set of n-tuples. Show that each element f in E possesses a best approximation of the form $\sum c_i g_i$ with $[c_1,\ldots,c_n] \in S$.

7. Convex Functions

In order to determine a point on a finite-dimensional subspace M (in a normed linear space) of minimum distance from a given point g we may select a basis $\{f_1, \ldots, f_n\}$ for M and then seek the minimum of the expression

$$\Delta(c) = \Delta(c_1, \ldots, c_n) = \left\| \sum_{i=1}^{n} c_i f_i - g \right\|$$

This is a continuous real-valued function of n real variables, n being the dimension of the subspace M. The continuity is readily proved by using the triangle inequality. Another significant property of the function Δ is that it is *convex*. A function ϕ defined in a convex set is said to be *convex* if the inequality

$$\phi(\lambda x + \mu y) \leq \lambda\phi(x) + \mu\phi(y)$$

is satisfied whenever $\lambda \geq 0$, $\mu \geq 0$, and $\lambda + \mu = 1$. Convexity of the domain of ϕ guarantees that $\lambda x + \mu y$ is in the domain when x and y are. Geometrically, this condition is that between any two points on the graph of ϕ, the graph lies below the chord; for the chord joining $[x,\phi(x)]$ to $[y,\phi(y)]$ has at the point $\lambda x + \mu y$ the ordinate $\lambda\phi(x) + \mu\phi(y)$.

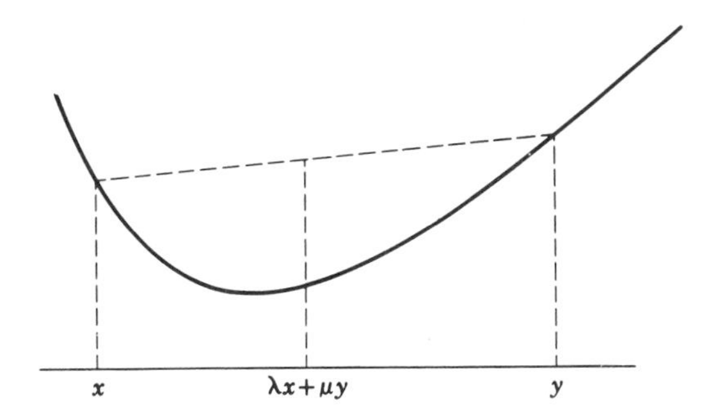

Let us prove that Δ is convex. We have

$$\begin{aligned}
\Delta(\lambda c + \mu d) &= \left\| \sum (\lambda c_i + \mu d_i) f_i - g \right\| \\
&= \left\| \lambda\left(\sum c_i f_i - g\right) + \mu\left(\sum d_i f_i - g\right) \right\| \\
&\leq \lambda \left\| \sum c_i f_i - g \right\| + \mu \left\| \sum d_i f_i - g \right\| \\
&= \lambda\,\Delta(c) + \mu\,\Delta(d)
\end{aligned}$$

Convex functions are especially tractable in minimization problems. This fact will emerge in Chap. 2, where algorithms are to be given for locating minima. One convenient property of convex functions is given in the following simple but important result.

Theorem. *A local minimum of a convex function is necessarily a global minimum.*

Proof. If possible, let ϕ be a convex function having a purely local minimum at the point x_1. This means that in some neighborhood N of x_1, say $N =$

$\{x: \| x - x_1 \| \leq \epsilon\}$, the inequality $\phi(x) \geq \phi(x_1)$ is valid, while at some other more distant point x_2, $\phi(x_2) < \phi(x_1)$. Clearly $\epsilon < \| x_1 - x_2 \|$, and consequently the number $\mu = \epsilon/\| x_1 - x_2 \|$ lies in the interval (0,1). Let $\lambda = 1 - \mu$. By the convexity of ϕ, $\phi(\lambda x_1 + \mu x_2) \leq \lambda\phi(x_1) + \mu\phi(x_2) < \lambda\phi(x_1) + \mu\phi(x_1) = \phi(x_1)$. But this is a contradiction since $\| (\lambda x_1 + \mu x_2) - x_1 \| = \mu \| x_2 - x_1 \| = \epsilon$, showing that $\lambda x_1 + \mu x_2 \in N$. ■

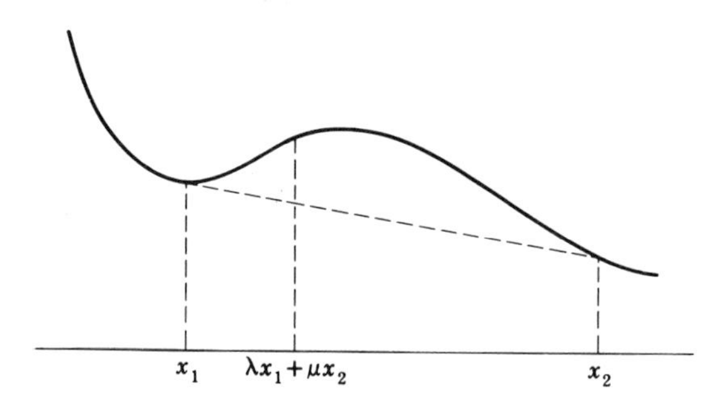

It is easy to give examples of discontinuous convex functions. For example, if $\phi(x) = 1$ when $| x | = 1$ and $\phi(x) = 0$ when $| x | < 1$, then ϕ is a discontinuous convex function on the set $\{x: | x | \leq 1\}$. In spite of this, the following result holds.

Theorem. *A convex function defined on an open convex set in n space must be continuous.*

Proof. Let ϕ be such a function and K its domain. Let x_0 be a point of K, at which the continuity of ϕ will be proved. Denote the vector $[0,\ldots,0,1,0,\ldots,0]$ with 1 as its ith coordinate by δ^i. Since K is open, there is a value of $\theta > 0$ for which the $2n$ points $x_0 \pm \theta\delta^i$ all belong to K. Let these vectors be denoted by $y_1, \ldots, y_{2n}$. An easy induction based upon the convexity of ϕ shows that if $\lambda_i \geq 0$ and $\sum \lambda_i = 1$, then $\phi(\sum \lambda_i y_i) \leq \sum \lambda_i \phi(y_i)$. Hence the function ϕ is bounded above on the convex hull of $\{y_1,\ldots,y_{2n}\}$ by the number $M = \max_i \phi(y_i)$. This convex hull contains a sphere about x_0 of radius, say, r. Now let z be an arbitrary vector of norm $\leq r$. Then

$$\phi(x_0) = \phi[\tfrac{1}{2}(x_0 + z) + \tfrac{1}{2}(x_0 - z)] \leq \tfrac{1}{2}\phi(x_0 + z) + \tfrac{1}{2}\phi(x_0 - z)$$

Hence $2\phi(x_0) \leq \phi(x_0 + z) + M$, showing that ϕ is bounded below on this sphere by $M' = 2\phi(x_0) - M$. Since x_0 and $x_0 + z$ lie in the sphere,

$$\phi[\lambda(x_0 + z) + (1 - \lambda)x_0] \leq \lambda\phi(x_0 + z) + (1 - \lambda)\phi(x_0)$$

whence

$$\phi(x_0 + \lambda z) - \phi(x_0) \leq \lambda[\phi(x_0 + z) - \phi(x_0)] \leq \lambda(M - M') \tag{1}$$

Similarly, since $x_0 - (1 - \lambda)z$ and $x_0 + \lambda z$ lie in the sphere, we have

$$\phi[\lambda(x_0 - z + \lambda z) + (1 - \lambda)(x_0 + \lambda z)] \le \lambda\phi(x_0 - z + \lambda z) + (1 - \lambda)\phi(x_0 + \lambda z)$$

whence

$$(2) \quad \phi(x_0) - \phi(x_0 + \lambda z) \le \lambda[\phi(x_0 - z + \lambda z) - \phi(x_0 + \lambda z)] \le \lambda(M - M')$$

Since z is an arbitrary vector of norm $\le r$, λz is an arbitrary vector of norm $\le \lambda r$. Hence inequalities (1) and (2) establish the continuity of ϕ at x_0. ∎

Problems

1. In a normed linear space, $\|f\|$ is a convex function of f.
2. If ϕ is a convex function, then every set $S_a = \{x: \phi(x) \le a\}$ is a convex set.
3. The converse of the assertion in Prob. 2 is false.
4. Is a convex function of a convex function necessarily convex?
5. A function ϕ is said to be *midpoint convex* if $f(\frac{1}{2}x + \frac{1}{2}y) \le \frac{1}{2}f(x) + \frac{1}{2}f(y)$. Show that a continuous midpoint convex function is convex.
6. A twice-differentiable function ϕ defined on the real line is convex if and only if $\phi'' \ge 0$.
7. Is the set of all convex functions defined on a convex set a convex set of functions?
8. A convex function defined on the real line has one-sided derivatives at every point. What is an appropriate generalization for functions of several variables?
9. If ϕ is a convex function on E, then its *supergraph* $\{(x,\lambda): \lambda \ge \phi(x)\}$ is a convex set in $E \oplus R$, and conversely.
10. If ϕ is convex, then $\lambda^{-1}[\phi(x + \lambda z) - \phi(x)]$ is nondecreasing for $\lambda > 0$.
11. If ϕ_i are convex functions, so are $\max_i \phi_i$ and $\sum \phi_i$.
12. If ϕ and $-\phi$ are convex, then $\phi(x) - \phi(0)$ is linear.

*13. If ϕ is convex on n space and if one set $\{x: \phi(x) \le a\}$ is bounded, then ϕ is bounded below.

14. In the proof of the second theorem, show that we may take $r = \theta/\sqrt{n}$. *Hint*: Show that the convex hull of $\{y_1, \ldots, y_{2n}\}$ is the set of points $x_0 + z$ where $\|z\|_1 = \sum |z_i| \le \theta$.
15. If f is a convex function defined on $(-\infty, \infty)$ and if there exist three points $a < b < c$ such that $f(b) \le \min\{f(a), f(c)\}$, then f attains its minimum on (a,c).

Chapter 2

The Tchebycheff solution of inconsistent linear equations

1. Introduction

In this chapter we shall consider some problems of approximation which arise from a system of linear equations

$$(1) \qquad \sum_{j=1}^{n} A_j^i x_j = b_i \qquad (i = 1, \ldots, m)$$

We suppose that the data A_j^i and b_i are prescribed, and that the unknowns x_j are to be determined. Thus, a system such as (1) presents us with an approximation problem: to determine its exact or approximate solutions. A system of linear equations may have *no* solution, exactly *one* solution, or *infinitely many* solutions, depending on the data. At first one might think that the case when no solution exists is the least interesting of the three and devoid of practical significance. But it turns out that the opposite is true: The calculation of an approximate solution to (1) when no exact solution exists is an important and nontrivial problem. Many practical problems of approximation reduce to one of this type, or sometimes to a sequence of such problems.

We begin with a theorem which indicates how the data of (1) determine the nature of the solutions. For each $j = 1, \ldots, n$ let A_j denote the vector $[A_j^1, A_j^2, \ldots, A_j^m]$. Also let b denote the vector $[b_1, b_2, \ldots, b_m]$. The system (1) may now be written in the form $\sum_{j=1}^{n} x_j A_j = b$. If this equation is consistent (i.e., possesses a solution), then b lies in the linear span of the vectors $A_1, \ldots, A_n$. With this remark in mind, the reader should have no difficulty in proving the following theorem.

Theorem. *The system of linear equations* (1) *is consistent iff b lies in the linear span of the vectors A_j. If the system is consistent, it has* exactly *one solution iff the A_j are linearly independent.*

If the system (1) is inconsistent (i.e., possesses no exact solution), then we may nevertheless seek to minimize the discrepancies between the numbers b_i and the numbers $\sum A_j{}^i x_j$. Looking at this in another light, we may ask that the vector $\sum x_j A_j - b$ (which cannot be zero) shall be in some sense *close* to zero. Paraphrased in still another way, the problem is to determine the x_j so that the vector $\sum x_j A_j$ will be as close as possible to b. From the general existence theorem of Chap. 1, Sec. 6, it follows that for any *norm* defined on R_m a solution to this problem exists. In other words, there exists a vector $x = [x_1, \ldots, x_n]$ for which the expression $\| \sum x_j A_j - b \|$ is a minimum. Such a vector may be termed a *best-approximate solution* of system (1). Generally there will be different best-approximate solutions for different choices of the norm, and even for a particular norm there may be a large set of best-approximate solutions. In this chapter we place special emphasis on the problem associated with the norm

$$\| y \|_T = \max_{1 \le i \le m} | y_i |$$

When this norm is employed, a best-approximate solution of (1) renders the expression

$$\Delta(x) = \max_{1 \le i \le m} | \sum_{j=1}^{n} A_j{}^i x_j - b_i | \tag{2}$$

a minimum. Such an x is therefore sometimes called a *minimax solution.* Since P. L. Tchebycheff was the first to carry out a systematic investigation of minimax approximations, x is also termed a *Tchebycheff approximation* and the norm the *Tchebycheff norm.* This explains the subscript T.

Our discussion will be simplified sometimes by considering a slightly different problem which encompasses the *Tchebycheff problem* described above. Instead of minimizing the expression (2), we minimize the expression

$$\delta(x) = \max_{i} \{ \sum_{j=1}^{n} A_j{}^i x_j - b_i \} \tag{3}$$

The way in which the Tchebycheff problem is subsumed by (3) is as follows. We double the number of data by setting

$$A^{i+m} = -A^i \qquad \text{and} \qquad b_{i+m} = -b_i$$

Thus if we put $r_i = \sum_{j=1}^{n} A_j{}^i x_j - b_i$, then $r_{i+m} = -r_i$ and $\max_{1 \le i \le m} | r_i | = \max_{1 \le i \le 2m} r_i$. Of course, the function $\delta(x)$ defined in (3) might not be bounded below. But if this function arises from a Tchebycheff problem in the manner just indicated, then it will be bounded below and achieve its minimum. It will be convenient to consider functions of the form δ which are bounded below by 0.

Problems

1. Prove the theorem of this section.
2. Prove that the function δ defined in equation (3) is convex.
3. For $i = 1, \ldots, m$ let $A^i = [A_1^i, \ldots, A_n^i]$. Prove that system (1) is consistent iff every linear dependence among the vectors A^i persists among the numbers b_i. That is, for any m-tuple c,

$$\sum c_i A^i = 0 \Rightarrow \sum c_i b_i = 0$$

4. Prove that if the inequalities $\langle A^i, x \rangle \leq b_i$ $(i = 1, \ldots, m)$ are consistent, then for all sufficiently large Q, every Tchebycheff solution of $\langle A^i, x \rangle = b_i - Q$ is a solution of the inequalities.

*5. Prove that if either of the numbers

$$\inf_x \max_{1 \leq i \leq m} r_i(x) \qquad \sup_x \min_{1 \leq i \leq m} r_i(x)$$

is finite, then so is the other, and the second does not exceed the first. Here $r_i(x) = \langle A^i, x \rangle - b_i$.

*6. If the function δ is bounded below then it achieves its infimum.

2. Systems of Equations with One Unknown

In this section we discuss several procedures for solving the Tchebycheff problem connected with a system of equations $a_i x = b_i$ $(i = 1, \ldots, m)$, this being the case $n = 1$ of system (1) considered above. The reason for dwelling upon this simple case is that our algorithms can be used as components of more sophisticated algorithms. Furthermore, familiarity with the simpler case will make the general case more transparent.

We begin with an idealized but concrete example which illustrates how a Tchebycheff problem may arise in practice.

Suppose that it is desired to estimate the so-called "spring constant" of a spring by measuring the elongation produced by various forces. Let us say that the results from experimentation are as follows:

x (force)	2.0	4.0	5.0	6.0
y (elongation)	1.2	2.1	2.6	3.1

By Hooke's law, $y = cx$. Thus *each* observation is capable of yielding a value of c, but because of errors in measurement, these values of c are not in agreement. We have in fact the following *inconsistent* system of linear equations for the determination of c:

$$\begin{aligned} 2.0c &= 1.2 \\ 4.0c &= 2.1 \\ 5.0c &= 2.6 \\ 6.0c &= 3.1 \end{aligned}$$

Let us agree (whether this be justified by statistical considerations or not) that c will be selected as a *Tchebycheff* solution of this system. We thus seek to minimize the expression

$$\Delta(c) = \max\{|2c - 1.2|, |4c - 2.1|, |5c - 2.6|, |6c - 3.1|\}$$

It is not difficult to graph this function. We begin by graphing the eight straight lines $v = 2c - 1.2$, $v = -2c + 1.2$, $v = 4c - 2.1$, etc. After this has been done, the topmost broken line is the graph of $\Delta(c)$. The result is shown in the figure.

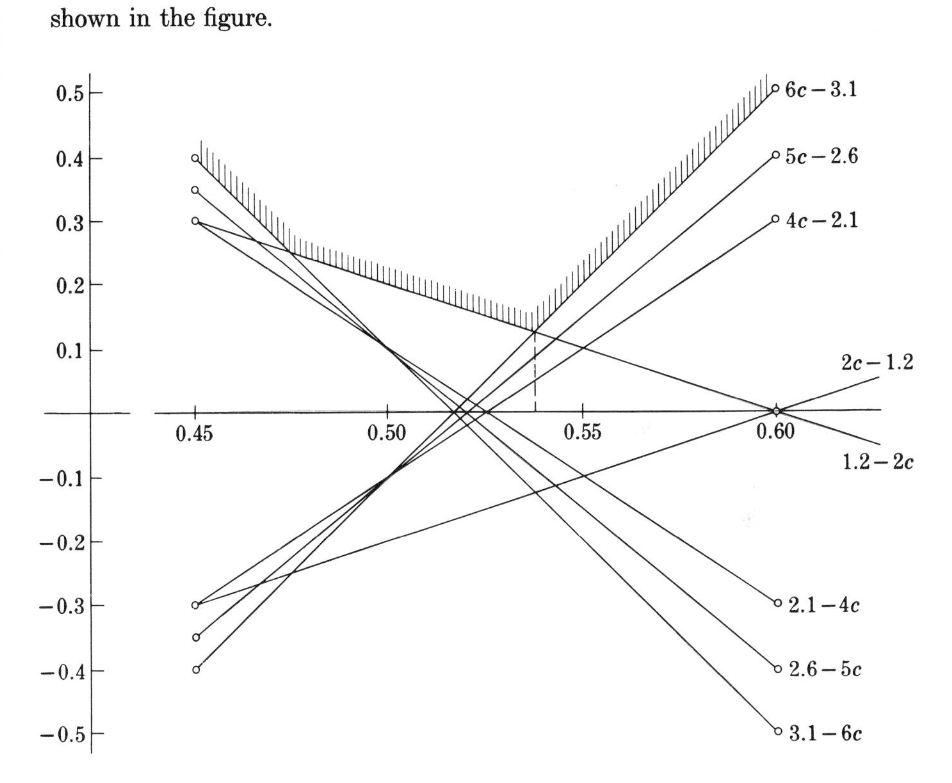

From the sketch we may estimate that the minimum occurs for $c = 0.54$ and is approximately 0.11. In fact, the minimum occurs at the intersection of two lines and is obtainable exactly by solving the equation

$$6c - 3.1 = 1.2 - 2c$$

Thus the correct value of c is 0.5375, and the corresponding value of Δ is 0.125.

Before leaving this example, we note from the graph of Δ its nondifferentiability and its *convexity*. A proof of convexity of such a function was given on page 25.

Also it is noteworthy that at the solution, $c = 0.5375$, two of the equations had errors that were equal in magnitude (viz., 0.125) while the remaining equations had smaller errors. A moment's reflection shows us that this would be expected in any similar problem, no matter how many equations there were. In the general problem, with n unknowns we will expect $n + 1$ errors equal in magnitude at the solution.

We now turn to the problem of constructing algorithms for these problems. The graphical method suggested by the above example is an algorithm to be sure, but it is not capable of being *automated.* Furthermore an extension to n variables appears to be quite difficult. As pointed out in Sec. 1, it is a more general problem to seek the minimum point for a function of the form

$$\delta(x) = \max_{1 \leq i \leq m} \{a_i x - b_i\}$$

As a matter of convenience, let $r_i(x) = a_i x - b_i$. These functions are called the *residuals.*

***Algorithm 1* (*Descent from Vertex to Vertex*).** Starting with any point x_0, define $M = \{i: r_i(x_0) = \delta(x_0)\}$. If M contains two indices j and k such that $a_j a_k < 0$, then x_0 is the solution. For no matter whether x is increased or decreased, one of $r_j(x)$ and $r_k(x)$ will increase, thus increasing $\delta(x)$. On the other hand, a local minimum is necessarily a global minimum, as we saw on page 25. If x_0 is not a solution, then select $j \in M$ for which $|a_j|$ is a minimum. In order to decrease $\delta(x)$ it is necessary and sufficient to decrease $r_j(x)$. The direction in which this may be accomplished is to the right if $a_j < 0$ and to the left if $a_j > 0$. We propose to move only as far as the first vertex on the graph of $\delta(x)$. Selecting j as we did ensures that $r_j(x)$ and $\delta(x)$ are identical along the line segment joining x_0 to the next point. This next point is among the points y_i defined by the equations $r_j(y_i) = r_i(y_i)$, and is the first to the right or left of x_0, depending on the sign of a_j.

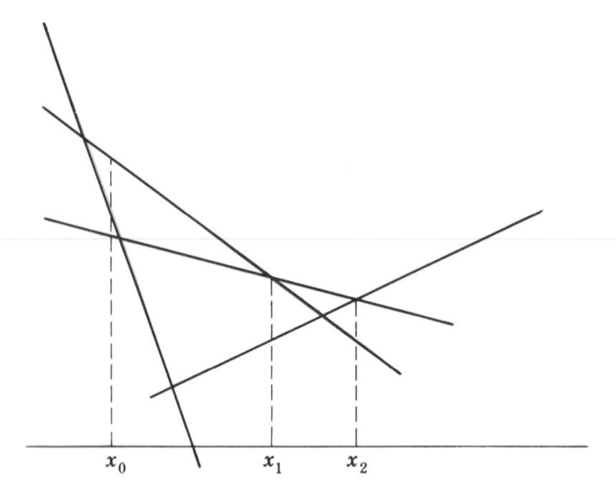

***Algorithm 2* (*Ascent from Vertex to Vertex*).** In each step of this algorithm we have a point x_0 and a pair of indices j and k such that $r_j(x_0) = r_k(x_0)$,

$a_j \leq 0 \leq a_k$, and $a_j \neq a_k$. It is easy to see that under these circumstances x_0 is a minimum point of the function max $\{r_j(x), r_k(x)\}$. The graph of δ suggests that at least one of the minimum points of δ will have the same property for an appropriate pair of indices. Now select i such that $r_i(x_0) = \delta(x_0)$. If $a_i < 0$, we proceed to the intersection of r_i and r_k. If $a_i > 0$, we proceed to the intersection of r_i and r_j. If $a_i = 0$, then we proceed to the intersection of r_i with whichever of r_j or r_k has a nonzero coefficient of x. In the first of these three cases, for example, we replace x_0 by the new point $x = (b_k - b_i)/(a_k - a_i)$. We also replace j by i and begin anew.

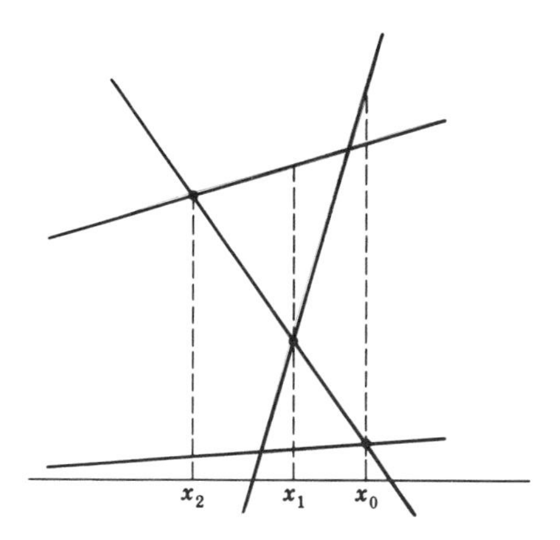

Algorithm 3 (***Searching***). In each step of this algorithm we have two points x and y lying on opposite sides of the minimum point. Say $x < y$ to be definite. This state of affairs will be evidenced as follows. Let $r_i(x) = \delta(x)$ and $r_j(y) = \delta(y)$. Then $a_i \leq 0 \leq a_j$. If $a_i = 0$, then x is a solution, and if $a_j = 0$, y is a solution. Put $z = \frac{1}{2}(x + y)$. Let $r_k(z) = \delta(z)$. If $a_k < 0$, replace x by z and i by k. If $a_k > 0$, replace y by z and j by k. Begin anew. In n steps the accuracy with which the minimum point is located will be improved by a factor of 2^{-n}.

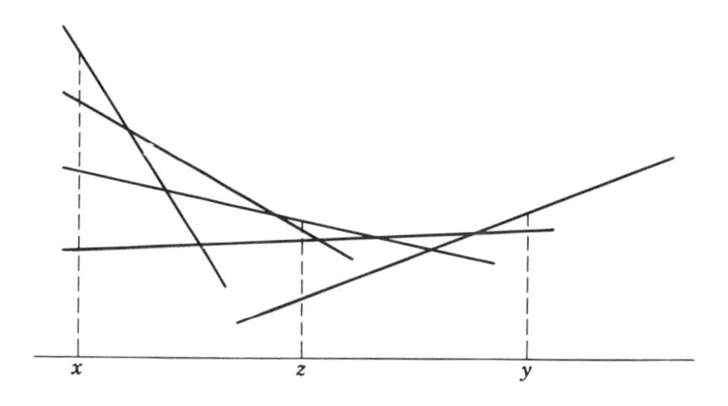

Problems

1. Invent starting procedures for Algorithms 2 and 3.
2. Locate minimum points for the following functions. (Each illustrates a different phenomenon.)

   (*a*) $\max \{2 - 2x, 1 - \frac{1}{2}x\}$ (*c*) $\max \{ |2 - 2x|, |1 - \frac{1}{2}x|, 1\}$

   (*b*) $\max \{ |2 - 2x|, |1 - \frac{1}{2}x| \}$ (*d*) $\max \{ |2 - 2x|, |1 - \frac{1}{2}x|, 1, x\}$

3. Locate the minimum of the function $\Delta(c) = \max_{1 \leq x \leq 2} |1/x - c|$. You will have found the best approximation to x^{-1} on $[1,2]$ by a constant. Draw a graph of the function Δ.
4. Locate the minimum of the function $\Delta(c_1,c_2) = \max_{0 \leq x \leq 1} |c_1x + c_2 - \sqrt{x}|$ You will have found the best approximation to $\sqrt{x}$ on $[0,1]$ by a linear function.
5. In the example concerning the spring constant, what happens if we regard x as the dependent and y as the independent variable?
6. In the same example, can you locate the line $y = cx$ so as to minimize the maximum of its distances from the data points?
7. (For those numerically inclined) Write computer programs for Algorithms 1 to 3, and compare their efficiency on some examples.
8. Write computer programs for the Tchebycheff problem in one variable that avoid the doubling of input data by means of appropriate logical calculations.

3. Characterization of the Solution

We consider again a system of linear equations

$$r_i(x) \equiv \sum_{j=1}^{n} A_j^i x_j - b_i = 0 \qquad (i = 1, \ldots, m)$$

and two functions which depend on it

$$\delta(x) = \max_i r_i(x)$$

$$\Delta(x) = \max_i |r_i(x)|$$

Our goal is to develop methods for locating minimum points of the functions δ and Δ. It is necessary first to find out what properties distinguish the solutions from all other points x. Let us denote the ith row of the matrix (A_j^i) by A^i. Then $r_i(x) = \langle A^i,x \rangle - b_i$.

Characterization Theorem. *A point z is a minimum point of the function δ if and only if the origin lies in the convex hull of the set $\{A^i: r_i(z) = \delta(z)\}$.*

Proof. Suppose that z is not a minimum point of δ. Then for some vector h, $\delta(z - h) < \delta(z)$. Let $M = \{i: r_i(z) = \delta(z)\}$. Then for $i \in M$ we have $r_i(z - h) \leq \delta(z - h) < \delta(z) = r_i(z)$ and $\langle A^i, z - h \rangle - b_i < \langle A^i,z \rangle - b_i$.

Thus

$$\langle A^i, h\rangle > 0 \qquad (i \in M)$$

By the theorem on linear inequalities (Chap. 1, Sec. 5) the origin does not lie in the convex hull of the set $\{A^i : i \in M\}$. For the converse, suppose that 0 does not lie in the convex hull of $\{A^i : i \in M\}$. By the cited theorem, there exists an h such that $\langle A^i, h\rangle > 0$ for $i \in M$. Hence the number $\alpha = \min_{i \in M} \langle A^i, h\rangle$ is positive. The residuals $r_i(z)$, for $i \in M$, will decrease in the direction $-h$ because for $\lambda > 0$,

$$r_i(z - \lambda h) = r_i(z) - \lambda\langle A^i, h\rangle \leq \delta(z) - \lambda\alpha$$

The residuals $r_i(z)$, for $i \notin M$, are less than $\delta(z)$, and by continuity remain so in a neighborhood of z. Thus there are points near z yielding lower values of δ. The details of this argument may be arranged as follows. Let $\beta = \max_{i \notin M} r_i(z)$, and $\gamma = \min_{1 \leq i \leq m} \langle A^i, h\rangle$. Then for $i \notin M$

$$r_i(z - \lambda h) = r_i(z) - \lambda\langle A^i, h\rangle \leq \beta - \lambda\gamma$$

We can make all residuals less than $\delta(z)$ by selecting $\lambda > 0$ in such a way that

$$\lambda\gamma > \beta - \delta(z)$$ ■

For the function Δ a similar theorem holds. We give it without proof. It is convenient to use the function sgn, which is defined by

$$\operatorname{sgn} x = \begin{cases} 1 & \text{if } x > 0 \\ 0 & \text{if } x = 0 \\ -1 & \text{if } x < 0 \end{cases}$$

Characterization Theorem. *Given a point $z \in R_n$, let $\sigma_i = \operatorname{sgn} r_i(z)$ and $M = \{i : |r_i(z)| = \Delta(z)\}$. The point z minimizes Δ if and only if the origin of R_n lies in the convex hull of the set $\{\sigma_i A^i : i \in M\}$.*

We now come to the n-dimensional formulation of a fact that has already been perceived in the case $n = 1$. If the reader will refer to the sketch on page 31, he will observe that at the minimum point, a certain *pair* of residual functions determines the graph of Δ. The precise result for n space is as follows.

Theorem. *If z is a minimum point of the function $\delta(x) = \max_{1 \leq i \leq m} r_i(x)$, then z is a minimum point of $\max_{i \in J} r_i(x)$, where J is a certain subset of $\{1, \ldots, m\}$ comprising at most $n + 1$ indices.*

Proof. By the characterization theorem (page 34) we know that the origin of R_n lies in the convex hull of the set $\{A^i: i \in M\}$, where $M = \{i: r_i(z) = \delta(z)\}$. If M contains $n + 1$ or fewer elements, we let $J = M$. Otherwise by Carathéodory's theorem (Chap. 1, Sec. 5), we select a subset J of M having at most $n + 1$ elements and such that $0 \in \mathcal{H}\{A^i: i \in J\}$. By the characterization theorem, z is also a minimum point of $\max_{i \in J} r_i(x)$. ∎ For the function Δ, a similar result is valid, and takes the following form.

Theorem. *Every minimax solution of the system* $\sum_{j=1}^{n} A_j{}^i x_j = b_i$ $(i = 1, \ldots, m > n)$ *is a minimax solution of an appropriate subsystem comprising* $n + 1$ *equations.*

Once such an "appropriate subsystem" is known, it is relatively easy to obtain its minimax solution. In several practical methods for computing minimax solutions of systems of equations the principal expenditure of effort is in locating this appropriate subsystem. We will postpone the discussion of this problem in order to see first how to obtain the minimax solution of $n + 1$ equations in n unknowns.

Problems

1. Determine whether the point $z = (1,1)$ is a minimum point of the function $\delta(x) = \max\{(x_1 + 2x_2 - 4), (-x_1 + 2x_2 - 3), (-x_1 - x_2 + 1), (x_1 - x_2 - 1)\}$. If it is not, determine an h such that $\delta(z - h) < \delta(z)$.
2. Determine whether the point $z = (2,1)$ is a minimum point of the function $\Delta(x) = \max\{|3x_1 + x_2 - 4|, |6x_1 - x_2 + 5|, |x_1 + x_2 + 2|, |-x_1 + 2x_2 - 5|\}$. If it is not, determine an h such that $\Delta(z - h) < \Delta(z)$. Then do the same problem for the point $z = (-1,3)$.
3. What can be said about the problem of minimizing the function δ when $b_i = 0$ for all i?
4. Give an example of a function of the form δ which has no minimum point. Give an example showing that the minimum point is not always unique.

*5. Try to resolve the difficulties that arise in proving the first characterization theorem when m is infinite.

6. The theorem of page 35 may also be proved with the aid of Helly's theorem and Prob. 14 of Chap. 1, Sec. 5. *Hint*: For each set J of $n + 1$ indices in $\{1,\ldots,m\}$ define

$$\epsilon(J) = \min_x \max_{i \in J} |\langle A^i, x\rangle - b_i|$$

Let $\epsilon = \max \epsilon(J)$. Define convex sets $K_i = \{x: |\langle A^i,x\rangle - b_i| \le \epsilon\}$. Every family of $n + 1$ of the sets K_i has a nonvoid intersection. Use Helly's theorem. See [Rademacher and Schoenberg, 1950].

4. The Special Case, $m = n + 1$

In this section we consider the problem of computing the minimax solution (or solutions) of a system of $n + 1$ equations in n unknowns:

$$r_i(x) = \sum_{j=1}^{n} A_j^i x_j - b_i = \langle A^i, x\rangle - b_i = 0 \qquad (i = 1, \ldots, n + 1)$$

Probably the most satisfactory way of solving such a system is the method of de La Vallée Poussin. Suppose that by some means we are able to discover a point x, signs $\sigma_i = \pm 1$, and a number ϵ such that

$$(1) \qquad r_i(x) = \sigma_i \epsilon \qquad (i = 1, \ldots, n + 1)$$

$$(2) \qquad 0 \in \mathcal{H}\{\sigma_1 A^1, \ldots, \sigma_{n+1} A^{n+1}\}$$

Then we claim that x is a minimax solution of the system. Indeed, by (1) we have $|r_i(x)| = |\epsilon|$. Then by the characterization theorem (page 35) and by property (2) above, x is a solution. We shall therefore set about securing conditions (1) and (2) above.

First we find a nontrivial solution of the linear equations $\sum_{i=1}^{n+1} \lambda_i A^i = 0$. This is possible because the $n + 1$ vectors $A^1, \ldots, A^{n+1}$ are necessarily linearly dependent, being elements of n space. If we define $\sigma_i = 1$ when $\lambda_i \geq 0$ and $\sigma_i = -1$ when $\lambda_i < 0$, then condition (2) is already secured because $0 = \sum (\sigma_i \lambda_i)(\sigma_i A^i)$. In order to complete the discussion we assume that the matrix is of rank n. Thus some set of n of its rows is linearly independent, and by renumbering the rows if necessary, we may take this set to be $\{A^1, \ldots, A^n\}$. Now if condition (1) is to be met, then $\langle A^i, x\rangle - b_i = \epsilon\sigma_i$. Multiplying this equation by λ_i and summing for $i = 1, \ldots, n + 1$ yields $\langle \sum \lambda_i A^i, x\rangle - \sum \lambda_i b_i = \epsilon \sum \sigma_i \lambda_i$. In view of what we already know, this reduces to $-\sum \lambda_i b_i = \epsilon \sum |\lambda_i|$, and this equation may be taken as the definition of ϵ since $\sum |\lambda_i| > 0$. It remains to be shown that with this definition of ϵ, equation (1) is actually *consistent*. If we leave out the equation corresponding to $i = n + 1$, the resulting system may be solved for a unique x because of our assumption that $\{A^1, \ldots, A^n\}$ is linearly independent. Thus we have $r_i(x) = \sigma_i \epsilon$ for $i = 1, \ldots, n$. But we have already seen that $\sum_{i=1}^{n+1} \lambda_i r_i(x) = \epsilon \sum_{i=1}^{n+1} \sigma_i \lambda_i$. Hence $\lambda_{n+1} r_{n+1}(x) = \epsilon\sigma_{n+1}\lambda_{n+1}$. If $\lambda_{n+1} = 0$, then the equation $\sum \lambda_i A^i = 0$ would represent a linear dependence among $A^1, \ldots, A^n$, which is impossible. Hence $\lambda_{n+1} \neq 0$ and $r_{n+1}(x) = \epsilon\sigma_{n+1}$.

Essentially the same method may be described with the use of determinants. Suppose, to take a concrete example, that we are confronted with the system

$$\begin{aligned} x_1 - x_2 &= 7 \\ 2x_1 + 3x_2 &= 5 \\ 3x_1 + x_2 &= -1 \end{aligned}$$

Conditions (1) in this case would read

$$\begin{aligned} x_1 - x_2 - \sigma_1\epsilon &= 7 \\ 2x_1 + 3x_2 - \sigma_2\epsilon &= 5 \\ 3x_1 + x_2 - \sigma_3\epsilon &= -1 \end{aligned}$$

By Cramer's rule,

$$\epsilon = \begin{vmatrix} 1 & -1 & 7 \\ 2 & 3 & 5 \\ 3 & 1 & -1 \end{vmatrix} \div \begin{vmatrix} 1 & -1 & -\sigma_1 \\ 2 & 3 & -\sigma_2 \\ 3 & 1 & -\sigma_3 \end{vmatrix} = \frac{-74}{7\sigma_1 + 4\sigma_2 - 5\sigma_3}$$

In order that ϵ shall have a minimum positive value, we must take $\sigma_1 = -1$, $\sigma_2 = -1$, and $\sigma_3 = 1$, thereby getting $\epsilon = \frac{37}{8}$. Knowing σ_1, σ_2, σ_3, and ϵ, we easily calculate $x_1 = \frac{3}{2}$ and $x_2 = -\frac{7}{8}$. Observe that 0 lies in the convex hull of the vectors $\sigma_i A^i$:

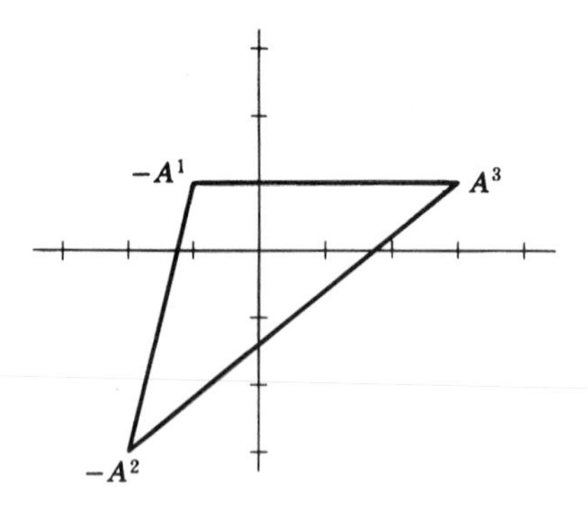

The two methods that we have just described for obtaining the signs σ_i are not necessarily the same from the computational viewpoint but are the same theoretically. In fact, in the second method the σ_i were chosen as the signs of the numbers -7, -4, 5, and the latter may serve as the numbers λ_i of the first method. Thus if the λ_i are taken equal to their own cofactors in the determinant

$$\begin{vmatrix} 1 & -1 & \lambda_1 \\ 2 & 3 & \lambda_2 \\ 3 & 1 & \lambda_3 \end{vmatrix}$$

then $$\lambda_1 \begin{pmatrix} 1 \\ -1 \end{pmatrix} + \lambda_2 \begin{pmatrix} 2 \\ 3 \end{pmatrix} + \lambda_3 \begin{pmatrix} 3 \\ 1 \end{pmatrix} = \begin{pmatrix} 0 \\ 0 \end{pmatrix}$$

for if the elements of one column of a determinant are multiplied by the cofactors of a different column and summed, the result is zero.

A noteworthy fact that emerges from the preceding discussion is that for an $(n + 1) \times n$ system of equations with rank n, once the signs σ_i are known, the minimax solution can be obtained by solving a consistent set of $n + 1$ linear equations in $n + 1$ unknowns. In problems of this type which arise from the approximation of continuous functions on an interval, the signs can usually be determined easily without solving any linear equations. In the general case that we are discussing here, however, the signs require a computation separate from that for the x_i and ϵ.

It is interesting that the signs of the residuals are the same for the minimax and least-squares solutions when $m = n + 1$. (We continue to assume that our matrix is of rank n.) Since this is true for a wide class of norms besides the least-squares norm, we shall dwell briefly upon this assertion. The discussion requires the notion of a hyperplane. A *hyperplane* in a vector space E is a set of points of the form

$$H = \{x \in E : f(x) = c\}$$

where c is a constant and f is a real-valued nonzero linear function. (In the case of a Banach space we require also that f be continuous.) Thus in the space R_{n+1} a hyperplane consists of all $(n + 1)$-tuples u for which $\langle u,f \rangle = c$, f being a fixed $(n + 1)$-tuple and c a constant. Now when we solve approximately a system of equations

$$\sum_{j=1}^{n} A_j^i x_j = b_i \qquad (i = 1, \ldots, n + 1)$$

we are attempting to make the vector

$$r = \sum_{j=1}^{n} x_j A_j - b$$

as small as possible in norm. The points of the form r lie on a hyperplane, a fact which we state formally as follows.

Lemma. *If the set of vectors $\{A_1, \ldots, A_n\}$ is independent in R_{n+1} and if b is a fixed element of R_{n+1}, then the set of points*

$$\{\sum_{j=1}^{n} x_j A_j - b : x_j \text{ real}\}$$

is a hyperplane.

Proof. By the Gram-Schmidt theorem (Chap. 1, Sec. 4) we may select a nonzero vector u orthogonal to $A_1, \ldots, A_n$. Put $c = -\langle u,b \rangle$. If $r = \sum x_j A_j - b$ then clearly $\langle u,r \rangle = c$. On the other hand, suppose that $\langle u,r \rangle = c$. Since $\{A_1, \ldots, A_n, u\}$ is a basis for R_{n+1}, an equation $r + b = \sum x_j A_j + x_0 u$ is possible. Taking the inner product of both sides of this equation with u yields $0 = x_0 \langle u,u \rangle$, whence $x_0 = 0$. ∎

Another proof of this theorem may be based upon the following idea. By the theorem on existence of best approximations (Chap. 1, Sec. 6), there exists a vector y for which the Euclidean norm $\| r(y) \|$ is a minimum. We can show that $r(y)$ is orthogonal to each vector A_j. Indeed, from the definition of y, $\| r(y) - \lambda A_j \|^2 \geq \| r(y) \|^2$, whence $-2\lambda \langle r(y), A_j \rangle + \lambda^2 \| A_j \|^2 \geq 0$. If this is to be true for all real λ, then $\langle r(y), A_j \rangle = 0$. Since $r(x) + b$ is a linear combination of the vectors A_j, it follows that $\langle r(y), r(x) + b \rangle = 0$ for all x. Thus for all x,

$$\langle r(y), r(x) \rangle = -\langle r(y), b \rangle$$

If $r(y) \neq 0$, then this shows that all vectors $r(x)$ lie on a certain hyperplane, proving half of the theorem. Since the above equation is true when $x = y$, we find that $-\langle r(y), b \rangle = \| r(y) \|^2$.

In the next theorem we require the concept of a *monotone norm* on R_n. Such a norm has the property that $\| x \| \leq \| y \|$ whenever the vectors $x = [x_1, \ldots, x_n]$ and $y = [y_1, \ldots, y_n]$ are related by the inequalities $| x_i | \leq | y_i |$ for $i = 1, \ldots, n$. All the norms

$$\| x \|_p = \sqrt[p]{\sum | x_i |^p} \qquad (1 \leq p \leq \infty)$$

have this property. But the norm in R_2

$$\| x \| = \max \{2 | x_1 + x_2 |, | x_1 - x_2 |\}$$

is a simple example of one which is not monotone.

Theorem. *Let H be a hyperplane in R_n. The points of H which minimize two different monotone norms have components which agree in sign (or may be so chosen in the event of nonunicity).*

Proof. If $0 \in H$, the theorem is trivial, and we therefore assume the contrary. Let $H = \{x \in R_n : \langle x,u \rangle = c\}$. By changing u to $-u$ if necessary, we may assume that $c > 0$. Let x be a point of H which minimizes a monotone norm $\| x \|$. By putting $x'_i = | x_i | \operatorname{sgn} u_i$ we obtain a point x' whose components agree in sign with those of u. Thus $\langle x',u \rangle \geq \langle x,u \rangle = c > 0$, and the number $\theta = c/\langle x',u \rangle$ lies in the interval $(0,1]$. By the monotonicity of the norm, $\| \theta x' \| \leq \| x' \| \leq \| x \|$. Since $\langle \theta x',u \rangle = c$, $\theta x'$ lies on H and minimizes the norm. ■

From the preceding theorem we see that the signs σ_i needed to solve the minimax problem for $n + 1$ equations in n unknowns can be obtained by solving the system in the least-squares sense and using for σ_i the sign of the ith least-squares residual. (This statement remains true for other monotone norms besides the Euclidean, but is of less practical significance.) Carrying this idea one step further, we find that the number ϵ in system (1) can also be determined. We summarize our findings as follows.

Theorem. *Let y be the least-squares solution of a system of $n + 1$ linear equations in n unknowns, $r_i(x) = 0$. Assume that the system is of rank n. Then the minimax solution is the exact solution of the system $r_i(x) = \sigma_i\epsilon$, where $\sigma_i = \operatorname{sgn} r_i(y)$ and $\epsilon = \sum r_i^2(y)/\sum | r_i(y) |$.*

Proof. From an earlier lemma (page 40) the points $r(x) = \sum x_jA_j - b$ fill out a hyperplane H in R_{n+1}. Actually the remarks following that lemma showed that $H = \{z\colon \langle z,r(y)\rangle = \langle r(y),r(y)\rangle\}$. If we define a point z with components $z_i = \sigma_i\epsilon$, then $z \in H$ because

$$\langle z,r(y)\rangle = \sum z_ir_i(y) = \epsilon\sum \sigma_ir_i(y) = \epsilon\sum | r_i(y) | = \sum r_i^2(y)$$

On the other hand, no point of H is closer to 0 than z in the Tchebycheff norm. For if $\| u \|_T < \| z \|_T$, then $\langle u,r(y)\rangle = \sum u_ir_i(y) \leq \max | u_i | \sum | r_i(y) | < \max | z_i | \sum | r_i(y) | = \epsilon \sum | r_i(y) | = \sum r_i^2(y)$, so that u does not lie on H. ■

As an illustration of this theorem, we consider the system

$$\begin{aligned} x_1 - x_2 &= 7 \\ 2x_1 + 3x_2 &= 5 \\ 3x_1 + x_2 &= -1 \end{aligned}$$

which we solved earlier, using the method of de La Vallée Poussin. In order to obtain the least-squares solution we may minimize the function

$$\phi(x_1,x_2) = (x_1 - x_2 - 7)^2 + (2x_1 + 3x_2 - 5)^2 + (3x_1 + x_2 + 1)^2$$

Setting the partial derivatives of ϕ equal to zero, we have

$$\begin{aligned} 14x_1 + 8x_2 &= 14 \\ 8x_1 + 11x_2 &= 7 \end{aligned}$$

whence $x_1 = \frac{49}{45}$ and $x_2 = -\frac{7}{45}$. The residual vector corresponding to this least-squares solution turns out to be $(-\frac{259}{45}, -\frac{148}{45}, \frac{185}{45})$, and the number ϵ turns out to be $\frac{37}{8}$. The minimax solution of our system is therefore the exact solution of the system

$$\begin{aligned} x_1 - x_2 - 7 &= -\tfrac{37}{8} \\ 2x_1 + 3x_2 - 5 &= -\tfrac{37}{8} \\ 3x_1 + x_2 + 1 &= \tfrac{37}{8} \end{aligned}$$

Of course, the minimax residual vector will be $(-\frac{37}{8}, -\frac{37}{8}, \frac{37}{8})$. The minimax solution itself is $(\frac{3}{2}, -\frac{7}{8})$.

Problems

1. A *Tchebycheff point* on the hyperplane $\{x: \langle u,x \rangle = c\}$ is any point for which $\| x \|_T$ is a minimum. Show that the point whose components are $x_i = \epsilon \operatorname{sgn} u_i$ with $\epsilon = c/\sum | u_i |$ is such a point.
2. The Tchebycheff point on the hyperplane in Prob. 1 is unique if and only if $u_i \neq 0$ for all i.
3. Prove that the point on the hyperplane $\{x: \langle u,x \rangle = c\}$ for which $\| x \|_1 = \sum | x_i |$ is a minimum may be defined by $x_j = c/u_j$ if j is the index for which $| u_j |$ is a maximum, and $x_i = 0$ for all other indices.
4. The point on the hyperplane of Prob. 3 for which $\| x \|_1$ is minimum is unique if and only if for some j, $| u_j | > | u_i |$ for all $i \neq j$.

*5. Describe the point on the hyperplane of Prob. 1 for which $\| x \|_p = (\sum | x_i |^p)^{1/p}$ is a minimum.

6. The Tchebycheff solution of an $(n + 1) \times n$ system of equations with rank n is unique if and only if the least-squares residual vector has no zero components.
7. If the Tchebycheff solution of the system $\sum_{j=1}^{n} A_j{}^i x_j = b_i$ $(i = 1, \ldots, n + 1)$ is not unique, then some set of n vectors selected from $\{A^1, \ldots, A^{n+1}\}$ is linearly dependent.

5. Pólya's Algorithm

In this section we consider the first of several methods for solving the general inconsistent system of equations

$$\sum_{j=1}^{n} A_j{}^i x_j = b_i \qquad (i = 1, \ldots, m)$$

in the *minimax* or *Tchebycheff* sense. It was pointed out in Sec. 3 that for this purpose we may discard all but $n + 1$ of the given equations, the $n + 1$ to be retained not generally being known at the start. The algorithm now to be described rests upon an idea of Pólya and may be used for determining this crucial set of $n + 1$ equations.

We recall the family of norms which were defined by the equation

$$\| v \|_p = \Big(\sum_{i=1}^{m} | v_i |^p\Big)^{1/p} \qquad (p \geq 1)$$

for any vector $v = (v_1, \ldots, v_m)$ of R_m. For a fixed vector v, the numbers $\| v \|_p$ converge to $\| v \|_T = \max_i | v_i |$ as $p \to \infty$. A proof of this fact is outlined in Prob. 2. Actually the convergence is monotonically downward. Now let us denote by $v^{(p)}$ the point of the form $\sum x_j A_j - b$ for which $\| v \|_p$ is a minimum, and similarly $v^{(T)}$. By this definition and the monotonicity just mentioned we have

$$(1) \qquad \| v^{(T)} \|_T \leq \| v^{(p)} \|_T \leq \| v^{(p)} \|_p \leq \| v^{(T)} \|_p$$

Letting $p \uparrow \infty$, we have $\| v^{(T)} \|_p \downarrow \| v^{(T)} \|_T$, and consequently $\| v^{(p)} \|_T \to \| v^{(T)} \|_T$. Thus for large values of p, $v^{(p)}$ is a good substitute for $v^{(T)}$. Note that we are *not* saying that $v^{(p)} \to v^{(T)}$. Without some further clarification this statement would be meaningless anyway, because $v^{(T)}$ need not be unique. We can easily see that if $v^{(T)}$ is unique, then $v^{(p)} \to v^{(T)}$. Indeed, the points $v^{(p)}$ are bounded since, for example, $\| v^{(p)} \|_T \leq \| v^{(T)} \|_1$. Thus from the family $\{v^{(p)}: p \geq 1\}$ we can extract a convergent sequence of points $v^{(p_1)}, v^{(p_2)}, \ldots$ $(p_k \to \infty)$. Now for the limit point v of such a sequence we have, using the continuity of $\| \cdot \|_T$ and inequality (1) above,

$$\| v \|_T = \| \lim_k v^{(p_k)} \|_T = \lim_k \| v^{(p_k)} \|_T = \| v^{(T)} \|_T$$

Since $v^{(T)}$ is unique, $v = v^{(T)}$. Since this is true for *any* convergent sequence from $\{v^{(p)}\}$, we have $\lim v^{(p)} = v^{(T)}$.

It is true, but we do not stop to prove it, that $v^{(p)}$ converges to one of the points $v^{(T)}$, even if the latter is not unique.

One algorithm which is suggested by these remarks, then, is simply to calculate for successively higher values of p the points $v^{(p)}$, and to extrapolate numerically for the limit vector. In practice, we can usually do much better by using the points $v^{(p)}$ only for determining some qualitative information about $v^{(T)}$, and then solving precisely for $v^{(T)}$. In fact, for large p, we may expect (because of the results of Sec. 3) that $v^{(p)}$ will exhibit $n + 1$ components nearly equal and maximum in magnitude. This tells us which $n + 1$ equations to retain from the original set of m and also what the signs σ_i of the residuals should be. We would then be in a position to use the methods of the preceding section. The method breaks down if by accident there are more than $n + 1$ equal maximum residuals in magnitude at the Tchebycheff solution.

Throughout the discussion we have referred to the residual vectors $v \in R_m$ rather than to the coefficient vectors $x \in R_n$. If the matrix of coefficients A_j^i

is of full rank (rank n), then x is uniquely determined by v via the equations

$$v_i = \sum_{j=1}^{n} A_j^i x_j - b_i$$

For the numerical work, it is convenient to minimize the function

$$\phi(x) = \sum_{i=1}^{m} \left| \sum_{j=1}^{n} A_j^i x_j - b_i \right|^{2p}$$

since the minima of $\| v \|_{2p}$ and $\| v \|_{2p}^{2p}$ occur at the same place. The even exponent is used to facilitate differentiation. The minimum may be sought by the method of steepest descent, or by solving the algebraic equations $\partial\phi/\partial x_j = 0$ by Newton's method.

As an illustration, we give a system of six equations in two unknowns, and show the approximate solutions for $p = 2, 4, 6, 40, 100, 400$ and the Tchebycheff solution, $p = \infty$. The example has been constructed to indicate that the critical set of $n + 1$ rows may not be discernible for the very lowest values of p.

$$\begin{aligned} x_1 + x_2 &= 3 \\ x_1 - x_2 &= 1 \\ x_1 + 2x_2 &= 7 \\ 2x_1 + 4x_2 &= 11.1 \\ 2x_1 + x_2 &= 6.9 \\ 3x_1 + x_2 &= 7.2 \end{aligned}$$

	$p = 1$	$p = 2$	$p = 3$	$p = 20$	$p = 50$	$p = 200$	$p = \infty$
x_1	2.0741	2.0466	2.0390	2.0141	2.0055	2.0014	2.0000
x_2	1.8078	1.9207	1.9498	1.9938	1.9977	1.9994	2.0000
r_1	0.88196	0.96730	0.98887	1.0078	1.0032	1.0008	1.0000
r_2	−0.73373	−0.87411	−0.91082	−0.97955	−0.99214	−0.99809	−1.0000
r_3	−1.3102	−1.1120	−1.0613	−0.99855	−0.99911	−0.99977	−1.0000
r_4	0.27961	0.67601	0.77742	0.90290	0.90178	0.90045	0.90000
r_5	−0.94392	−0.88610	−0.87211	−0.87810	−0.89125	−0.89786	−0.90000
r_6	0.83020	0.86049	0.86692	0.83602	0.81428	0.80349	0.80000
$\phi^{1/2p}$	2.1659	1.4452	1.2610	1.0257	1.0102	1.0025	1.0000

Problems

1. *Jensen's theorem.* For fixed $v = (v_1,\ldots,v_m)$, the function $\| v \|_p = \sqrt[p]{\sum_{i=1}^{m} | v_i |^p}$ is a decreasing function of p. *Hint*: If $\| v \|_p = 1$, then $| v_i | \leq 1$, and consequently $| v_i |^q \leq | v_i |^p$ when $q \geq p$. Use homogeneity of the norm to remove the restriction $\| v \|_p = 1$.
2. For fixed v, $\| v \|_p \downarrow \| v \|_T = \max | v_i |$ as $p \uparrow \infty$. *Hint*: If $\| v \|_T = 1$, then $1 \leq \sqrt[p]{\sum | v_i |^p} \leq \sqrt[p]{n} \downarrow 1$. Use Prob. 1 and the homogeneity of the norm.
3. Prove that $\langle u,v \rangle \leq \| u \|_1 \| v \|_T$. This inequality and the Cauchy-Schwarz inequality (page 13) are special cases of the inequality in the next problem.
4. *Hölder's inequality.* If $p \geq 1$ and $1/p + 1/q = 1$, then $\langle u,v \rangle \leq \| u \|_p \| v \|_q$. *Hint*: Assume $\| u \|_p = \| v \|_q = 1$, and remove this restriction at the end by homogeneity. Prove first that for $x > 0$, $x^{1/p} \leq x/p + 1/q$. Then set $x = | u_i |^p | v_i |^{-q}$, multiply through by $| v_i |^q$, and sum.
5. In a linear space E, let $\phi_1, \phi_2, \ldots$ be a sequence of norms such that $\lim_{n\to\infty} \phi_n(x)$ exists for every x. Is either $\lim_{n\to\infty} \phi_n$ or $\sup_n \phi_n$ a norm on E?
6. Let $\phi_0, \phi_1, \ldots$ be a sequence of norms such that for each x, $\phi_n(x) \downarrow \phi_0(x)$. If $x^{(n)}$ denotes a point on a fixed finite-dimensional manifold for which $\phi_n(x^{(n)})$ is least, what can you say about $\lim \phi_n(x^{(n)})$ and $\lim x^{(n)}$?

6. The Ascent Algorithm

In this section we consider the general ascent method which was illustrated with $n = 1$ in Sec. 2 (page 31). In order to make the calculations proceed smoothly, we are going to assume about the matrix $(A_j{}^i)$ that its rows satisfy a rather strong requirement of nondegeneracy called the *Haar condition*: A set of vectors in n space is said to satisfy the Haar condition if every set of n of them is linearly independent. Expressed otherwise, each selection of n vectors from such a set is a basis for n space.

Exchange Theorem. *Let $\{A^0,\ldots,A^{n+1}\}$ be a set of vectors in n space satisfying the Haar condition. If 0 lies in the convex hull of $\{A^0,\ldots,A^n\}$, then there is an index $j \leq n$ such that this condition remains true when A^j is replaced by A^{n+1}.*

Proof. By hypothesis there exist constants $\theta_i \geq 0$ such that $0 = \sum_{i=0}^{n} \theta_i A^i$ and $\sum_{i=0}^{n} \theta_i = 1$. The Haar condition would be violated if any θ_i were zero; hence $\theta_i > 0$. We may therefore solve for any A^j, obtaining $A^j = \sum_{\substack{i=0\\ i\neq j}}^{n} \frac{-\theta_i}{\theta_j} A^i$.

Since $\{A^0,\ldots,A^n\}$ spans n space, it is possible to write $A^{n+1} = \sum_{i=0}^{n} \lambda_i A^i$ for appropriate λ_i. Hence

$$\begin{aligned} 0 &= A^{n+1} - \sum_{i=0}^{n} \lambda_i A^i \\ &= A^{n+1} - \lambda_j A^j - \sum_{\substack{i=0 \\ i\neq j}}^{n} \lambda_i A^i \\ &= A^{n+1} - \lambda_j \sum_{\substack{i=0 \\ i\neq j}}^{n} \frac{-\theta_i}{\theta_j} A^i - \sum_{\substack{i=0 \\ i\neq j}}^{n} \lambda_i A^i \\ &= A^{n+1} + \sum_{\substack{i=0 \\ i\neq j}}^{n} \left(\frac{\lambda_j \theta_i}{\theta_j} - \lambda_i \right) A^i \end{aligned}$$

Now if j is selected so that $\lambda_j\theta_i/\theta_j - \lambda_i \geq 0$, then our final equation above will express 0 as a nonnegative linear combination of $A^0, \ldots, A^{n+1}$ with A^j not appearing, which would suffice to prove that 0 is in the convex hull of these points. Our requirement on j is that $\lambda_j/\theta_j \geq \lambda_i/\theta_i$; in other words we must select j so that λ_j/θ_j is the largest possible of these ratios. (This index j is unique, for if there were two such largest ratios λ_j/θ_j, then one of the coefficients in the above equation would vanish, contradicting the Haar condition.) ■

We now proceed to a description of the algorithm. We seek a point where the function

$$\Delta(x) = \max_{1\leq i\leq m} | r_i(x) | = \max_{1\leq i\leq m} | \langle A^i, x\rangle - b_i |$$

achieves its minimum value. It is assumed that the Haar condition is satisfied by the set of vectors $\{A^1, \ldots, A^m\}$. The basic idea of this algorithm is to calculate the minimax solutions of a succession of subsystems, each comprising $n + 1$ equations. By the theorem on page 36, the solution of one of these subsystems is the point sought. On the other hand, there can be but a finite number of these subsystems, and this simple observation will be the basis for a proof that the algorithm is effective.

In each computing cycle, we will have a set of $n + 1$ indices $J = \{i_0, \ldots, i_n\}$ and a vector of signs $\sigma = \{\sigma_0, \ldots, \sigma_n\}$ such that

(1) $$0 \in \mathcal{H}\{\sigma_0 A^{i_0}, \ldots, \sigma_n A^{i_n}\}$$

We then solve the following system of $n + 1$ linear equations to determine a vector $y = [y_1, \ldots, y_n]$ and a number e:

(2) $$\sigma_j r_{i_j}(y) = e \qquad (j = 0, \ldots, n)$$

In order to ensure that $e > 0$, we may change the signs of all σ_j without losing property (1). As we have already seen on page 37, conditions (1) and (2) imply that y is a minimax solution of the system

$$\langle A^{i_j}, y \rangle = b_{i_j} \qquad (j = 0, \ldots, n)$$

If $e = \Delta(y)$, then by the characterization theorem of page 35, y is a minimax solution of the original system of m equations. In the contrary case, there exists at least one index α (not in J) such that $|r_\alpha(y)| > e$. Ordinarily we would take α so that $|r_\alpha(y)| = \Delta(y)$, but this is not necessary. Now let $\mu = \operatorname{sgn} r_\alpha(y)$. Using the exchange theorem, we replace one of the vectors $\sigma_0 A^{i_0}, \ldots, \sigma_n A^{i_n}$ by μA^α in such a way that the origin remains in the convex hull of the set—property (1). The situation presented now is the same as that in the beginning, and we proceed as before.

The necessity for calculating y and e forces us to make some assumption about the given data, a convenient one being that the matrix

$$\begin{bmatrix} \sigma_0 & A_1{}^{i_0} & \cdots & A_n{}^{i_0} \\ \cdots & \cdots & \cdots & \cdots \\ \sigma_n & A_1{}^{i_n} & \cdots & A_n{}^{i_n} \end{bmatrix}$$

shall be nonsingular. We will encounter cases later when this condition can be verified by *a priori* considerations.

The computations of the algorithm break down only when $e = \Delta(y)$, and this equation signifies that y is a solution. We have already remarked that only a finite number of sets J exist. All that remains to be shown, then, is that the computations do not "cycle," i.e., return infinitely many times to the *same* subset J. This proof will be accomplished by showing that the number e is a function of J which strictly increases from step to step. To this end, suppose for simplicity of notation that at a certain step J is $\{1, \ldots, n + 1\}$ and that α is $n + 2$. Suppose further that in the next step J becomes $J' = \{2, \ldots, n + 2\}$. Let y' and e' be the values of y and e corresponding to the new set J'. By the choice of α, $|r_2(y)| < |r_{n+2}(y)|$, while $|r_2(y')| = |r_{n+2}(y')|$. Hence $y - y' \neq 0$. Since $\langle \sigma_i A^i, y - y' \rangle = \sigma_i r_i(y) - \sigma_i r_i(y') = e - e'$ for $i = 2, \ldots, n + 1$, the Haar conditions imply that $e - e' \neq 0$. Now $\langle \sigma_{n+2} A^{n+2}, y - y' \rangle = \sigma_{n+2} r_{n+2}(y) - \sigma_{n+2} r_{n+2}(y') > e - e'$. If $e - e' > 0$, then $\langle \sigma_i A^i, y - y' \rangle > 0$ for $i = 2, \ldots, n + 2$, contradicting the fact that $0 \in \mathcal{H}\{\sigma_i A^i : 2 \leq i \leq n + 2\}$. [Recall the theorem on linear inequalities (Chap. 1, Sec. 5).] Hence we may conclude that $e - e' < 0$, as was to be proved.

One small point may require clarification: How are the initial set J and vector σ determined? J may be taken arbitrarily, and then we may find a nontrivial solution to the equation

$$\sum_{j=0}^{n} \theta_j A^{i_j} = 0$$

putting then $\sigma_j = \operatorname{sgn} \theta_j$.

A convenient arrangement of the computations is as follows. Observe first that the equations (2), viz., $e = \sigma_j r_{i_j}(y) = \sigma_j[\langle A^{i_j}, y\rangle - b_{i_j}]$, may be written as $-\sigma_j e + \langle A^{i_j}, y\rangle = b_{i_j}$ and thence in matrix notation as

$$\begin{bmatrix} \sigma_0 & A_1^{i_0} & \cdots & A_n^{i_0} \\ \cdot & \cdot & \cdot & \cdot \\ \sigma_n & A_1^{i_n} & \cdots & A_n^{i_n} \end{bmatrix} \begin{bmatrix} -e \\ y_1 \\ \cdot \\ \cdot \\ \cdot \\ y_n \end{bmatrix} = \begin{bmatrix} b_{i_0} \\ \cdot \\ \cdot \\ \cdot \\ b_{i_n} \end{bmatrix}$$

If we assume that the matrix A_J on the left has an inverse $C = (C_j^i)$, then we may write

$$\begin{bmatrix} -e \\ y_1 \\ \cdot \\ \cdot \\ \cdot \\ y_n \end{bmatrix} = \begin{bmatrix} C_0^0 & \cdots & C_n^0 \\ \cdot & \cdot & \cdot \\ C_0^n & \cdots & C_n^n \end{bmatrix} \begin{bmatrix} b_{i_0} \\ \cdot \\ \cdot \\ \cdot \\ b_{i_n} \end{bmatrix}$$

Since C is the inverse of A_J, we have

$$\begin{bmatrix} C_0^0 & \cdots & C_n^0 \\ \cdot & \cdot & \cdot \\ C_0^n & \cdots & C_n^n \end{bmatrix} \begin{bmatrix} \sigma_0 & A_1^{i_0} & \cdots & A_n^{i_0} \\ \cdot & \cdot & \cdot & \cdot \\ \sigma_n & A_1^{i_n} & \cdots & A_n^{i_n} \end{bmatrix} = \begin{bmatrix} 1 & \cdots & 0 \\ \cdot & \cdot & \cdot \\ 0 & \cdots & 1 \end{bmatrix}$$

From this it is evident that $\sum_{j=0}^{n} \sigma_j C_j^0 = 1$ and $\sum_{j=0}^{n} C_j^0 A^{i_j} = 0$. Thus the numbers $\sigma_j C_j^0$ are the coefficients needed to express the fact that 0 lies in the convex hull of the points $\sigma_j A^{i_j}$. These coefficients enter into the calculations relating to the exchange theorem. From the proof of that theorem we see that we must express μA^α as a linear combination of $\sigma_0 A^{i_0}, \ldots, \sigma_n A^{i_n}$. If we set $A^\alpha = \sum_{j=0}^{n} \lambda_j A^{i_j}$, then the coefficients λ_j may be obtained by solving the matrix equation

$$(\lambda_0, \ldots, \lambda_n) \begin{bmatrix} \sigma_0 & A_1^{i_0} & \cdots & A_n^{i_0} \\ \cdot & \cdot & \cdot & \cdot \\ \sigma_n & A_1^{i_n} & \cdots & A_n^{i_n} \end{bmatrix} = (\mu, A_1^\alpha, \ldots, A_n^\alpha)$$

of which the solution is

$$(\lambda_0, \ldots, \lambda_n) = (\mu, A_1{}^\alpha, \ldots, A_n{}^\alpha) \begin{bmatrix} C_0{}^0 & \cdots & C_n{}^0 \\ \cdot\cdot\cdot & \cdot\cdot\cdot & \cdot\cdot\cdot \\ C_0{}^n & \cdots & C_n{}^n \end{bmatrix}$$

Since $\mu A^\alpha = \sum_{j=0}^{n} (\mu\sigma_j\lambda_j)(\sigma_j A^{i_j})$, the ratios to be computed in the exchange theorem are $\mu\sigma_j\lambda_j/\sigma_j C_j{}^0 \equiv \mu\lambda_j/C_j{}^0$. The number β is chosen as the index of the largest of these ratios.

Before summarizing the algorithm in a flow diagram, we should observe that in proceeding from one cycle of the calculation to the next, only one row of the matrix A_J changes. The effect on C may be predicted by use of the following theorem.

Theorem. *Let A be a nonsingular matrix and $C_1, \ldots, C_n$ the columns of its inverse. Let $\bar{A}$ be the matrix obtained by replacing the βth row of A by a vector v. If $\lambda \equiv \langle v, C_\beta \rangle \neq 0$, then $\bar{A}$ is nonsingular, and the columns of its inverse are given by the rules $\bar{C}_\beta = \lambda^{-1} C_\beta$ and $\bar{C}_j = C_j - \langle v, C_j \rangle \bar{C}_\beta$ $(j \neq \beta)$.*

Proof. In order to verify that $\bar{A}\bar{C} = I$, we compute the inner product of $\bar{A}^i$ (the ith row of $\bar{A}$) with $\bar{C}_j$. There are four cases. In case 1, $i = \beta$ and $j = \beta$. Then $\langle \bar{A}^\beta, \bar{C}_\beta \rangle = \langle v, \lambda^{-1} C_\beta \rangle = \lambda^{-1} \langle v, C_\beta \rangle = 1$. In case 2, $i \neq \beta$ and $j = \beta$. Then $\langle \bar{A}^i, \bar{C}_\beta \rangle = \langle A^i, \lambda^{-1} C_\beta \rangle = \lambda^{-1} \langle A^i, C_\beta \rangle = 0$. In case 3, $i = \beta$ and $j \neq \beta$. Then $\langle \bar{A}^i, \bar{C}_j \rangle = \langle v, C_j - \langle v, C_j \rangle \bar{C}_\beta \rangle = \langle v, C_j \rangle - \langle v, C_j \rangle \lambda^{-1} \langle v, C_\beta \rangle = 0$. In case 4, $i \neq \beta$ and $j \neq \beta$. Then $\langle \bar{A}^i, \bar{C}_j \rangle = \langle A^i, C_j - \langle v, C_j \rangle \bar{C}_\beta \rangle = \langle A^i, C_j \rangle - \langle v, C_j \rangle \lambda^{-1} \langle A^i, C_\beta \rangle = \langle A^i, C_j \rangle = \delta_{ij}$. ■

In the present algorithm, we shall replace the row $(\sigma_\beta, A_1{}^{i_\beta}, \ldots, A_n{}^{i_\beta})$ by a new row $(\mu, A_1{}^\alpha, \ldots, A_n{}^\alpha)$. We shall need the number λ which is the inner product of the new row with the βth column of C, viz., $\mu C^0 + \sum_{j=1}^{n} A_j{}^\alpha C_\beta{}^j$. But this is the number λ_β previously computed. In the flow diagram it simplifies matters to let y_0 denote the number $-e$. Also, we have not given the details for steps (2) and (4) in the first box, these being standard problems in linear algebra. Finally, the shorthand $x \rightarrow w$ denotes that the quantity x replaces the quantity w, or that the quantity x is stored in the memory at the cell labeled w.

The flow diagram for minimizing the function $\Delta(x) = \max_{1 \leq i \leq m} | \langle A^i, x \rangle - b_i |$ by the ascent method is shown on page 50.

Getting started:

(1) Read A_j^i, b_i, $J = \{i_0, \ldots, i_n\}$.

(2) Solve $\sum_{j=0}^{n} \theta_j A^{i_j} = 0$, $\sum_{j=0}^{n} \theta_j = 1$.

(3) $\operatorname{sgn} \theta_j \to \sigma_j \qquad (j = 0, \ldots, n)$.

(4)
$$\begin{bmatrix} \sigma_0 & A_1^{i_0} & \cdots & A_n^{i_0} \\ \cdots & \cdots & \cdots & \cdots \\ \sigma_n & A_1^{i_n} & \cdots & A_n^{i_n} \end{bmatrix}^{-1} \to \begin{bmatrix} C_0^0 & \cdots & C_n^0 \\ \cdots & \cdots & \cdots \\ C_0^n & \cdots & C_n^n \end{bmatrix}.$$

(1) $\sum_{j=0}^{n} C_j^k b_{i_j} \to y_k \qquad (k = 0, \ldots, n)$.

(2) $\sum_{k=1}^{n} A_k^i y_k - b_i \to r_i \qquad (i = 1, \ldots, m)$.

(3) Select α so that $| r_\alpha |$ is maximum.

(4) Print $y_0, \ldots, y_n, r_1, \ldots, r_m, i_0, \ldots, i_n, \alpha$.

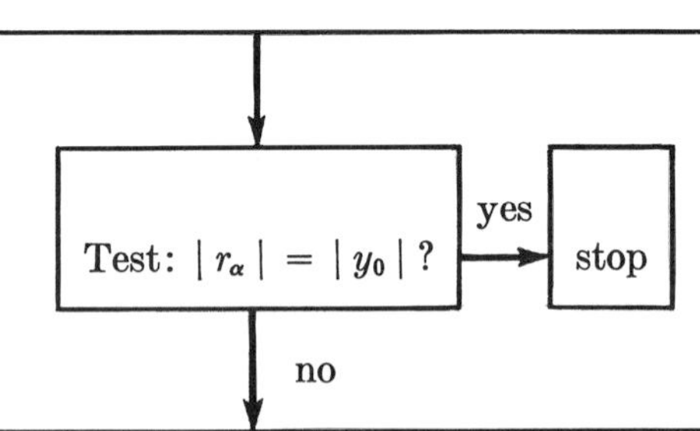

(1) $\operatorname{sgn} r_\alpha \to \mu$.

(2) $\mu C_s^0 + \sum_{j=1}^{n} A_j^\alpha C_s^j \to \lambda_s \qquad (s = 0, \ldots, n)$.

(3) Select β so that $\mu\lambda_\beta / C_\beta^0$ is a maximum.

(4) $C_\beta^k / \lambda_\beta \to C_\beta^k \qquad (k = 0, \ldots, n)$.

(5) $C_j^k - \lambda_j C_\beta^k \to C_j^k \qquad (k = 0, \ldots, n; j = 0, \ldots, n; j \neq \beta)$.

(6) $\alpha \to i_\beta$.

Problems

1. Prove that the index j in the statement of the exchange theorem is unique.
2. Prove, under the hypotheses of the exchange theorem, that there is an index k such that $-A^k$ is a nonnegative linear combination of $A^0, \ldots, A^{k-1}, A^{k+1}, \ldots, A^{n+1}$.
3. If $\{A^1, \ldots, A^n\}$ is a dependent set in R_n, then for all sufficiently small $\epsilon \neq 0$, the set $\{A^1 + \epsilon\delta^1, \ldots, A^n + \epsilon\delta^n\}$ is independent. Here δ^i is the vector $(0,\ldots,1,0,\ldots,0)$ with 1 as the ith component.
4. How can the idea in Prob. 3 be exploited to carry out the ascent algorithm when the Haar conditions fail?
5. Prove or disprove: If f has a minimum and if g is convex, then $g(f)$ has a minimum at the same location.

7. The Descent Algorithm

In this section we consider the general descent method which was illustrated with $n = 1$ on page 32. For variety, the discussion is directed toward minimizing the function

$$\delta(x) = \max_{1 \leq i \leq m} \{\langle A^i, x\rangle - b_i\}$$

The manner in which this includes the Tchebycheff problem was discussed on page 29. The general idea of the method is to proceed downward from vertex to vertex on the hypersurface in $(n + 1)$ space whose equation is

$$z = \delta(x)$$

A special technique is used to arrive at a first vertex.

Let x^0 be any initial vector. By renumbering the residuals for convenience in the description, we may assume that

$$\delta(x^0) = r_1(x^0) = r_2(x^0) = \cdots = r_k(x^0) > r_{k+i}(x^0) \qquad (i \geq 1)$$

We seek a direction to move from x^0 in which the residuals $r_1, \ldots, r_k$ will decrease at an equal rate. We proceed in this direction until a $(k + 1)$st residual function rises to meet the first k. By reapplying this technique we arrive in at most n steps at a *vertex*, where $n + 1$ equal maximum residuals occur. Now, the rate of change in r_i as we move from x^0 in the direction y is readily seen to be the number $\langle A^i, y\rangle$. Indeed,

$$\frac{d}{d\lambda} r_i(x^0 + \lambda y) = \frac{d}{d\lambda} [\langle A^i, x^0 + \lambda y\rangle - b_i] = \langle A^i, y\rangle$$

Hence a direction y in which residuals $r_1, \ldots, r_k$ remain equal by decreasing at a common rate may be determined by solving the equations

$$\langle A^i, y\rangle = -1 \qquad (i = 1, \ldots, k) \tag{1}$$

If we assume the Haar conditions (i.e., every set of n vectors A^i is independent), then this condition on y is easily met as long as $k \leq n$. In fact, we can take y to be of the form

$$y = \sum_{j=1}^{k} c_j A^j$$

the equations then being

$$\sum_{j=1}^{k} c_j \langle A^i, A^j \rangle = -1 \qquad (i = 1, \ldots, k)$$

We shall not stop to prove the nonsingularity of the *Gram matrix*, $\langle A^i, A^j \rangle$, but refer the reader to page 103, where this fact is proved.

Having determined a vector y with properties (1), we take our next point to be of the form $x^1 = x^0 + \lambda y$ where λ is the smallest positive coefficient for which $k + 1$ equal maximum residuals occur.

Let us assume now that a point x^0 is known where $n + 1$ equal maximum residuals occur, say,

$$\delta(x^0) = r_1(x^0) = \cdots = r_{n+1}(x^0) > r_{n+i}(x^0) \qquad (i > 1)$$

It is conceivable that x^0 is a solution. This will be the case if and only if the system of linear inequalities

$$\langle A^i, y \rangle < 0 \qquad (i = 1, \ldots, n + 1)$$

is inconsistent. If this system is consistent, then by the theorem on linear inequalities (Chap. 1, Sec. 5) 0 is *not* in the convex hull of $\{A^1, \ldots, A^{n+1}\}$. Consequently if we solve the following equations for $\theta_1, \ldots, \theta_{n+1}$,

$$\sum_{i=1}^{n+1} \theta_i A^i = 0 \qquad \sum_{i=1}^{n+1} \theta_i = 1$$

then at least one coefficient θ_i will be negative. Now, issuing from the vertex associated with the point x^0 there are a number of "edges," which are one-dimensional linear manifolds along which n residuals remain equal to one another. Not all these edges actually lie on the surface $z = \delta(x)$, however. For each set of n indices selected from $\{1, \ldots, n + 1\}$ there is an edge along which those n residuals are equal. The one remaining residual, say r_j, will generally change at a different rate. The direction of this edge will therefore be obtained by solving a system such as the following for the vector y^j:

$$\langle A^i, y^j \rangle = -1 \qquad (i = 1, \ldots, n + 1; i \neq j)$$

In the direction y^j, residual r_j will change at a rate $\langle A^j, y^j \rangle$, and this may be greater than or less than -1. If $\langle A^j, y^j \rangle < -1$, then the corresponding edge

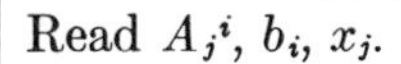

Select $M = \{i_0, \ldots, i_k\}$ such that

$r_i(x) = \delta(x)$ for $i \in M$, and

$r_i(x) < \delta(x)$ for $i \notin M$.

$k \geq n$?

yes

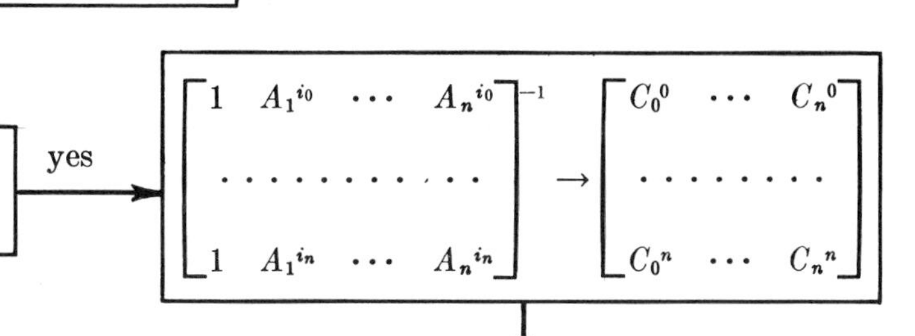

no

Solve $\sum_{p=0}^{k} c_p \langle A^{i_p}, A^{i_q} \rangle = -1$.

$\sum_{q=0}^{k} c_q A^{i_q} \rightarrow y$.

Let j be the index of the smallest

positive ratio $\lambda_j = \dfrac{\delta(x) - r_j(x)}{\langle A^j, y \rangle + 1}$.

$x + \lambda_j y \rightarrow x$.

$j \rightarrow i_{k+1}$.

$k + 1 \rightarrow k$.

$C_j^0 > 0$?

$j = 0, \ldots, n$.

yes

Print x, $\delta(x)$,

r_i, M. Stop.

no

Select p so that C_p^0 is a minimum.

$C_p^j / C_p^0 \rightarrow y_j$.

Let α be the index of smallest positive ratio:

$$t_\alpha = \frac{\delta(x) - r_\alpha(x)}{\langle A^\alpha, y \rangle + 1}.$$

$x + t_\alpha y \rightarrow x$.

$\sum_{j=0}^{n} A_j^\alpha C_s^j \rightarrow \lambda_s \qquad (s = 0, \ldots, n)$.

Select β so that $\lambda_\beta / C_\beta^0$ is a maximum

$C_\beta^k / \lambda_\beta \rightarrow C_\beta^k \qquad (k = 0, \ldots, n)$.

$C_j^k - \lambda_j C_\beta^k \rightarrow C_j^k \qquad (k = 0, \ldots, n;\ j = 0, \ldots, n;\ j \neq \beta)$.

$\alpha \rightarrow i_\beta$.

lies on the surface. To compute $\langle A^j,y^j\rangle$ we write

$$0 = \langle 0,y^j\rangle = \sum_{i=1}^{n+1} \theta_i\langle A^i,y^j\rangle = \sum_{\substack{i=1\\ i\neq j}}^{n+1} -\theta_i + \theta_j\langle A^j,y^j\rangle$$

Thus
$$\langle A^j,y^j\rangle = \frac{1}{\theta_j}\sum_{\substack{i=1\\ i\neq j}}^{n+1} \theta_i = \frac{1-\theta_j}{\theta_j} = \frac{1}{\theta_j} - 1$$

Certainly one of these ratios will be less than -1 because at least one of the θ_j is negative. If j is such an index, the next point is of the form $x^0 + \lambda y^j$, where we take λ to be the smallest positive coefficient for which $n + 1$ equal maximum residuals occur. From this stage on, the computing can be streamlined by using the theorem on page 49.

The flow diagram for minimizing the function $\delta(x) = \max_{1\leq i\leq m} \{\langle A^i,x\rangle - b_i\}$ by the descent method is shown on page 53.

8. Convex Programming

The name *convex programming* has come to signify the problem of locating the minimum points of a convex function defined on a convex set. We consider one such problem of a general nature in n space. Let the convex function which is to be minimized be defined by

$$F(x) = \max_{1\leq i\leq k} r_i(x) = \max_{1\leq i\leq k} \left(\sum_{j=1}^{n} A_j{}^i x_j - b_i\right)$$

Let the convex set be defined as $K = \{x\colon G(x) \leq 0\}$ where

$$G(x) = \max_{k< i\leq m} r_i(x) = \max_{k< i\leq m} \left(\sum_{j=1}^{n} A_j{}^i x_j - b_i\right)$$

We assume that the set $\{A^1,\ldots,A^m\}$ satisfies the Haar condition and that K is nonempty. In each step of the algorithm a set J of $n+1$ indices is given, $J\subset \{1,\ldots,m\}$. At least one element of J must be in the range $\{k+1, \ldots, m\}$, and the origin must lie in the convex hull of the set $\{A^i\colon i \in J\}$. Now we compute a point x and a number λ such that

$$i \leq k \quad\text{and}\quad i \in J \quad\Rightarrow\quad r_i(x) = \lambda$$

$$i > k \quad\text{and}\quad i \in J \quad\Rightarrow\quad r_i(x) = 0$$

If $G(x) \leq 0$ and $F(x) = \lambda$, then we may stop, for x is a solution; i.e., x is a point of K for which F is a minimum. If $G(x) > 0$, then $x \notin K$, and we select an index $\alpha > k$ for which $r_\alpha(x) > 0$. If $G(x) \leq 0$ and $F(x) > \lambda$, then select an index $\alpha \leq k$ for which $r_\alpha(x) > \lambda$. Now, using the exchange theorem, we replace an element $\beta \in J$ by α in such a way that the origin lies in the convex hull of $\{A^i: i \in J'\}$ where $J' = J \cup \{\alpha\} \sim \{\beta\}$. Then we start anew with J' in place of J.

Theorem. *The above algorithm is effective in the convex programming problem: in a finite number of steps a point of K is produced for which $F(x)$ is a minimum.*

Proof. Suppose that $J = \{i_0,\ldots,i_n\}$ with $\{i_0,\ldots,i_j\} \subset \{1,\ldots,k\}$ and $\{i_{j+1},\ldots,i_n\} \subset \{k+1, \ldots, m\}$. The algorithm requires us to solve the system of linear equations:

$$\begin{bmatrix} A_1^{i_0} & \cdots & A_n^{i_0} & 1 \\ \cdots & \cdots & \cdots & \cdots \\ A_1^{i_j} & \cdots & A_n^{i_j} & 1 \\ A_1^{i_{j+1}} & \cdots & A_n^{i_{j+1}} & 0 \\ \cdots & \cdots & \cdots & \cdots \\ A_1^{i_n} & \cdots & A_n^{i_n} & 0 \end{bmatrix} \begin{bmatrix} x_1 \\ \cdot \\ \cdot \\ \cdot \\ x_n \\ \lambda \end{bmatrix} = \begin{bmatrix} b_{i_0} \\ \cdot \\ \cdot \\ \cdot \\ b_{i_n} \end{bmatrix}$$

We shall prove that the matrix on the left—call it B—is nonsingular. Suppose, on the contrary, that a vector $v = (x,\lambda) = (x_1,\ldots,x_n,\lambda) \neq 0$ exists for which $Bv = 0$. By the Haar conditions $\lambda \neq 0$. Since $0 \in \mathcal{H}\{A^i: i \in J\}$, we may write $0 = \sum_{i\in J} \theta_i A^i$ with $\theta_i \geq 0$ and $\sum \theta_i = 1$. By the Haar conditions, $\theta_i > 0$. Thus $0 = \langle 0,v\rangle = \sum \theta_i\langle A^i,x\rangle$. But this is not possible since $\langle A^i,x\rangle = -\lambda$ when $k \geq i \in J$, and $\langle A^i,x\rangle = 0$ when $k < i \in J$. The nonsingularity of B ensures that x and λ can be computed as specified in the algorithm. Now we must prove that x is a solution if $F(x) = \lambda$ and $G(x) \leq 0$. If not, then there is a point y such that $G(y) \leq 0$ and $F(y) < \lambda$. The vector $z = y - x$ satisfies then the system of inequalities

$$\langle A^i,z\rangle < 0 \qquad (k \geq i \in J)$$

$$\langle A^i,z\rangle \leq 0 \qquad (k < i \in J)$$

This leads to a contradiction by the same reasoning as above. Next we must verify that the set of indices J' has the properties of J. In particular, we must see that J' contains an index from the range $\{1,\ldots,k\}$. If not, then J had exactly one element, β, from $\{1,\ldots,k\}$, and the index α was in the range $\{k+1, \ldots, m\}$. Let y be an arbitrary point of K. Then $\langle A^i, y - x\rangle \leq 0$ when $k < i \in J$, and $\langle A^\alpha, y - x\rangle < 0$. This contradicts the property $0 \in \mathcal{H}\{A^i: i \in J'\}$. Finally, we shall prove that the value of λ increases from one step to the next. (Thus a particular subset J cannot recur, and the calculations must terminate in a finite number of steps.) With the set J', we determine x' and λ' such that

$$\langle A^i,x' \rangle - b_i = \lambda' \qquad (k \geq i \in J')$$

$$\langle A^i,x' \rangle - b_i = 0 \qquad (k < i \in J')$$

Subtraction of the analogous quantities involving x and λ yields

$$\langle A^i, x' - x \rangle = \lambda' - \lambda \qquad (k \geq i \in J \cap J')$$

$$\langle A^i, x' - x \rangle = 0 \qquad (k < i \in J \cap J')$$

$$\langle A^\alpha, x' - x \rangle < \lambda' - \lambda \qquad (\text{if } \alpha \leq k)$$

$$\langle A^\alpha, x' - x \rangle < 0 \qquad (\text{if } \alpha > k)$$

Since $0 \in \mathcal{H}\{A^i : i \in J'\}$, we conclude that $\lambda' - \lambda \geq 0$. The Haar conditions imply $\lambda' - \lambda > 0$. ∎

Problems

1. Let $\{A^0,\ldots,A^n\}$ denote a set of vectors in n space satisfying the Haar condition. Prove that $0 \in \mathcal{H}\ \{A^0,\ldots,A^n\}$ if and only if the system of inequalities $\langle A^i,z \rangle \geq 0$ has no nontrivial solution.
2. In the algorithm for convex programming, show that it is not necessary to assume that K is nonvoid, for if $K = \phi$, this fact will be disclosed in the computation.
3. Devise an algorithm for determining a best, nonnegative, minimax solution to a system of equations $\langle A^i,x \rangle = b_i$ $(i = 1,\ldots,m)$.
4. Let $h_i = \{x: \langle A^i,x \rangle \leq b_i\}$ with $\bigcap_{i=1}^{m} h_i$ empty. Prove that $\bigcap_{i \in J} h_i$ is empty for some set J of at most $n + 1$ indices. How does this relate to Prob. 2?

Chapter 3

Tchebycheff approximation by polynomials and other linear families

1. Introduction

In this chapter we consider the problem of approximating a continuous function f defined on an interval $[a,b]$ by a polynomial $P(x) = c_n x^n + c_{n-1}x^{n-1} + \cdots + c_0$. Our interest centers on approximations which minimize expressions of the form

$$(1) \qquad \max_{a \le x \le b} | f(x) - P(x) |$$

or

$$(2) \qquad \max_{1 \le i \le m} | f(x_i) - P(x_i) |$$

Some of the discussion can be given, however, for the more general approximation problem in which the monomials $1, x, x^2, \ldots, x^n$ are replaced by other fixed functions $g_0, g_1, \ldots, g_n$. In this connection we may speak of *generalized polynomials*, by which we would mean functions of the form $\sum_{i=0}^{n} c_i g_i$. Thus we shall try to encompass in our theory an approximation problem of the bizarre form

$$f(x) \approx c_1 \log x + c_2 \cos x + c_3 e^x + c_4 (x - 2)^{-1}$$

A very special case of the theory occurs when in the expression (2) above the number of points is taken equal to $n + 1$. In this case the approximation problem has an immediate explicit solution. We discuss this topic, interpolation, first; the techniques developed here will be useful in other types of approximation.

2. Interpolation

We know that a straight line having the equation $y = ax + b$ can be passed through any two points having distinct abscissas. Similarly a parabola $y =$

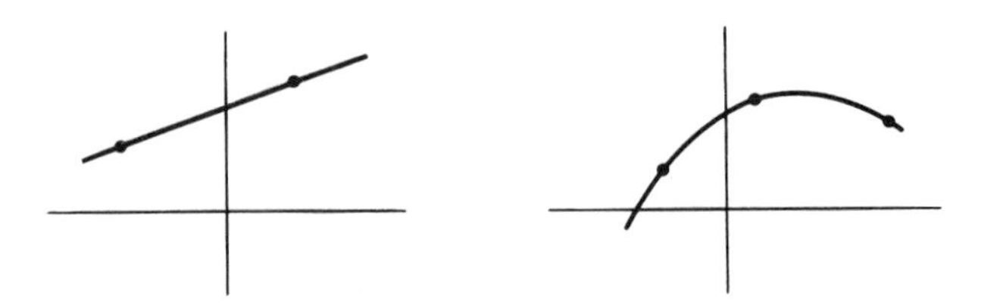

$ax^2 + bx + c$ can be made to pass through any three points having distinct abscissas. The general result along these lines is easily surmised, and takes the following form.

Interpolation Theorem. *There exists a unique polynomial of degree $\leq n$ which assumes prescribed values at $n + 1$ distinct points.*

Proof 1. Let $x_0, \ldots, x_n$ be the points and $y_0, \ldots, y_n$ the prescribed values. We seek a polynomial P such that $P(x_i) = y_i$ $(i = 0, \ldots, n)$. Since the polynomial is of degree $\leq n$, it may be expressed as $P(x) = \sum_{j=0}^{n} c_j x^j$. Hence our requirement reads now $\sum_{j=0}^{n} c_j x_i^j = y_i$ $(i = 0, \ldots, n)$. Written out in matrix form, this becomes

$$\begin{bmatrix} 1 & x_0 & x_0^2 & \cdots & x_0^n \\ \cdot & \cdot & \cdot & \cdots & \cdot \\ 1 & x_n & x_n^2 & \cdots & x_n^n \end{bmatrix} \begin{bmatrix} c_0 \\ \cdot \\ \cdot \\ \cdot \\ c_n \end{bmatrix} = \begin{bmatrix} y_0 \\ \cdot \\ \cdot \\ \cdot \\ y_n \end{bmatrix}$$

In this equation, of course, the c's are the unknowns while the x matrix and right-hand side are known. This equation has a unique solution because the coefficient matrix is nonsingular. The determinant of this matrix, known as *Vandermonde's determinant*, has the value

$$D = \prod_{0 \leq j < i \leq n} (x_i - x_j)$$

(A proof of this identity is outlined in Prob. 1.) The right-hand side of this formula denotes the *product* of *all* factors $(x_i - x_j)$ for which the pair (i,j) satisfies $0 \leq j < i \leq n$. From this formula for D, it is clear that $D \neq 0$ if and only if the points x_i are distinct. ■

Proof 2. Our polynomial could be written down at once if there existed polynomials l_i of degree $\leq n$ with the property $l_i(x_j) = \delta_{ij}$. Indeed, we would write $P(x) = \sum y_i l_i(x)$, whence $P(x_j) = \sum_i y_i l_i(x_j) = \sum_i y_i \delta_{ij} = y_j$. A moment's reflection shows that l_i is

$$l_i(x) = \prod_{\substack{j=0 \\ j \neq i}}^{n} \frac{x - x_j}{x_i - x_j}$$

Another way to define l_i is to start with $W(x) = \prod_{j=0}^{n} (x - x_j)$. Then

$$l_i(x) = a_i \frac{W(x)}{x - x_i}$$

Now the requirement $l_i(x_i) = 1$ leads, via L'Hospital's rule to

$$1 = a_i W'(x_i)$$

The formula that results is known as the *Lagrange interpolation formula*:

$$P(x) = \sum_{i=0}^{n} y_i l_i(x) \qquad l_i(x) = \prod_{\substack{j=0 \\ j \neq i}}^{n} \frac{x - x_j}{x_i - x_j} = \frac{W(x)}{(x - x_i) W'(x_i)}$$

There remains now the unicity of P. Suppose then that P and Q are two polynomials of degree $\leq n$ which have the property $Q(x_i) = P(x_i) = y_i$. Then $P - Q$ is a polynomial of degree $\leq n$ which vanishes at the $n + 1$ distinct points x_i. Hence $P - Q = 0$. ∎

Proof 3. Let us attempt to determine the desired polynomial in the form

$$P(x) = a_0 + a_1(x - x_0) + a_2(x - x_0)(x - x_1) + \cdots + a_n(x - x_0) \cdots (x - x_{n-1})$$

Setting $x = x_0$, we find that $P(x_0) = a_0$. Since $P(x_0)$ is prescribed to be y_0, we must take $a_0 = y_0$. With a_0 now known, we proceed to put $x = x_1$, getting $P(x_1) = a_0 + a_1(x_1 - x_0)$. Solving for the coefficients one by one in this manner yields

$$a_0 = y_0$$

$$a_1 = \frac{y_1 - a_0}{x_1 - x_0}$$

$$a_2 = \frac{y_2 - a_0 - a_1(x_2 - x_0)}{(x_2 - x_0)(x_2 - x_1)}$$

$$a_3 = \frac{y_3 - a_0 - a_1(x_3 - x_0) - a_2(x_3 - x_0)(x_3 - x_1)}{(x_3 - x_0)(x_3 - x_1)(x_3 - x_2)}$$

etc.

The existence of P, then, follows from the fact that in these formulas the denominators do not vanish. We do not stop to give a unicity proof based on the above formula. ■

Each of the three proofs just given suggests a different numerical procedure for obtaining the *interpolating polynomial.* In order to illustrate this let us determine the quadratic $P(x) = c_0 + c_1x + c_2x^2$ passing through the three points (1,2), (2,−1), (3,1), using each method in turn.

In the first method we solve the equations

$$\begin{aligned} c_0 + c_1 + c_2 &= 2 \\ c_0 + 2c_1 + 4c_2 &= -1 \\ c_0 + 3c_1 + 9c_2 &= 1 \end{aligned}$$

obtaining as a result $P(x) = 10 - \frac{21}{2}x + \frac{5}{2}x^2$.

Using Lagrange's formula, P comes out in the form

$$P(x) = (x-2)(x-3) + (x-1)(x-3) + \tfrac{1}{2}(x-1)(x-2)$$

Finally, using the method of the third proof, we obtain P in the form

$$P(x) = 2 - 3(x-1) + \tfrac{5}{2}(x-1)(x-2)$$

We turn next to the question of assessing the interpolation process as an instrument of approximation. At the beginning of this chapter, two related measures of the discrepancy between two functions f and P were introduced, viz.,

$$\max_{a \le x \le b} |f(x) - P(x)| \qquad \text{and} \qquad \max_{1 \le i \le m} |f(x_i) - P(x_i)|$$

The polynomial P of degree $\le n$ which interpolates to f at $n+1$ points x_i clearly solves the problem of minimizing the second expression when $m = n+1$. We ask, will the first expression also be small when P is chosen in this way? The answer is certainly "no" if the behavior of f between the interpolating points is not somehow controlled. It turns out that such control is possible for functions which possess $n+1$ continuous derivatives.

Theorem. *If f possesses n continuous derivatives on $[a,b]$, if P is the polynomial of degree $<n$ which interpolates to f at n nodes x_i in $[a,b]$, and if $W(x) = \prod(x - x_i)$, then in terms of the Tchebycheff norm,*

$$\|f - P\| \le \frac{1}{n!}\|f^{(n)}\|\,\|W\|$$

Proof. We shall prove somewhat more, namely, that to each x in $[a,b]$ there corresponds a $\xi \in (a,b)$ such that

$$f(x) - P(x) = \frac{1}{n!}f^{(n)}(\xi)W(x) \tag{1}$$

This formula is obvious if x is one of the nodes. Otherwise, we put $\phi = f - P - \lambda W$ where λ is chosen to make $\phi(x) = 0$. It is clear that ϕ vanishes also at all the nodes x_i. Thus ϕ vanishes in at least $n + 1$ points of $[a,b]$. By Rolle's theorem, ϕ' vanishes at least once between any two zeros of ϕ and thus vanishes in at least n points. Continuing this argument, we see that $\phi^{(n)}$ has at least one root on the interval, say at the point ξ. But $\phi^{(n)} = f^{(n)} - \lambda n!$ since P is a polynomial of degree $<n$ and $W(z) = z^n + \cdots$. Thus $f^{(n)}(\xi) = \lambda n!$ Since the value of λ is $[f(x) - P(x)]/W(x)$, the proof is complete. ∎

Let us now answer a question raised in a natural way by the foregoing theorem. How can we situate the nodes so as to optimize the error bound? Since the nodes enter this formula only in the function W, we must attempt to minimize the norm of W.

Theorem. *The uniform norm of* $W(x) = \prod_{i=1}^{n} (x - x_i)$ *is minimized on* $[-1,1]$ *when* $x_i = \cos [(2i - 1]\pi/2n]$.

Proof. It is known (see Prob. 2 below) that $\cos n\theta$ can be expressed in the form $\sum_{k=0}^{n} a_k \cos^k \theta$ with appropriate coefficients a_k, the leading one, a_n, being 2^{n-1}. Letting $T_n(x) = \sum_{k=0}^{n} a_k x^k$, we have $T_n(\cos \theta) = \cos n\theta$. The n roots of T_n are therefore the points x_i given above. The polynomial $W = 2^{1-n}T_n$ is of the form contemplated above since its leading coefficient is unity. The maximum of $| W(x) |$ on $[-1,1]$ occurs then at the points $y_i = \cos i\pi/n$ since $T_n(y_i) = \cos i\pi = (-1)^i$. Now, if possible, let V be another polynomial of the same form as W for which $\| V \| < \| W \|$. Then $V(y_0) < W(y_0)$, $V(y_1) > W(y_1)$, etc., from which it follows that $W - V$ must vanish at least once in each interval (y_1,y_0), (y_2,y_1), ... for a total of n times. But this is not possible because both V and W have leading coefficient unity, and their difference is therefore of degree $<n$. ∎

***Theorem* (Hermite Interpolation).** *There exists a unique polynomial* P *of degree* $\leq 2n - 1$ *such that* P *and its derivative* P' *take on prescribed values at* n *points.*

Proof. Let the conditions on P be that $P(x_i) = y_i$ and $P'(x_i) = y_i'$ for $i = 1, \ldots, n$. As in the case of ordinary interpolation, our polynomial can be written in an explicit formula (Hermite's interpolation formula)

$$P(x) = \sum_{i=1}^{n} [y_i A_i(x) + y_i' B_i(x)]$$

where A_i and B_i are polynomials of degree $\leq 2n - 1$ with the properties

$A_i(x_j) = \delta_{ij}$, $B_i(x_j) = 0$, $A'_i(x_j) = 0$, and $B'_i(x_j) = \delta_{ij}$. In terms of the functions $l_i(x)$, A_i and B_i take the form

$$A_i(x) = [1 - 2(x - x_i)l'_i(x_i)]l_i^2(x)$$

$$B_i(x) = (x - x_i)l_i^2(x)$$

It is left to the reader to verify, using the equation $l_i(x_j) = \delta_{ij}$, that the polynomial P so defined has the desired interpolating properties. In order to prove the uniqueness of P, we suppose on the contrary that another polynomial Q of degree $\leq 2n - 1$ exists having the properties $Q(x_i) = y_i$ and $Q'(x_i) = y'_i$. Then $P - Q$ is a polynomial which has roots of at least multiplicity 2 at each point x_i, since $(P - Q)'(x_i) = 0$. Since $P - Q$ is of degree $\leq 2n - 1$, it must be zero. ■

Problems

1. *Vandermonde's determinant.* By induction show that

$$\begin{vmatrix} 1 & x_0 & x_0^2 & \cdots & x_0^n \\ \cdots & \cdots & \cdots & \cdots & \cdots \\ 1 & x_n & x_n^2 & \cdots & x_n^n \end{vmatrix} = \prod_{0 \leq j < i \leq n} (x_i - x_j)$$

To reduce the given determinant to one of order n, subtract x_0 times each column from its successor, proceeding from right to left. Then factor, and use the induction hypothesis.

2. *Tchebycheff polynomials.*
   (*a*) Prove that $\cos (n + 1)\theta = 2 \cos \theta \cos n\theta - \cos (n - 1)\theta$. The formula $\cos (A \pm B) = \cos A \cos B \mp \sin A \sin B$ will be helpful.
   (*b*) Prove by induction that $\cos n\theta$ is expressible in the form $T_n(\cos \theta) = \sum_{k=0}^{n} a_{nk} \cos^k \theta$, and that $\cos^n \theta$ is expressible in the form $\sum_{k=0}^{n} b_{nk} \cos k\theta$.
   (*c*) If $T_n(x) = \sum_{k=0}^{n} a_{nk}x^k$, then $T_{n+1}(x) = 2xT_n(x) - T_{n-1}(x)$.
   (*d*) For $x \in [-1,1]$, $T_n(x) = \cos (n \arccos x)$.
   (*e*) Show that $T_0(x) = 1$, $T_1(x) = x$, $T_2(x) = 2x^2 - 1$, $T_3(x) = 4x^3 - 3x$, $T_4(x) = 8x^4 - 8x^2 + 1$, $T_5(x) = 16x^5 - 20x^3 + 5x$.
   (*f*) Prove that when n is even, T_n is an even function: i.e., $T_n(-x) = T_n(x)$. When n is odd, T_n is odd: i.e., $T_n(-x) = -T_n(x)$.
   (*g*) The leading coefficient in $T_n(x)$ is 2^{n-1} when $n \geq 1$.
   *(*h*) Show that

$$T_n(x) = 2^{n-1}\left[x^n - \frac{n}{2^2}x^{n-2} + \frac{n(n-3)}{2!2^4}x^{n-4} - \frac{n(n-4)(n-5)}{3!2^6}x^{n-6} + \frac{n(n-5)(n-6)(n-7)}{4!2^8}x^{n-8} - \cdots\right]$$

3. Let $x_0, \ldots, x_n$ be fixed "nodes" (points) in $[a,b]$. For any function $f \in C[a,b]$ let $L_n f$ denote the polynomial of degree $\leq n$ which agrees with f at $x_0, \ldots, x_n$. Show that L_n is a linear operator: i.e., $L_n(\lambda f) = \lambda L_n f$ and $L_n(f+g) = L_n f + L_n g$.

4. Show that $L_n f = f$ if and only if f is a polynomial of degree $\leq n$.

5. Show that the Lagrange interpolating polynomial for nodes at the zeros of T_n is

$$P(x) = \frac{1}{n}\sum_{i=1}^{n} f(x_i)\,\frac{T_n(x)(-1)^{i-1}\sin\theta_i}{x - x_i}$$

where $x_i = \cos\theta_i$ and $\theta_i = (2i-1)\pi/2n$.

6. Show that, in the notation of page 59, $\sum_{i=0}^{n} l_i(x) \equiv 1$, and more generally, $\sum_{i=0}^{n} x_i^k l_i(x) \equiv x^k$ for $k = 0, \ldots, n$. *Hint*: The polynomial which interpolates to 1 is 1 itself.

7. Prove that on any interval the successive Lagrangian interpolating polynomials for e^x (with arbitrary nodes) converge uniformly to e^x. Generalize to arbitrary functions for which $\|f^{(n)}\| \leq M^n$ on $[a,b]$.

8. *More about Tchebycheff polynomials*. Prove the following properties of T_n.

(*a*) $(1-x^2)T_n''(x) - xT_n'(x) + n^2 T_n(x) = 0$.

(*b*) $\dfrac{1-tx}{1-2tx+t^2} = \sum_{n=0}^{\infty} T_n(x)t^n$ for $-1 < t < 1$.

Hint: The series on the right is the real part of $\sum_{n=0}^{\infty} (te^{i\theta})^n$ when $\cos\theta = x$.

(*c*) $2^{1-n}T_n(x) = x^n - \binom{n}{2}x^{n-2}(1-x^2) + \binom{n}{4}x^{n-4}(1-x^2)^2 - \cdots$. *Hint*: The polynomial on the right is the real part of the binomial expansion of $(\cos\theta + i\sin\theta)^n$ when $x = \cos\theta$.

(*d*) $T_{2n}(x) = T_n(2x^2 - 1)$.

(*e*) The polynomial $P_n(x) = x^n + c_1x^{n-1} + c_2x^{n-2} + \cdots$ for which $\max_{a\leq x\leq b} |P_n(x)|$ is a minimum is given by $P_n(x) = 2\left(\dfrac{b-a}{4}\right)^n T_n\left(\dfrac{2x-a-b}{b-a}\right)$.

(*f*) For $|x| > 1$, $T_n(x) = \cosh nz$ where $\cosh z = x$.

(*g*) $\int T_n = T_{n+1}/(2n+2) - T_{n-1}/(2n-2)$.

(*h*) $T_n(T_m) = T_{nm}$.

9. If P is any polynomial of degree $\leq 2n-1$, then $P = \sum_{i=1}^{n} [P(x_i)A_i + P'(x_i)B_i]$, where A_i and B_i are the polynomials defining the Hermite interpolation formula. *Hint*: Use the uniqueness part of the Hermite interpolation theorem.

10. In the theorem which governs the error in interpolation, $f^{(n)}(\xi)$ is a continuous function of x. *Hint*: Solve for this term, and use L'Hospital's rule.

11. Vandermonde's determinant formula may be proved as follows. By the theory of determinants, the determinant in question must be a polynomial in the letters $x_0, x_1, \ldots, x_n$ of degree n in each. Since this polynomial vanishes whenever two of the variables coincide, each factor $x_i - x_j$ with $i \neq j$ must occur once. Complete the argument.

12. For a function f possessing $2n$ continuous derivatives, the error in the Hermite interpolation process may be expressed as follows:

$$f(x) - P(x) = \frac{f^{(2n)}(\xi)W^2(x)}{(2n)!}$$

Hint: Proceed as in the similar theorem for Lagrange interpolation. Define $\phi = f - P - \lambda W^2$.

13. *Inverse of Vandermonde's matrix.* The inverse A of the Vandermonde matrix $V = (x_i{}^j)$ $(i = 0, \ldots, n)$ is obtained as follows. Put $W = \prod_{i=0}^{n} (x - x_i)$ and $l_j(x) = W(x)[(x - x_j)W'(x_j)]^{-1}$. Let the jth column of A consist of the coefficients in the polynomial l_j. *Hint*: Use the Lagrange interpolation formula.

14. The polynomial P of degree $\leq n$ which takes values $y_0, \ldots, y_n$ at arguments $x_0, \ldots, x_n$ may be written

$$P(x) = c \begin{vmatrix} 0 & 1 & x & x^2 & \cdots & x^n \\ y_0 & 1 & x_0 & x_0{}^2 & \cdots & x_0{}^n \\ y_1 & 1 & x_1 & x_1{}^2 & \cdots & x_1{}^n \\ \cdots & \cdots & \cdots & \cdots & \cdots & \cdots \\ y_n & 1 & x_n & x_n{}^2 & \cdots & x_n{}^n \end{vmatrix}$$

where c is a certain constant. Determine c, and prove the formula.

15. The result of Prob. 14 may be generalized as follows. If there exists a linear combination P of the functions $g_1, \ldots, g_n$ which assumes values $y_1, \ldots, y_n$ at points $x_1, \ldots, x_n$, then it is

$$P(x) = c \begin{vmatrix} 0 & g_1(x) & \cdots & g_n(x) \\ y_1 & g_1(x_1) & \cdots & g_n(x_1) \\ \cdots & \cdots & \cdots & \cdots \\ y_n & g_1(x_n) & \cdots & g_n(x_n) \end{vmatrix}$$

What assumption about $\{g_1, \ldots, g_n\}$ is necessary if such P is always to exist? (Cf. Prob. 10, Sec. 4.)

16. Prove that $\sum (x - x_i)^2 l_i{}^2(x) l_i'(x_i) = 0$. *Hint*: If Hf denotes the Hermite interpolating polynomial for f, then $H1 = 1$ and $Hx = x$.

17. Consider a set of n equally spaced points x_i in $[0,1]$, and estimate the norm of $W(x) = \Pi(x - x_i)$. *Hint*: If $x_i = (2i - 1)/2n$, then $\| W \| = | W(0) |$. Use the Stirling inequality

$$1 < \frac{n!e^n}{n^n \sqrt{2\pi n}} < 1 + \frac{1}{4n}$$

to establish that $(1 + 1/4n)^{-1} \sqrt{2}\ e^{-n} \leq \| W \| \leq (1 + 1/8n)^{-1} \sqrt{2}\ e^{-n}$.

18. Using Probs. 3 and 6 show that the error in Lagrange interpolation is

$$(L_n f - f)(x) = \sum [f(x_i) - f(x)] l_i(x)$$

19. If P is any polynomial of degree $\leq n$ such that $|P(x)| \leq 1$ when $-1 \leq x \leq 1$, then for any y *outside* $[-1,1]$, $|P(y)| \leq |T_n(y)|$. *Hint*: If not, let $\lambda = P(y)/T_n(y)$, and consider the polynomial $Q = \lambda T_n - P$. It changes sign n times in $[-1,1]$ and vanishes at y. [Tchebycheff, 1881]

20. In the complex plane, let P_n be the polynomial of degree $<n$ which interpolates to the function $f(z) = 1/z$ at the n nth roots of unity. Show that $P_n(z) = z^{n-1}$. Show that $\| P_n - f \| \not\to 0$ as $n \to \infty$, where the sup-norm is computed over the set of z for which $|z| = 1$. [Méray, 1884]

21. Consider the Lagrange formula $(Lf)(x) = \sum f(x_i) l_i(x)$ with $a \leq x_1 < \cdots < x_n \leq b$. Show that $\| Lf \| \leq \lambda \| f \|$ with $\lambda = \max_x \sum |l_i(x)|$. Show that no smaller λ has this property for all $f \in C[a,b]$.

***22.** Show that for $-1 \leq x_1 < x_2 < x_3 \leq 1$ the minimum value of λ in Prob. 21 is $\frac{5}{4}$ and is attained when $-x_1 = x_3 \geq \frac{2}{3}\sqrt{2}$ and $x_2 = 0$. If x_1, x_2, and x_3 are roots of T_3, then $\lambda = \frac{5}{3}$. Thus placing the nodes at the roots of T_n generally does not lead to a minimum value of λ. [Bernstein, 1931]

3. The Weierstrass Theorem

In the preceding section we saw how to construct a polynomial P_n of degree $\leq n$ which agreed with a given function f at certain $n + 1$ points. It remains a matter of speculation whether $P_n(x) \to f(x)$ as $n \to \infty$ for all points x. It would seem reasonable, in view of the theorem on page 60, that for a very smooth function f (say one possessing derivatives of all orders), $\| P_n - f \| \to 0$. However, this is not to be expected in general. Specifically, if we take $f(x) = (x^2 + 1)^{-1}$ and compute the Lagrangian interpolating polynomials P_n for equally spaced nodes in $[-5,5]$, we find that $\| P_n - f \|$ becomes *arbitrarily large*. This situation is rather startling since the function f is regular at all points other than $\pm i$. This example is due to Runge. For a simpler example of Méray see Prob. 20 of the preceding section.

Another example is due to Bernstein: The polynomials P_n of degree n which interpolate the function $f(x) = |x|$ at $n + 1$ equally spaced points on $[-1,1]$ converge to $f(x)$ *only* at $+1$, 0, and -1.

It might be suspected that the equal spacing of the nodes was somehow to blame for this unfortunate state of affairs. Indeed, we shall see later that a clustering of the nodes near the extremities of the interval is usually advisable. Nevertheless, Faber showed that no matter how the nodes for interpolation are prescribed, there will be some continuous function whose interpolating polynomials fail to converge uniformly. (See Chap. 6, Sec. 5.)

Against the background of these "negative" results, the Weierstrass theorem appears the more remarkable. According to it, there exists *some* sequence of polynomials converging to a prescribed continuous function, uniformly on a closed bounded interval. The examples above indicate that such a sequence of polynomials cannot be obtained by interpolating at a fixed set of nodes. On the other hand, it need hardly be pointed out that Taylor series are not available for this purpose either, except for a very small class of continuous functions. This class does not even include the elementary function $|x|$. In order to appreciate the difficulties involved, let us consider this function on the interval $[-1,1]$. We start with the Taylor series for $\sqrt{1-z}$, which converges uniformly for $0 \leq z \leq 1$:

$$\sqrt{1-z} = 1 - \frac{1}{2}z - \frac{1}{2\cdot 4}z^2 - \frac{1\cdot 3}{2\cdot 4\cdot 6}z^3 - \frac{1\cdot 3\cdot 5}{2\cdot 4\cdot 6\cdot 8}z^4 - \cdots$$

Then we replace z by $1 - x^2$ to obtain

$$|x| = \sqrt{x^2} = \sqrt{1-(1-x^2)}$$
$$= 1 - \frac{1}{2}(1-x^2) - \frac{1}{2\cdot 4}(1-x^2)^2 - \frac{1\cdot 3}{2\cdot 4\cdot 6}(1-x^2)^3 - \cdots$$

It is to be especially noted that this is *not* a Taylor series in x. Lebesgue's proof of the Weierstrass theorem uses this series in an essential way.

Weierstrass Approximation Theorem. *Let f be a continuous function defined on $[a,b]$. To each $\epsilon > 0$ there corresponds a polynomial P such that $\|f - P\| < \epsilon$. Thus $|f(x) - P(x)| < \epsilon$ for all $x \in [a,b]$.*

We are going to derive this theorem presently as a consequence of another, more powerful theorem. By way of introducing this theorem, let us consider in outline the proof given by Bernstein of Weierstrass' theorem. Bernstein constructed, for a given $f \in C[0,1]$, a sequence of polynomials (now called *Bernstein* polynomials) $B_n f$ by means of the formula

$$(B_n f)(x) = \sum_{k=0}^{n} f\left(\frac{k}{n}\right)\binom{n}{k} x^k (1-x)^{n-k} \tag{1}$$

Here $\binom{n}{k}$ is the binomial coefficient $\dfrac{n!}{(n-k)!k!}$. The formula (1) also defines for each n a linear operator, B_n. We mean by this that to each element f in $C[0,1]$ there corresponds another element $B_n f$ of $C[0,1]$ in such a way that the condition of linearity is met.

$$B_n(af + bg) = aB_n f + bB_n g \tag{2}$$

The operators B_n are readily seen to have a further property expressed by the implication

$$f \geq g \Rightarrow B_n f \geq B_n g \tag{3}$$

We mean by an inequality $f \geq g$ that $f(x) \geq g(x)$ for all x (in the domain of f). An operator for which (3) is true is said to be a *monotone* operator. An examination of Bernstein's proof reveals that the crux of the matter is the verification that these operators have the convergence properties

$$B_n f \to f \qquad \text{for } f(x) = 1, x, x^2 \tag{4}$$

The conclusion of the proof is, of course, that $B_n f \to f$ for all $f \in C[0,1]$, convergence here being in the sense of the uniform norm.

A striking generalization of this theorem of Bernstein has recently been given by Bohman and by Korovkin. This result is that properties (2), (3), and (4) are sufficient for *any* sequence of operators L_n to have the property $L_n f \to f$ for all $f \in C[0,1]$.

Theorem on Monotone Operators. *For a sequence of monotone linear operators L_n on $C[a,b]$ the following conditions are equivalent:*

(*i*) $L_n f \to f$ *(uniformly) for all* $f \in C[a,b]$
(*ii*) $L_n f \to f$ *for the three functions* $f(x) = 1, x, x^2$
(*iii*) $L_n 1 \to 1$ *and* $(L_n\phi_t)(t) \to 0$ *uniformly in* t *where* $\phi_t(x) \equiv (t-x)^2$

Proof. The implication $(i) \Rightarrow (ii)$ is trivial.

For the proof that $(ii) \Rightarrow (iii)$ define $f_i(x) = x^i$. Since $\phi_t(x) = t^2 - 2tx + x^2$, we have $\phi_t = t^2 f_0 - 2tf_1 + f_2$ and $L_n\phi_t = t^2 L_n f_0 - 2tL_n f_1 + L_n f_2$. Thus

$$\begin{aligned}(L_n\phi_t)(t) &= t^2[(L_n f_0)(t) - 1] - 2t[(L_n f_1)(t) - t] + [(L_n f_2)(t) - t^2]\\ &\leq t^2 \| L_n f_0 - f_0 \| + |2t| \, \| L_n f_1 - f_1 \| + \| L_n f_2 - f_2 \|\end{aligned}$$

Since t^2 and $|2t|$ are bounded on $[a,b]$, we see that $(L_n\phi_t)(t)$ converges uniformly to zero, thus proving (*iii*).

For the proof that $(iii) \Rightarrow (i)$, let f be an arbitrary element of $C[a,b]$. Given $\epsilon > 0$, we shall obtain the inequality $\| L_n f - f \| < 3\epsilon$ for all sufficiently large n. Begin by selecting $\delta > 0$ such that

$$|x - y| < \delta \Rightarrow |f(x) - f(y)| < \epsilon$$

Now put $\alpha = 2 \|f\| \delta^{-2}$, and let t be an arbitrary but fixed point of $[a,b]$. If $|t - x| < \delta$, then $|f(t) - f(x)| < \epsilon$, whereas if $|t - x| \geq \delta$, then $|f(t) - f(x)| \leq 2 \|f\| \leq 2 \|f\| (t-x)^2/\delta^2 = \alpha\phi_t(x)$. Thus for *all* x, the following inequality is satisfied:

$$-\epsilon - \alpha\phi_t(x) \leq f(t) - f(x) \leq \epsilon + \alpha\phi_t(x)$$

In order to write an inequality on the *functions*, let $f_0(x) = 1$. Then we have

$$-\epsilon f_0 - \alpha\phi_t \leq f(t) f_0 - f \leq \epsilon f_0 + \alpha\phi_t$$

By the linearity and monotonicity of L_n we have

$$-\epsilon(L_n f_0)(t) - \alpha(L_n\phi_t)(t) \leq f(t)(L_n f_0)(t) - (L_n f)(t) \leq \epsilon(L_n f_0)(t) + \alpha(L_n\phi_t)(t)$$

This yields $|f(t)(L_n f_0)(t) - (L_n f)(t)| \leq \epsilon \| L_n f_0 \| + \alpha(L_n\phi_t)(t)$. Since $L_n f_0 \to f_0$ and $(L_n\phi_t)(t) \to 0$, it is clear that this inequality essentially finishes the proof. In order to see exactly how large n must be, write

$$\begin{aligned} |f(t) - (L_n f)(t)| &\leq |f(t) - f(t)(L_n f_0)(t)| + |f(t)(L_n f_0)(t) - (L_n f)(t)| \\ &\leq |f(t)|\,|1 - (L_n f_0)(t)| + \epsilon \| L_n f_0 \| + \alpha(L_n\phi_t)(t) \\ &\leq \|f\|\,\|f_0 - L_n f_0\| + \epsilon(1 + \|f_0 - L_n f_0\|) \\ &\qquad + \alpha(L_n\phi_t)(t) \end{aligned}$$

Thus we should select N so that whenever $n \geq N$ it will follow that $(\|f\| + \epsilon)\,\|f_0 - L_n f_0\| < \epsilon$ and $\alpha(L_n\phi_t)(t) < \epsilon$. ■

Proof of the Weierstrass Theorem. We are going to prove the theorem for the interval $[0,1]$, leaving the extension to an arbitrary interval for the problems. It will be shown that for any $f \in C[0,1]$ the Bernstein polynomials $B_n f$ (defined on page 66) converge to f. The linearity and monotonicity of B_n have already been mentioned. By the theorem on monotone operators, it will suffice to show that $B_n f \to f$ for $f(x) = 1, x$, and x^2. That $B_n 1 = 1$ follows from the binomial theorem:

$$(B_n 1)(x) = \sum_{k=0}^{n} \binom{n}{k} x^k (1 - x)^{n-k} = [x + (1 - x)]^n = 1$$

For the function $f(x) = x$ we have

$$\begin{aligned} (B_n f)(x) &= \sum_{k=0}^{n} \frac{k}{n} \binom{n}{k} x^k (1 - x)^{n-k} \\ &= \sum_{k=1}^{n} \frac{kn!}{n(n - k)!k!} x^k (1 - x)^{n-k} \\ &= x \sum_{k=1}^{n} \binom{n - 1}{k - 1} x^{k-1} (1 - x)^{n-k} \\ &= x \sum_{k=0}^{n-1} \binom{n - 1}{k} x^k (1 - x)^{n-1-k} \\ &= x[x + (1 - x)]^{n-1} = x \end{aligned}$$

For the function $f(x) = x^2$ we have

$$(B_n f)(x) = \sum_{k=0}^{n} \left(\frac{k}{n}\right)^2 \binom{n}{k} x^k (1-x)^{n-k}$$

$$= \sum_{k=1}^{n} \frac{k}{n} \binom{n-1}{k-1} x^k (1-x)^{n-k}$$

$$= \frac{n-1}{n} \sum_{k=1}^{n} \frac{k-1}{n-1} \binom{n-1}{k-1} x^k (1-x)^{n-k}$$

$$+ \frac{1}{n} \sum_{k=1}^{n} \binom{n-1}{k-1} x^k (1-x)^{n-k}$$

$$= \frac{n-1}{n} x^2 + \frac{1}{n} x \to x^2 \qquad \blacksquare$$

It was remarked earlier that the Lagrange interpolation procedure, for a fixed array of nodes, fails to furnish approximations of arbitrarily high precision to all continuous functions. Thus the Weierstrass theorem is not a simple consequence of the Lagrange interpolation formula. We should expect this to be true all the more for the Hermite interpolation formula since in its usual application this involves the *derivatives* of the function. In 1930 Fejér made the startling discovery that the Hermite interpolation procedure can be arranged so as to yield a proof of the Weierstrass theorem.

In the Hermite formula (page 61) let us take $y_i = f(x_i)$ and $y_i' = 0$. The resulting operator we shall call the *Fejér-Hermite* operator; it takes the form

$$(5) \qquad (Lf)(x) = \sum_{i=1}^{n} f(x_i) A_i(x)$$

$$= \sum_{i=1}^{n} f(x_i)[1 - 2(x - x_i) l_i'(x_i)] l_i^2(x)$$

For the purposes of Fejér's theorem it will be convenient to express the operator (5) in terms of the function $W(x) = \prod (x - x_i)$. We have, as in the previous section,

$$l_i(x) = \frac{W(x)}{(x - x_i) W'(x_i)}$$

If we differentiate and then use L'Hospital's rule to evaluate the indeterminate form, we obtain $l_i'(x_i) = \frac{1}{2} W''(x_i)/W'(x_i)$. Thus (5) takes the alternative form

$$(6) \qquad (Lf)(x) = \sum f(x_i) \left[1 - \frac{(x - x_i) W''(x_i)}{W'(x_i)}\right] \left[\frac{W(x)}{(x - x_i) W'(x_i)}\right]^2$$

Theorem. [*Fejér*] *Let L_n denote the Fejér-Hermite operator with nodes at the zeros of the nth Tchebycheff polynomial T_n. Then $L_n f \to f$ for all $f \in C[-1,1]$.*

Proof. [Korovkin, 1959] We begin by establishing a formula for L_n in this special case:

$$(L_n f)(x) = \frac{1}{n^2} T_n^2(x) \sum_{i=1}^{n} f(x_i) \frac{1 - xx_i}{(x - x_i)^2} \tag{7}$$

where the points x_i are the zeros of T_n, viz., $x_i = \cos (2i - 1)\pi/2n$. A comparison of this formula with the general formula (6) shows that we must establish the equality

$$\frac{1}{n^2}(1 - xx_i) = \left[1 - (x - x_i)\frac{T_n''(x_i)}{T_n'(x_i)}\right]\frac{1}{[T_n'(x_i)]^2} \tag{8}$$

Since $T_n(x) = \cos (n \cos^{-1} x)$, we have $T_n'(x) = n \sin (n \cos^{-1} x)(1 - x^2)^{-1/2}$ and $[T_n'(x_i)]^2 = n^2 \sin^2 [(2i - 1)\pi/2](1 - x_i^2)^{-1} = n^2(1 - x_i^2)^{-1}$. On the other hand, starting with the differential equation

$$(1 - x^2) T_n''(x) - xT_n'(x) + n^2 T_n(x) = 0$$

(Prob. 8*a*, Sec. 2), we may set $x = x_i$ to deduce that $(1 - x_i^2) T_n''(x_i) = x_i T_n'(x_i)$ and $T_n''(x_i)/T_n'(x_i) = x_i/(1 - x_i^2)$. With these replacements it is easy to verify equation (8). To complete the proof we use the theorem on monotone operators. That L_n is monotone is immediate from (7) since $1 - xx_i \geq 0$. By the uniqueness part of the Hermite interpolation theorem (Sec. 2) it follows that $L_n 1 = 1$. In accordance with the theorem on monotone operators, we need only show that $(L_n \phi_x)(x) \to 0$ uniformly in x, where $\phi_x(t) = (x - t)^2$. We have

$$(L_n \phi_x)(x) = \frac{1}{n^2} T_n^2(x) \sum_{i=1}^{n} (x - x_i)^2 \frac{1 - xx_i}{(x - x_i)^2}$$

The sum cannot exceed $2n$ since $0 \leq 1 - xx_i \leq 2$. Hence, the entire expression converges to zero uniformly as $n \to \infty$. ∎

Problems

1. Prove that $\frac{k}{n}\binom{n}{k} = \binom{n-1}{k-1}$. Give a direct proof for the function $f(x) = x^3$ that $B_n f \to f$.

2. Consider an operator of the form $(Lf)(x) = \sum_{i=1}^{n} f(x_i) g_i(x)$ where $a \leq x_1 < x_2 < \cdots < x_n \leq b$ and $g_i \in C[a,b]$. Prove that L is monotone iff $g_i(x) \geq 0$ for all $x \in [a,b]$ and for all i.

3. The identity operator I, defined by $If = f$, is an example of a monotone linear operator. Use the monotone-operator theorem to show that the identity is the *only* monotone linear operator L defined on $C[a,b]$ such that $Lf = f$ for all quadratic functions, $f(x) = ax^2 + bx + c$. (Thus monotone operators suffer from a severe drawback in practical approximation.)

***4.** Let $h_n(x) = \sum_{k=0}^{n} a_{nk}x^k$. Discuss the operators $(L_nf)(x) = (1-x)^{-n}\left[h_n\left(\frac{x}{1-x}\right)\right]^{-1}$

$\times \sum_{k=0}^{n} f(k/n)a_{nk}x^k(1-x)^{n-k}$, which generalize the Bernstein operators. When are L_n

monotone? When is it true that $L_nf \to f$ for all $f \in C[0,1]$? When is it true that $L_nf = f$ for the function $f(x) = x$?

***5.** If f is a polynomial of degree $\leq k$, then so is B_nf, *for all* n. *Hint*: Use induction on k. The case $k = 0$ is done in the text. In the inductive step (from $k - 1$ to k) it will suffice to consider only the special function $f(x) = x^k$.

6. Prove that if $f \in C[a,b]$ and $\phi(x) = f[a + x(b-a)]$, then $\phi \in C[0,1]$. Use this fact to establish the Weierstrass theorem for an arbitrary closed interval. If f is continuous on $(-\infty,+\infty)$, then polynomials P_n exist such that for each x, $P_n(x) \to f(x)$.

7. Let $B_n^*f = B_nf + \phi_nf$, where ϕ_n are fixed elements of $C[0,1]$. What are suitable hypotheses on ϕ_n such that $B_n^*f \to f$ for all $f \in C[0,1]$?

8. For the function $f(x) = x^2$, how high must n be to obtain $\| B_nf - f \| < 10^{-8}$? Do you think the Bernstein polynomials afford a practical means of getting good polynomial approximations?

9. Show that the general Fejér-Hermite operator of equation (5) is monotone in $C[-1,1]$ if and only if the nodes satisfy the conditions

$$(1 - x_i)^{-1} \geq 2 \sum_{\substack{j=1 \\ j \neq i}}^{n} (x_i - x_j)^{-1} \geq -(1 + x_i)^{-1}$$

10. Show that the Fejér-Hermite operator of equation (5) is a monotone operator if and only if the *conjugate nodes*, $\bar{x}_i = x_i + \frac{1}{2}/l_i'(x_i)$ lie outside $[a,b]$. [Fejér, 1934]

11. In the case of nodes taken at the zeros of T_n, the conjugate nodes (Prob. 10) are $\bar{x}_i = x_i^{-1}$. [Fejér, 1934]

***12.** The Fejér-Hermite operator with nodes at the zeros of the nth Legendre polynomial P_n is also a monotone operator. Prove this with the aid of the differential equation $(1 - x^2)P_n''(x) - 2xP_n'(x) + n(n+1)P_n(x) = 0$. (The Legendre polynomials are defined in Chap. 4, Sec. 2.)

13. If $f \in C[a,b]$ and $0 < a < b < 1$, then there exists a sequence of polynomials P_nf *with integer coefficients* converging uniformly to f. *Hint*: Let $[u]$ denote the largest

integer $\leq u$, and let $(P_nf)(x) = \sum_{k=0}^{n} \left[\binom{n}{k} f\left(\frac{k}{n}\right)\right] x^k(1-x)^{n-k}$. Since $B_nf \to f$, it suffices

to prove $\| B_nf - P_nf \| \to 0$.

14. An operator L on $C[a,b]$ is said to be *nonnegative* if $f \geq 0 \Rightarrow Lf \geq 0$. Show that for linear operators, nonnegativity and monotonicity are equivalent.

15. If f and its derivative f' belong to $C[a,b]$, then to each $\epsilon > 0$ there corresponds a polynomial P satisfying $\| f - P \| < \epsilon$ and $\| f' - P' \| < \epsilon$. [Painlevé, 1898]

16. Prove that $\| B_nf \| \leq \| f \|$.

***17.** What happens if for a given f and n we form the sequence B_nf, $B_n^2f = B_n(B_nf)$, $B_n^3f = B_n(B_n^2f)$, etc.? Investigate $\lim_{k\to\infty} B_n^kf$, $\lim_{n\to\infty} B_n^kf$, and $\lim_{n\to\infty} B_n^nf$.

18. Show that if we define $f_{n0}(x) = 1$ and

$$f_{nm}(x) = x\left(x - \frac{1}{n}\right)\left(x - \frac{2}{n}\right)\cdots\left(x - \frac{m-1}{n}\right) \qquad (m \geq 1)$$

then $(B_nf_{nm})(x) = f_{nm}(1)x^m$.

19. Using the result of Probs. 16 and 18, we can show directly that $B_nf \to f$ for all polynomials by writing $g_m(x) = x^m$, $g_m = \lim_n f_{nm} = \lim_n B_nf_{nm}$ and

$$\| B_ng_m - g_m \| \leq \| B_n(g_m - f_{nm}) \| + \| B_nf_{nm} - g_m \|$$

20. An alternative proof of the result in Prob. 13 goes as follows. Take n so that $\sum_{i\geq n} b^i < \epsilon$, and then take a polynomial $Q(x) = \sum c_ix^i$ such that $| Q(x) - f(x)x^{-n} | < \epsilon$. Let $P(x) = \sum [c_i]x^{i+n}$, and show that $\| P - f \| < 2\epsilon$.

***21.** If $f' \in C[0,1]$, then $(B_nf)' \to f'$. *Hint*: Show by calculating that $(B_nf)' = B_{n-1}\phi_n$, where ϕ_n is the function $\phi_n(x) = n\left[f\left(\frac{n-1}{n}x + \frac{1}{n}\right) - f\left(\frac{n-1}{n}x\right)\right]$. Then prove that $\phi_n(x) \to f'(x)$. Problem 15 will be helpful. [Wigert, 1930]

22. Using Probs. 16 and 19 and the Weierstrass theorem, prove that $B_nf \to f$ for all $f \in C[0,1]$.

23. If $g(x) = \int_0^x f''(t)(x - t)\,dt$, then $f - B_nf = g - B_ng$. [Stancu]

4. General Linear Families

Up to this point we have been considering the approximation of functions by ordinary polynomials of degree $\leq n$, and these are of course simply linear combinations of the functions

$$1, x, x^2, \ldots, x^n$$

It is natural to generalize the concept of a polynomial to include linear combinations of other prescribed functions, say

$$g_1, g_2, \ldots, g_n$$

We shall always assume that such functions are continuous on some fixed compact metric space X; their linear combinations $\sum_{i=1}^{n} c_ig_i$ will be termed *generalized polynomials*. The existence theorem of Chap. 1, Sec. 6, guarantees that to each $f \in C[X]$ there corresponds at least one generalized polynomial $\sum c_ig_i$ which best approximates f. We turn now to the problem of characterizing these best approximations.

Characterization Theorem. *In order that the coefficients $c_1, \ldots, c_n$ shall render the (uniform) norm of $r \equiv \sum c_i g_i - f$ a minimum, it is necessary and sufficient that the origin of n space shall lie in the convex hull of the point set $\{r(x)\hat{x}: |r(x)| = \|r\|\}$, where $\hat{x}$ denotes the n-tuple $[g_1(x), \ldots, g_n(x)]$.*

Proof. Suppose that $\|r\|$ is not a minimum. Then for some n-tuple d, we have $\|\sum (c_i - d_i) g_i - f\| < \|\sum c_i g_i - f\|$. Now let $X_0 = \{x \in X: |r(x)| = \|r\|\}$. For $x \in X_0$ we have then

$$[r(x) - \sum d_i g_i(x)]^2 < [r(x)]^2$$

from which we obtain the following system of linear inequalities to be satisfied by d:

$$r(x) \sum d_i g_i(x) \equiv r(x) \langle d, \hat{x} \rangle > 0 \qquad (x \in X_0) \tag{1}$$

By the theorem on linear inequalities (Chap. 1, Sec. 5), 0 lies outside the convex hull of $\{r(x)\hat{x}: x \in X_0\}$. It is left for the reader to verify that this set of n-tuples is compact so that the cited theorem is applicable.

For the converse, suppose that 0 lies outside the aforementioned convex hull. By the same theorem, there exists an n-tuple d such that inequality (1) above is true for $x \in X_0$. Since X_0 is compact, the number $\epsilon = \min_{x \epsilon X_0} r(x) \langle d, \hat{x} \rangle$ is positive (theorem on page 5). Define $X_1 = \{x \in X: r(x) \langle d, \hat{x} \rangle \leq \epsilon/2\}$. It is easy to see that X_1 is a closed set containing no points of X_0. It follows that $|r(x)|$ achieves its supremum E on X_1 and that $E < \|r\|$. We are going to prove that for some $\lambda > 0$, $\|r - \lambda \sum d_i g_i\| < \|r\|$. First take $x \in X_1$ and assume that $0 < \lambda < (\|r\| - E)/\|\sum d_i g_i\|$. Then

$$|r(x) - \lambda \sum d_i g_i(x)| \leq |r(x)| + \lambda |\sum d_i g_i(x)|$$
$$\leq E + \lambda \|\sum d_i g_i\| < \|r\|$$

Next take $x \notin X_1$ and assume that $0 < \lambda < \epsilon/\|\sum d_i g_i\|^2$. Then $[r(x) - \lambda \sum d_i g_i(x)]^2 = [r(x)]^2 - 2\lambda r(x) \langle d, \hat{x} \rangle + \lambda^2 [\sum d_i g_i(x)]^2 < \|r\|^2 + \lambda(-\epsilon + \lambda \|\sum d_i g_i\|^2) < \|r\|^2$. ∎

The reader should observe the similarity between this theorem and the one on page 34, which dealt with the discrete version of the same approximation problem. It would be easy, by using the complex analogue of the theorem on page 19, to extend the characterization to the complex case. The only change would be that 0 would lie in the convex hull of $\{\overline{r(x)} \cdot \hat{x}: |r(x)| = \|r\|\}$, where $\overline{r(x)}$ denotes the complex conjugate of $r(x)$. Actually, no new ideas would be necessary to deal with functions which took instead of *scalar* values, values in *Hilbert space.*

For certain types of generalized polynomials, the characterization of best approximations can be given in a much more convenient form. The additional property that we require is the *Haar condition*, which was encountered in Chap. 2, page 45. A system of functions $\{g_1,\ldots,g_n\}$ is said to satisfy the *Haar condition* if each g_i is continuous and if every set of n vectors of the form

$$\hat{x} = [g_1(x),\ldots,g_n(x)]$$

is independent. Expressed otherwise, each determinant

$$D[x_1,\ldots,x_n] = \begin{vmatrix} g_1(x_1) & \cdots & g_n(x_1) \\ \cdots & \cdots & \cdots \\ g_1(x_n) & \cdots & g_n(x_n) \end{vmatrix}$$

made up from n distinct points $x_1, \ldots, x_n$ is *nonzero*. The nonvanishing of Vandermonde's determinant (page 58) for distinct points implies that $\{1,x,x^2,\ldots,x^n\}$ satisfies the Haar condition on any interval (and for any n). A system of functions satisfying the Haar condition is sometimes termed a *Tchebycheff* system. We note (Prob. 1) that $\{g_1,\ldots,g_n\}$ satisfies the Haar condition if and only if 0 is the only function of the form $\sum_{i=1}^{n} c_i g_i$ which has n or more roots on $[a,b]$. We establish next a preliminary result.

Lemma. *Let $\{g_1,\ldots,g_n\}$ be a system of elements from $C[a,b]$ satisfying the Haar condition. Let $a \leq x_0 < x_1 < \cdots < x_n \leq b$, and let $\lambda_0, \ldots, \lambda_n$ be nonzero constants. In order that 0 lie in the convex hull of the n-tuples $\lambda_0\hat{x}_0, \ldots, \lambda_n\hat{x}_n$ it is necessary and sufficient that the λ's alternate in sign: $\lambda_i\lambda_{i-1} < 0$ for $i = 1, \ldots, n$.*

Proof. We note first that all the determinants $D[x_1,\ldots,x_n]$ with $x_1 < \cdots < x_n$ have the same sign. Indeed, suppose that $D[x_1,\ldots,x_n] < 0 < D[y_1,\ldots,y_n]$. Then for some $\lambda \in (0,1)$, $D[\lambda x_1 + (1-\lambda)y_1,\ldots, \lambda x_n + (1-\lambda)y_n] = 0$. From the Haar condition it follows that for some distinct i and j, $\lambda x_i + (1-\lambda)y_i = \lambda x_j + (1-\lambda)y_j$. Hence $x_i - x_j$ and $y_i - y_j$ have opposite signs.

Now the origin lies in the hull of the points $\lambda_i\hat{x}_i$ if and only if the equation $\sum \theta_i\lambda_i\hat{x}_i = 0$ has a positive solution, $[\theta_0,\ldots,\theta_n]$. Writing this in the form

$$\hat{x}_0 = \sum_{i=1}^{n} \frac{-\theta_i\lambda_i}{\theta_0\lambda_0}\hat{x}_i$$

and solving by Cramer's rule, we have

$$\frac{-\theta_i\lambda_i}{\theta_0\lambda_0} = \frac{D[x_1,\ldots,x_{i-1},x_0,x_{i+1},\ldots,x_n]}{D[x_1,\ldots,x_n]}$$

Since it requires $i-1$ column interchanges in the numerator determinant to restore the natural order of the arguments, and since each such interchange alters the sign of a determinant, we see that sgn $(-\theta_i\lambda_i/\theta_0\lambda_0) = (-1)^{i-1}$, whence sgn $\lambda_i = (-1)^i$ sgn λ_0. ■

Alternation Theorem. *Let $\{g_1,\ldots,g_n\}$ be a system of elements of $C[a,b]$ satisfying the Haar condition, and let X be any closed subset of $[a,b]$. In order that a certain generalized polynomial $P = \sum c_i g_i$ shall be a best approximation on X to a given $f \in C[X]$ it is necessary and sufficient that the error function $r = f - P$ exhibit on X at least $n + 1$ "alternations" thus: $r(x_i) = -r(x_{i-1}) = \pm \| r \|$, with $x_0 < \cdots < x_n$ and $x_i \in X$. Here $\| r \| = \max_{x \in X} | r(x) |$.*

Proof. By the characterization theorem on page 73, $\| r \|$ is a minimum if and only if the origin lies in the convex hull of the set $\{r(x)\hat{x}: | r(x) | = \| r \|\}$. By Carathéodory's theorem (Chap. 1, Sec. 5), we may express 0 as a convex linear combination of some of these elements, $0 = \sum_{i=0}^{k} \lambda_i r(x_i)\hat{x}_i$, with $k \leq n$. By the Haar condition, $k \geq n$, and thus $k = n$. Now let the x_i be arranged in order, $a \leq x_0 < x_1 < \cdots < x_n \leq b$. By the lemma our equation for 0 is possible if and only if the numbers $\lambda_i r(x_i)$ alternate in sign. ■

The theorem just proved is very important for the numerical determination of best approximations. For example, if we wish to approximate $\sin \frac{1}{2}\pi x$ by a polynomial of first degree, $P(x) = c_0 + c_1 x$ over $[0,1]$, then the error function $f - P$ must alternate at least three times. From the sketch we surmise that

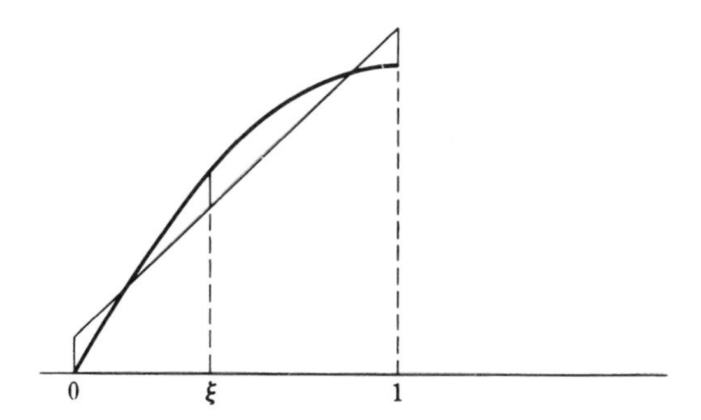

the points of alternation are 0, ξ, and 1, where ξ is not known. Let $\epsilon = \| f - P \|$. Then

$$f(0) - P(0) = -\epsilon$$

$$f(\xi) - P(\xi) = +\epsilon$$

$$f(1) - P(1) = -\epsilon$$

Since $f(x) = \sin \frac{1}{2}\pi x$ and $P(x) = c_0 + c_1 x$, these equations become

$$c_0 = \epsilon$$

$$c_0 + c_1\xi = \sin \tfrac{1}{2}\pi\xi - \epsilon$$

$$c_0 + c_1 = \epsilon + 1$$

If we knew ξ, we could solve for c_0, c_1, and ϵ. At ξ, the error is to be a maximum. Thus $f'(\xi) - P'(\xi) = 0$, whence $(\pi/2) \cos \frac{1}{2}\pi\xi = c_1$. Since $c_1 = 1$, we obtain $\xi = (2/\pi) \cos^{-1} (2/\pi) = 0.56050850$ and then $\epsilon = c_0 = \frac{1}{2}(-\xi + \sin \frac{1}{2}\pi\xi) = 0.10533467$. This situation is typical: If we know the location of the extrema of the error curve, we can easily obtain the polynomial by solving some *linear* equations. The location of these extrema, however, almost always involves the solution of some *nonlinear* equations. The reader should recall a similar observation made on page 36 in connection with overdetermined systems of linear equations.

As an interesting application of the alternation theorem, we can prove a certain "nonexistence theorem." First we define a *Markoff system.* A finite or an infinite ordered system of continuous functions $\{g_1,g_2,\ldots\}$ on an interval $[a,b]$ is termed a Markoff system if every initial segment $\{g_1,\ldots,g_n\}$ satisfies the Haar condition. For example, the monomials $1, x, x^2, \ldots$ form a Markoff system on any interval.

Theorem. *Let $\{g_1,g_2,\ldots\}$ be an infinite Markoff system on $[a,b]$. Let M denote the closed linear subspace of $C[a,b]$ generated by the elements g_i. Then no point outside M has a closest point inside M.*

Proof. Suppose on the contrary that g is a point of M closest to f, and that $\|f - g\| = \epsilon > 0$. Then 0 is the nearest point of M to $f - g$. Let M_n be the finite-dimensional subspace generated by $g_1, \ldots, g_n$. Then 0 is the point of M_n closest to $f - g$. By the alternation theorem, $f - g$ exhibits at least $n + 1$ oscillations of amplitude ϵ. Since this is true for all n, $f - g$ cannot be continuous. ■

In another application of the alternation theorem, we take X to be any set of $n + 2$ points on the real line. We have then the following special case.

Corollary. *Let ordinary polynomials P and Q of degree $\leq n + 1$ be determined by the conditions $P(x_i) = f(x_i)$ and $Q(x_i) = (-1)^i$, where $x_0 < \cdots < x_{n+1}$. Then the polynomial of degree $\leq n$ which best approximates f on $\{x_i\}$ is $P - \lambda Q$, where λ is chosen so that the resulting polynomial is of degree $\leq n$. The error in the approximation is $|\lambda|$.*

Proof. The existence of P and Q is guaranteed by the interpolating property. Since Q alternates in sign at the points x_i, it has $n + 1$ roots, and is therefore of degree $n + 1$. Hence λ exists. Since $f(x_i) - P(x_i) + \lambda Q(x_i) = (-1)^i\lambda$, $P - \lambda Q$ is the best approximation of f, by the alternation theorem. ■

In the next theorem we have a system of continuous functions $\{g_1,\ldots,g_n\}$ satisfying the Haar condition. We denote by $E(f)$ the infimum of $\| P - f \|$ as P ranges over all generalized polynomials, $P = \sum c_i g_i$.

Theorem of de La Vallée Poussin. *If P is a generalized polynomial such that $f - P$ assumes alternately positive and negative values at $n + 1$ consecutive points x_i of $[a,b]$, then* $E(f) \geq \min_i |f(x_i) - P(x_i)|$.

Proof. If the conclusion is false, there exists a generalized polynomial P_0 such that $\|f - P_0\| < \min_i |f(x_i) - P(x_i)|$. Then the generalized polynomial $P_0 - P = (f - P) - (f - P_0)$ is alternately positive and negative at the consecutive points x_i, and consequently vanishes at n points. But this is not possible. (See Prob. 1.) ■

Problems

1. The system $\{g_1,\ldots,g_n\}$ satisfies the Haar condition iff no nontrivial generalized polynomial $\sum c_i g_i$ has more than $n - 1$ roots.
2. Determine whether the following systems of functions satisfy the Haar condition on the interval indicated. (Use Prob. 1.)
   (*a*) $\{1,x^2,x^4\}$ on $[0,1]$
   (*b*) $\{1,x^2,x^4\}$ on $[-1,1]$
   (*c*) $\{1,x^2,x^3\}$ on $[-1,1]$
   (*d*) $\{1/(x + 1), 1/(x + 2), 1/(x + 3)\}$ on $[0,1]$
   (*e*) $\{1,e^x,e^{2x}\}$ on $[0,1]$
3. State and prove some interesting generalizations suggested by Prob. 2. For part (*d*), Cauchy's lemma from Chap. 6, Sec. 2, will be helpful.
4. Using the alternation theorem, re-prove the theorem of page 61. *Hint*: When we minimize $\| W \|$, we are obtaining a best approximation of $-x^n$ by a polynomial of degree $<n$.
5. Using de La Vallée Poussin's theorem, establish the sufficiency half of the alternation theorem.
6. For any interval and any n there exists a system $\{g_1,\ldots,g_n\}$ satisfying the Haar condition which is not a Markoff system and cannot be rearranged to form a Markoff system. *Hint*: If $\{g_1,\ldots\}$ is a Markoff system, then g_1 can have no root.
7. For any interval and for any n there exists a Markoff system $\{g_1,\ldots g_n\}$ of which every rearrangement is a Markoff system. *Hint*: See Prob. 2.
8. If $f^{(n)}(x) > 0$ on $[a,b]$, then the system of functions $\{1,x,\ldots,x^{n-1},f\}$ satisfies the Haar condition. *Hint*: If not, then some function $\phi = \sum_{i=0}^{n-1} c_i x^i + c_n f(x)$ has $n + 1$ roots. Its derivative ϕ' has n roots, etc. [Bernstein, 1926]

9. If $\{g_1,\ldots,g_n\}$ satisfies the Haar condition on $[a,b]$ and if $x_2,\ldots,x_n$ are any distinct points in $[a,b]$, then there exists a generalized polynomial having these points for zeros, and it can be written $D[x,x_2,x_3,\ldots,x_n]$. (See page 74 for the definition of D.) Any two such generalized polynomials are multiples of each other.

10. Let $\{g_1,\ldots,g_n\}$ satisfy the Haar condition on $[a,b]$. A generalized polynomial taking prescribed values y_i at n points x_i in $[a,b]$ is given by

$$P(x) = \sum y_i \frac{D[x_1,\ldots,x_{i-1},x,x_{i+1},\ldots,x_n]}{D[x_1,\ldots,x_n]}$$

11. If $\{g_1,\ldots,g_n\}$ satisfies the Haar condition, and if $A = (a_{ij})$ is a nonsingular $n \times n$ matrix, then the set of functions $h_i = \sum_{j=1}^{n} a_{ij}g_j$ also satisfies the Haar condition.

12. From Prob. 11 and the fact that $\{e^{\lambda_1 x},e^{\lambda_2 x},\ldots\}$ forms a Markoff system on any interval, deduce that $\{1, \cosh x, \sinh x, \cosh 2x, \sinh 2x, \ldots, \cosh nx, \sinh nx\}$ satisfies the Haar condition. [Remes, 1957]

13. If $\{g_1,\ldots,g_n\}$ satisfies the Haar condition on $[a,b]$ and if ϕ is monotonic on $[a,b]$, then the set of functions $h_i(x) = g_i(\phi^{-1}(x))$ satisfies the Haar condition on the interval $[\phi(a),\phi(b)]$. [Remes, 1957]

14. If the set of functions $\{g_1,\ldots,g_n\}$ satisfies the Haar condition on $[a,b]$, then every determinant of $n+1$ rows $A^i = [(-1)^i,g_1(x_i),\ldots,g_n(x_i)]$ is different from zero when $x_0 < x_1 < \cdots < x_n$. *Hint*: Expand by the elements of the first column.

15. In Prob. 10 suppose that $y_i = f(x_i)$ where $f, g_1, \ldots, g_n$ all possess continuous nth derivatives. Then to each x there corresponds ξ such that

$$f(x) - P(x) = [f^n(\xi) - P^{(n)}(\xi)] \prod_{i=1}^{n} \frac{x - x_i}{i}$$

16. Generalize the corollary to the case of generalized polynomials constructed from a Markoff system.

17. Define $E_n(f)$ as the infimum of $\|f - P\|$ as P ranges over all polynomials of degree $\leq n$. Prove that if f has a continuous $(n+1)$st derivative on $[0,1]$, then $E_n(f) \leq \|f^{(n+1)}\|/(n+1)!$. *Hint*: Use the error bound for Lagrange interpolation.

18. Show that to each $f \in C[a,b]$ there corresponds a system of points $\{x_{ni}\}$ with $n = 0,1,\ldots$ and $i = 0,\ldots,n$ such that if L_nf is the Lagrange interpolating polynomial to f on the nodes $x_{n0}, x_{n1}, \ldots, x_{nn}$, then $L_nf \to f$ uniformly. [Marcinkiewicz, 1937]

19. Prove, without assuming the Haar conditions, that if $P = \sum_{i=1}^{n} c_ig_i$ is a best approximation to f on a compact set X, then there is a finite subset X_0 containing at most $n+1$ points such that P is a best approximation to f on X_0.

20. The set of functions $\{1, \cos x, \ldots, \cos nx, \sin x, \ldots, \sin nx\}$ satisfies the Haar condition on any interval $[a, a+2\pi)$. *Hint*: $\sum_{k=0}^{n} (a_k \cos k\theta + b_k \sin k\theta)$ can be put into the complex form $e^{-in\theta} \sum_{k=0}^{2n} c_ke^{ik\theta}$. The polynomial $\sum c_kz^k$ can have at most $2n$ roots.

21. Let $E_n(f) = \min \|f - P\|$, where P ranges over all ordinary polynomials of degree $\leq n$. Given $x_0 < \cdots < x_{n+1}$, let λ and P be determined by the conditions $\lambda(-1)^i + P(x_i) = f(x_i)$. Write $\lambda = \lambda(x_0,\ldots,x_{n+1})$. Prove that

$$E_n(f) = \max_{x_0 < \cdots < x_{n+1}} |\lambda(x_0,\ldots,x_{n+1})|$$

22. Establish the following formula for the function λ defined in Prob. 21:

$$\lambda(x_0,\ldots,x_{n+1}) = \sum_{i=0}^{n+1} \frac{\alpha_i}{\alpha} f(x_i)$$

with $\alpha_i = \prod_{\substack{j=0 \\ j\neq i}}^{n+1} (x_i - x_j)^{-1}$ and $\alpha = \sum_{i=0}^{n+1} (-1)^i \alpha_i$.

23. $\{f_1,\ldots,f_n\}$ is independent iff there exist points $x_1,\ldots,x_n$ such that $\det f_i(x_j) \neq 0$. *Hint*: If $\{f_1,\ldots,f_n\}$ is independent while $\det f_i(x_j) = 0$ for all $x_1,\ldots,x_n$, then the function $\phi(x_n) = \det f_i(x_j)$ is identically zero for all $x_1,\ldots,x_{n-1}$. By expanding this determinant we obtain $\phi(x) = \sum_{i=1}^{n} \lambda_i f_i(x)$. By the independence, $\lambda_n = 0$. Hence $\det f_i(x_j) = 0$ $(1 \leq i, j \leq n - 1)$. Apply induction.

***24.** *Generalized de La Vallée Poussin theorem.* If we relax the Haar condition, we may still prove that

$$0 \in \mathcal{H}\,\{e(x_i)\hat{x}_i\} \Rightarrow E(f) \geq \min |\, e(x_i)\,|$$

Here $e = p - f$, p is selected from any prescribed linear subspace P of $C[X]$, and E is the distance from f to P. $\mathcal{H}$ denotes the convex hull, and x_i are any points from X.

***25.** Prove the alternation theorem for ordinary polynomial approximation by the following direct argument. Let x_0 be the first point of $[a,b]$ at which r attains the value $\pm \| r \|$, say for definiteness $r(x_0) = \| r \|$. Let x_1 be the first point of $[x_0,b]$ at which $r(x_1) = -\| r \|$. Continuing in this way, we define points $x_0 < \cdots < x_k$ such that $r(x_i) = (-1)^i \| r \|$. If $k > n$, we are finished. Otherwise define z_i (for $i = 1, \ldots, k$) to be the largest root of r in $[x_{i-1},x_i]$. Put $h(x) = \prod (z_i - x)$, and show that for some $\lambda > 0$, $P - \lambda h$ is a better approximation to f than is P.

26. Let $\{g_1,\ldots,g_n\}$ satisfy the Haar conditions on $[a,b]$. Let $\{x_0,\ldots,x_n\}$ be any set of $n + 1$ points in $[a,b]$. For each i let P_i be a generalized polynomial such that $P_i(x_j) = f(x_j)$ for $j = 0, \ldots, i - 1, i + 1, \ldots, n$. Then the generalized polynomial which *best* approximates f on $\{x_0,\ldots,x_n\}$ is a convex linear combination of $P_0, \ldots, P_n$. [de La Vallée Poussin, 1910]

27. Let $P(x) = a_0 + a_1 x + \cdots + a_n x^n$ be the polynomial of best approximation to $\sqrt{x}$ on $[0,1]$. Then $a_0 > 0$ and $\operatorname{sgn} a_i = (-1)^{i-1}$ for $i \geq 1$. *Hint*: Since $P(x) - \sqrt{x}$ alternates $n + 2$ times, it has $n + 1$ roots. So then has the polynomial $Q(t) = P(t^2) - t$. Use Descartes's rule of signs. [de La Vallée Poussin, 1910]

28. If $f^{(n+1)}(x) \geq 0$ throughout $[a,b]$, and if P is the polynomial of degree $\leq n$ which best approximates f, then there are *no more than* $n + 2$ points of alternation for $f - P$. *Hint*: Prove first that there must exist $n + 1$ points x_i where $P(x_i) = f(x_i)$. Then use the formula for the error in Lagrangian interpolation [equation (1), Sec. 2]. See [Shohat, 1941].

***29.** In order to discover the polynomial of degree $< n$ which best approximates x^n on $[-1,1]$, Tchebycheff reasoned as follows. The error function $P(x) = x^n + a_{n-1}x^{n-1} + \cdots$ must have $n + 1$ extrema of equal magnitude ϵ. These may be the points ± 1 or internal points, where $P'(x) = 0$. We conclude that the polynomials $P^2 - \epsilon^2$ and $(1 - x^2)(P')^2$ have the same roots and are therefore scalar multiples of each other. We are led to the differential equation

$$(1 - x^2)[P'(x)]^2 + n^2[P(x)]^2 - n^2\epsilon^2 = 0$$

[Tchebycheff, Collected Works I, p. 295]

5. The Unicity Problem

As in the preceding section we suppose that a certain linearly independent set of continuous functions $\{g_1, \ldots, g_n\}$ has been prescribed on an interval $[a,b]$. We contemplate the approximation of other functions f by "generalized polynomials" $\sum c_i g_i$ in such a way as to minimize the norm $\| f - \sum c_i g_i \| = \max_{a \leq x \leq b} | f(x) - \sum c_i g_i(x) |$. The question to be considered is: When will such a best approximation be unique?

Unicity Theorem. *If the functions $g_1, \ldots, g_n$ are continuous on $[a,b]$ and satisfy the Haar condition, then the best approximation of each continuous function by a generalized polynomial $\sum c_i g_i$ is unique.*

Proof. If possible, let P and Q be distinct generalized polynomials of best approximation to a given function f. By the triangle inequality we see easily that $\frac{1}{2}(P + Q)$ is also a best approximation to f. By the alternation theorem (Sec. 4), there exist points $x_0 < x_1 < \cdots < x_n$ in $[a,b]$ such that $f(x_i) - \frac{1}{2}(P + Q)(x_i) = (-1)^i \epsilon$, where $| \epsilon | = \| f - P \|$. Hence

$$\tfrac{1}{2}[f(x_i) - P(x_i)] + \tfrac{1}{2}[f(x_i) - Q(x_i)] = (-1)^i \epsilon$$

Since the bracketed numbers do not exceed $| \epsilon |$ in magnitude, we see that

$$(f - P)(x_i) = (f - Q)(x_i) = (-1)^i \epsilon$$

But then P and Q are equal at $n + 1$ points x_i, which is not possible. (See Prob. 1, Sec. 4.) ∎

There are two directions in which this theorem may be improved. In the first place, we can prove a converse theorem due to Haar: If the generalized polynomials of best approximation to f are unique for *all* choices of f, then the set of functions $\{g_1, \ldots, g_n\}$ satisfies the Haar condition. In the second place, we can obtain some detailed information about how fast $\| f - P \|$ increases as P recedes from the best approximation. We consider the latter now. Incidentally, we relax the condition that X be an interval. Any compact metric space is suitable here.

Strong Unicity Theorem. *Let the set of functions $\{g_1, \ldots, g_n\}$ satisfy the Haar condition. Let P_0 be the generalized polynomial of best approximation to a given continuous function f. Then there exists a constant $\gamma > 0$ depending on f such that for any generalized polynomial P, $\| f - P \| \geq \| f - P_0 \| + \gamma \| P_0 - P \|$.*

Proof. If $\| f - P_0 \| = 0$, then $\gamma = 1$ since $\| P - P_0 \| \leq \| P - f \|$. Let us assume therefore that $\| f - P_0 \| > 0$. By the characterization theorem of page 73 there exist points $x_0, \ldots, x_k$ and signs $\sigma_0, \ldots, \sigma_k$ such that $f(x_i) -$

$P_0(x_i) = \sigma_i \| f - P_0 \|$ and such that the origin of n space lies in the convex hull of the n-tuples $\sigma_i[g_1(x_i),\ldots,g_n(x_i)]$. Thus an equation $0 = \sum_{i=0}^{k} \theta_i \sigma_i g_j(x_i)$ is valid with $\theta_i > 0$ and $j = 1, \ldots, n$. By the Haar condition, $k \geq n$. By Carathéodory's theorem, we may assume that $k \leq n$. Hence $k = n$. Now let Q be any generalized polynomial of norm 1. We have $\sum_{i=0}^{n} \theta_i \sigma_i Q(x_i) = 0$. By the Haar condition, the numbers $\sigma_i Q(x_i)$ are not all zero. Since $\theta_i > 0$, we infer that at least one of $\sigma_i Q(x_i)$ is positive. Consequently the expression $\max_i \sigma_i Q(x_i)$ is a positive function of Q. Hence the number $\gamma = \min_{\|Q\|=1} \max_i \sigma_i Q(x_i)$ is positive, being the minimum of a positive continuous function on a compact set. Now let P be an arbitrary generalized polynomial. If $P = P_0$, the inequality to be proved is trivial. Otherwise, the generalized polynomial $Q = (P_0 - P)/\| P_0 - P \|$ is of unit norm. Consequently there is an index i for which $\sigma_i Q(x_i) \geq \gamma$. We have then

$$\begin{aligned} \| f - P \| &\geq \sigma_i(f - P)(x_i) \\ &= \sigma_i(f - P_0)(x_i) + \sigma_i(P_0 - P)(x_i) \\ &\geq \| f - P_0 \| + \gamma \| P_0 - P \| \end{aligned}$$ ■

Haar's Unicity Theorem. *The best approximation to a continuous function f by a generalized polynomial $\sum c_i g_i$ is unique for* all *choices of f if and only if $\{g_1,\ldots,g_n\}$ satisfies the Haar condition.*

Proof. Half of this theorem has already been established by the preceding theorem. For the other half, suppose that $\{g_1,\ldots,g_n\}$ does not satisfy the Haar condition. Then there exist points $x_1, \ldots, x_n$ such that the matrix $[g_i(x_j)]$ is singular. Let nonzero vectors $[a_1,\ldots,a_n]$ and $[b_1,\ldots,b_n]$ be selected orthogonal to the columns and rows, respectively, of this matrix. Thus $\sum_i a_i g_i(x_j) = 0$ and $\sum_j b_j g_i(x_j) = 0$. Put $Q = \sum_i a_i g_i$ so that $Q(x_j) = 0$. We may suppose that $\| Q \| < 1$. Select $f \in C[a,b]$ such that $\| f \| = 1$ and $f(x_j) = \operatorname{sgn} b_j$. Put $F(x) = f(x)[1 - | Q(x) |]$. Then $F(x_j) = f(x_j) = \operatorname{sgn} b_j$. Now for any generalized polynomial P, we must have $\| F - P \| \geq 1$. Indeed, if $\| F - P \| < 1$, then $\operatorname{sgn} P(x_j) = \operatorname{sgn} F(x_j) = \operatorname{sgn} b_j$, contradicting the equation $\sum b_j P(x_j) = 0$. Now for $0 \leq \lambda \leq 1$, λQ is a best approximation to F because $| F(x) - \lambda Q(x) | \leq | F(x) | + \lambda | Q(x) | = |f(x)| \times (1 - | Q(x) |) + \lambda | Q(x) | \leq 1 - | Q(x) | + \lambda | Q(x) | \leq 1$. ■

It might be conjectured that the inequality of the strong unicity theorem, viz., $\| f - P \| \geq \| f - P_0 \| + \gamma \| P - P_0 \|$, could be proved without assuming that the set $\{g_1, \ldots, g_n\}$ satisfies the Haar condition, but assuming instead that the best approximation P_0 was unique *for the particular f under consideration.* A counterexample is provided by taking $g_1(x) = x$ and $f(x) = x^2 - 1$ on the interval $[-1,1]$. The graph shows f and a typical approximation

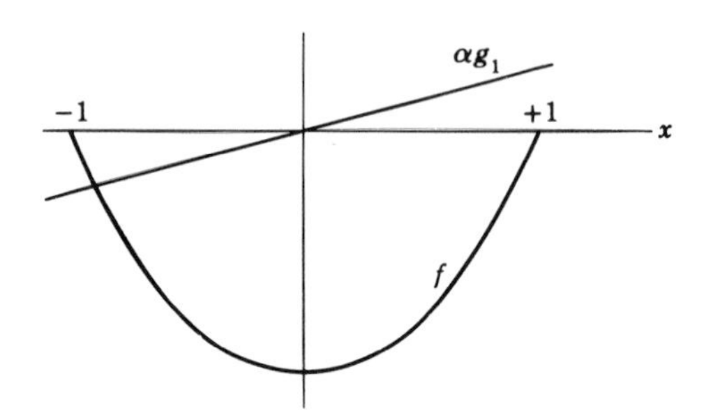

αg_1. If $\alpha \neq 0$, then the error of approximation near $x = 0$ will be greater than 1. In fact, if $|\alpha| < 1$, then the maximum of $|f(x) - \alpha g_1(x)|$ occurs at $\frac{1}{2}\alpha$ and has the value $1 + \frac{1}{4}\alpha^2$. Thus $\alpha = 0$ is the unique solution. However, if the inequality of the strong unicity theorem were valid, we would have for all α, $1 + \frac{1}{4}\alpha^2 \geq 1 + \gamma\alpha$, which is false for small α.

One of the interesting applications of the strong unicity theorem is in establishing that if a function f is altered slightly, its generalized polynomial of best approximation changes only slightly. To make this precise, suppose that a system of functions satisfying the Haar conditions on X is prescribed: $\{g_1, \ldots, g_n\}$. For each $f \in C[X]$ let $\mathfrak{J}f$ be the (unique) generalized polynomial of best approximation to f. Then $\mathfrak{J}$ is a continuous operator. In fact it satisfies at each point a Lipschitz condition:

Theorem. [*Freud*] *To each f_0 there corresponds a number $\lambda > 0$ such that for all f, $\|\mathfrak{J}f_0 - \mathfrak{J}f\| \leq \lambda \|f_0 - f\|$.*

Proof. By the strong unicity theorem there exists a constant γ depending only upon f_0 such that $\|f_0 - P\| \geq \|f_0 - \mathfrak{J}f_0\| + \gamma\|\mathfrak{J}f_0 - P\|$. Letting $P = \mathfrak{J}f$, we have

$$\begin{aligned}
\gamma\|\mathfrak{J}f_0 - \mathfrak{J}f\| &\leq \|f_0 - \mathfrak{J}f\| - \|f_0 - \mathfrak{J}f_0\| \\
&\leq \|f_0 - f\| + \|f - \mathfrak{J}f\| - \|f_0 - \mathfrak{J}f_0\| \\
&\leq \|f_0 - f\| + \|f - \mathfrak{J}f_0\| - \|f_0 - \mathfrak{J}f_0\| \\
&\leq \|f_0 - f\| + \|f - f_0\| + \|f_0 - \mathfrak{J}f_0\| - \|f_0 - \mathfrak{J}f_0\| \\
&= 2\|f_0 - f\|
\end{aligned}$$

Thus we may take $\lambda = 2\gamma^{-1}$. ∎

Problems

1. The operator $\mathfrak{J}$ defined on page 82 is not linear. But nevertheless $\mathfrak{J}(f+g) = \mathfrak{J}f + \mathfrak{J}g$ when g is a generalized polynomial. Also $\mathfrak{J}(\lambda f) = \lambda \mathfrak{J}f$. Other properties of $\mathfrak{J}$ are $\mathfrak{J}^2 = \mathfrak{J}$ and $\| \mathfrak{J}f \| \leq 2 \| f \|$.

2. Let $E(f) = \inf_{c_1,\ldots,c_n} \| f - \sum c_i g_i \|$. Prove the following properties of this functional:

   (*a*) $E(\lambda f) = |\lambda| E(f)$
   (*b*) $E(f+g) \leq E(f) + E(g)$
   (*c*) $E(f+g) = E(f)$ if g is a generalized polynomial
   (*d*) E is continuous
   (*e*) $E(f) \leq \| f \|$
   (*f*) $E(f + \lambda g) \to \infty$ as $\lambda \to \infty$ if g is not a generalized polynomial

   How many of these properties require the hypothesis that $\{g_1, \ldots, g_n\}$ satisfies the Haar condition?

3. Let a function $f \in C[a,b]$ and a positive number ϵ be given. Then there exists a number $\delta > 0$ such that $\| P - \mathfrak{J}f \| < \epsilon$ whenever P is a generalized polynomial satisfying $\| P - f \| < (1+\delta)E(f)$. [de La Vallée Poussin, 1919, p. 87]

4. If $\{g_1, \ldots, g_n\}$ does not satisfy the Haar condition, then generally a function f will have several best approximations of the form $\sum c_i g_i$. Let $\mathfrak{J}f$ denote the set of all such best approximations to f. Show that $\mathfrak{J}f$ is a convex set for all f. Is it possible for $\mathfrak{J}$ to be discontinuous in the sense that f and g may arbitrarily close while $\max_{P \in \mathfrak{J}f} \min_{Q \in \mathfrak{J}g} \| P - Q \|$ is large? What about $\min_{P \in \mathfrak{J}f} \min_{Q \in \mathfrak{J}g} \| P - Q \|$?

5. Systems of continuous functions satisfying the Haar condition do not exist on all compact metric spaces X. Let X denote a set in which there exists a closed curve having no double points, i.e., a subset of X of the form $\{\phi(t) : 0 \leq t \leq 1\}$, where ϕ is a one-to-one continuous function from $[0,1]$ to X satisfying $\phi(0) = \phi(1)$. Then there exists on X no system $\{g_1, g_2\}$ satisfying the Haar condition. Indeed, let x_1 and x_2 be two points lying on such a closed curve. By a continuous motion around this curve, x_1 and x_2 can be made to exchange positions without ever becoming coincident. In this process the determinant

   $$\begin{vmatrix} g_1(x_1) & g_2(x_1) \\ g_1(x_2) & g_2(x_2) \end{vmatrix}$$

   will change sign and therefore vanish for some pair of distinct points $(y_1 y_2)$. This contradicts the Haar condition.

6. The constant γ in the strong unicity theorem must satisfy $\gamma \leq 1$.

7. Let $\mathfrak{J}^{-1}[P] = \{ f \in C[a,b] : \mathfrak{J}f = P \}$ where $\mathfrak{J}$ is the operator defined on page 82. Describe $\mathfrak{J}^{-1}[P]$ geometrically.

8. Is the strong unicity theorem true for least-squares approximation, the inner product being $\langle f, g \rangle = \int_a^b f(x) g(x)\, dx$?

9. In the proof of the unicity theorem, it is only necessary to know that each best approximation reaches its maximum deviation from f in at least n points. This remark opens the way to unicity theorems when the domain for approximation is not an interval.

6. Discretization Errors: General Theory

For the practical problem of computing best approximations on an interval, it is often desirable to replace that interval by a finite set of points and to seek an approximation which is optimum on that set. It seems reasonable to expect that if this finite set of points somehow fills out the interval, leaving no wide gaps, then an approximation obtained in this way will be satisfactory. In numerical analysis, the maneuver of replacing the continuum by a discrete set is called *discretization*. An extreme case of this occurs in interpolation, where the number of points in the finite set is taken equal to the number of parameters in the approximating function. We have already seen (page 61) in the case of ordinary polynomials on $[-1,1]$, that for interpolation with n points one disposition of points is more favorable than others, viz., $x_i = \cos[(2i - 1)/2n]$. Similar conclusions will be reached in the next section for sets of arbitrarily many points. The results of this section are not limited to ordinary polynomials nor to intervals.

The errors due to discretization here are of two types. In the first place, an approximation which is optimum on a discrete set may involve large discrepancies outside the discrete set. In the second place, even if the maximum discrepancies are nearly a minimum, the approximating function may be far from the best one. The latter situation is less serious, since two approximations which are far from each other may be equally satisfactory. We shall see that in the case of approximation families satisfying the Haar condition, even the errors of the second type will go to zero as the discrete set fills up the interval.

It is necessary to establish some notation. Let us consider a compact metric space (X,d). We shall be particularly interested in the case when X is the interval $[-1,1]$ and $d(x,y)$ is taken to be $|x - y|$ or $(2/\pi)|\cos^{-1} x - \cos^{-1} y|$. For any subset Y of X we require a measure of how nearly Y fills up X. For this purpose we define the *density* of Y in X by the equation

$$|Y| = \max_{x \in X} \inf_{y \in Y} d(x,y)$$

The word *density* is abused here since $|Y| \to 0$ as Y becomes "dense." For example, a set Y of $n + 1$ equally spaced points in $[-1,1]$ has the density $1/n$ if the end points are included in Y and if $d(x,y) = |x - y|$.

In correspondence with any subset Y, we may define a seminorm in $C[X]$ by writing

$$\|f\|_Y = \sup_{y \in Y} |f(y)|$$

With this notation, the original norm in $C[X]$ is $\| \quad \|_X$. Since there will generally exist continuous nonzero functions which vanish on Y, we can have

$\| f \|_Y = 0$ and $f \neq 0$ simultaneously. Thus it is not possible to have a general inequality of the form $\| f \|_X \leq k \| f \|_Y$. In special cases, however, such an inequality is possible, as in the following lemma.

Lemma 1. *Let $\{g_1, \ldots, g_n\}$ be any set of continuous functions on the compact metric space X. To each $\alpha > 1$ there corresponds a positive δ such that $\| P \| < \alpha \| P \|_Y$ for all generalized polynomials $P = \sum c_i g_i$ and for all sets Y such that $| Y | < \delta$.*

Proof. If the set $\{g_1, \ldots, g_n\}$ is not linearly independent, we may replace it by a smaller set without altering the concept of a generalized polynomial. Therefore we may assume the independence of $\{g_1, \ldots, g_n\}$. Let θ denote the minimum of $\| \sum c_i g_i \|$ on the compact set defined by the equation $\sum | c_i | = 1$. Since $\{g_1, \ldots, g_n\}$ is independent, $\theta > 0$. Now define a function Ω on the positive reals as follows:

$$\Omega(\delta) = \max_{1 \leq i \leq n} \max_{d(x,y) \leq \delta} | g_i(x) - g_i(y) |$$

We may call this the *joint modulus of continuity* of the family $\{g_1, \ldots, g_n\}$. It clearly has the property $\Omega(\delta) \downarrow 0$ as $\delta \downarrow 0$ because of the (uniform) continuity of the functions g_i. Now let δ be small enough to ensure that $\Omega(\delta) < \theta$, and let Y be sufficiently "fine" that $| Y | < \delta$. Consider an arbitrary generalized polynomial $P = \sum c_i g_i$. Let x be a point of X where $| P(x) |$ is a maximum, and let y be a point of Y within distance δ from x. Then

$$\begin{aligned} \theta \sum | c_i | \leq \| P \| &\equiv | P(x) | \\ &\leq | P(x) - P(y) | + | P(y) | \\ &\leq \sum | c_i | \, | g_i(x) - g_i(y) | + \| P \|_Y \\ &\leq \Omega(\delta) \sum | c_i | + \| P \|_Y \end{aligned}$$

Hence
$$\sum | c_i | \leq \frac{\| P \|_Y}{\theta - \Omega(\delta)}$$

and
$$\| P \| \leq \Omega(\delta) \frac{\| P \|_Y}{\theta - \Omega(\delta)} + \| P \|_Y$$

$$= \| P \|_Y \left[1 + \frac{\Omega(\delta)}{\theta - \Omega(\delta)} \right]$$

Since the expression in brackets approaches 1 as $\delta \to 0$, and is independent of P, the proof is complete. ∎

If f is a bounded real-valued function defined on a metric space X, then a measure of its continuity is given by the function ω, defined for $\delta \geq 0$ by the equation

$$\omega(\delta) = \sup_{d(x,y) \leq \delta} | f(x) - f(y) |$$

This function ω is termed the *modulus of continuity* of f. For example, if the graph of f has a jump of magnitude 1, then $\omega(\delta)$ can never be less than 1. On the other hand, if f is *uniformly continuous*, then $\omega(\delta) \downarrow 0$ as $\delta \downarrow 0$. Indeed, given $\epsilon > 0$, there exists by uniform continuity a positive δ such that *for all* x and y,

$$d(x,y) \leq \delta \Rightarrow | f(x) - f(y) | \leq \epsilon$$

This is equivalent to the inequality $\omega(\delta) \leq \epsilon$.

Lemma 2. *Let X be a compact metric space and $f, g_1, \ldots, g_n$ arbitrary elements of $C[X]$. Then for any generalized polynomial $P = \sum c_i g_i$ the seminorms $\| f - P \|_Y$ converge to $\| f - P \|_X$, if $| Y | < \delta \to 0$, according to the inequality $\| f - P \| < \| f - P \|_Y + \omega(\delta) + \beta \| P \| \Omega(\delta)$, where β is independent of f and P, ω is the modulus of continuity of f, and Ω is the joint modulus of continuity defined in the proof of Lemma 1.*

Proof. As in Lemma 1, there is no loss of generality in assuming that the set $\{g_1, \ldots, g_n\}$ is independent. Let θ be defined as in Lemma 1 so that $\| \sum c_i g_i \| \geq \theta \sum | c_i |$ for arbitrary c_i. Now let $P = \sum c_i g_i$ be fixed. Select a point $x \in X$ where $| f(x) - P(x) |$ is a maximum. Select a point $y \in Y$ such that $d(x,y) < \delta$. Then $\| f - P \| = | f(x) - P(x) | \leq | f(x) - f(y) | + | f(y) - P(y) | + | P(y) - P(x) | \leq \omega(\delta) + \| f - P \|_Y + \sum | c_i | \, | g_i(y) - g_i(x) |$. The last term does not exceed $\sum_i | c_i | \max_i | g_i(y) - g_i(x) | \leq \theta^{-1} \| P \| \Omega(\delta)$. ■

Theorem 1. *Let X be a compact metric space and $f, g_1, \ldots, g_n$ elements of $C[X]$. For each $Y \subset X$ let P_Y denote a generalized polynomial $\sum c_i g_i$ of best approximation to f on Y. Then $\| f - P_Y \| \to \| f - P_X \|$ as $| Y | \to 0$.*

Proof. The inequality $\| f - P_Y \|_Y \leq \| f - P_X \|_Y \leq \| f - P_X \| \leq \| f - P_Y \|$ is obvious. By Lemma 2, $\| f - P_Y \| - \| f - P_X \| \leq \| f - P_Y \| - \| f - P_Y \|_Y \leq \omega(\delta) + \beta \| P_Y \| \Omega(\delta)$. To complete the proof we estimate $\| P_Y \|$ as follows: $\| P_Y \|_Y \leq \| P_Y - f \|_Y + \| f \|_Y \leq \| 0 - f \|_Y + \| f \|_X \leq 2 \| f \|$. Hence by Lemma 1, $\| P_Y \| \leq 2\alpha \| f \|$. Note that we have established the inequality

$$\| f - P_Y \| - \| f - P_X \| \leq \omega(\delta) + 2\alpha\beta \| f \| \Omega(\delta)$$

■

In order to state our next result, we require the concept of a *fundamental* set. A subset G of $C[X]$ is said to be fundamental if each element of $C[X]$ can be arbitrarily well approximated by linear combinations of elements of G. Formally, given $f \in C[X]$ and $\epsilon > 0$, there must exist $g_i \in G$ and coefficients c_i such that $\| f - \sum_{i=1}^{n} c_i g_i \| < \epsilon$. For example, the set of functions $\{1, x, x^2, \ldots\}$ is fundamental in $C[a,b]$ by the Weierstrass theorem.

Theorem 2. *Let X be a compact metric space and $\{g_1, g_2, \ldots\}$ a fundamental set in $C[X]$. Then it is possible to prescribe a sequence of finite subsets Y_n of X in such a way that for all $f \in C[X]$ the generalized polynomials $P_n = \sum_{i=1}^{n} c_i^{(n)} g_i$ of best approximation to f on Y_n converge uniformly to f as $n \to \infty$.*

Proof. For each set $\{g_1, \ldots, g_n\}$, a constant β_n may be computed as in Lemma 2. Also a function α_n and a joint modulus of continuity Ω_n may be computed as in Lemma 1 (they depend on δ). Let the finite sets Y_n be so chosen that if $\delta_n = | Y_n |$, then

$$\alpha_n \beta_n \Omega_n(\delta_n) \to 0 \qquad (\text{as } n \to \infty)$$

(The proof that finite sets in a compact metric space can have arbitrarily small density is outlined in Prob. 9.) From the fundamentality of $\{g_1, \ldots\}$ it follows that the distance $E_n(f)$ from f to the subspace spanned by $\{g_1, \ldots, g_n\}$ tends to zero as $n \to \infty$. Hence from the final inequality in the preceding proof we have

$$\| f - P_n \| \leq E_n(f) + \omega(\delta_n) + 2\alpha_n \beta_n \| f \| \Omega_n(\delta_n)$$

Since the right member goes to zero as $n \to \infty$, we have $P_n \to f$. ∎

In the preceding theorems we have not found it necessary to assume that our system of functions $\{g_1, \ldots, g_n\}$ satisfied the Haar condition. On the other hand, we have not yet proved that the discrete approximations P_Y converge to P_X (and indeed the latter need not be unique when the Haar condition fails). It turns out that unicity of P_X is all that is necessary for such a result.

Theorem 3. *Let X be a compact metric space and $f, g_1, \ldots, g_n$ elements of $C[X]$. If f possesses a unique generalized polynomial $P_X = \sum c_i g_i$ of best approximation on X, then its best approximations P_Y on subsets Y converge to P_X as $| Y | \to 0$.*

Proof. For every $\epsilon > 0$ define

$$\delta(\epsilon) = \inf_{\|P-P_X\|\geq\epsilon} [\| f - P \| - \| f - P_X \|]$$

where P denotes a generic linear combination of $g_1, \ldots, g_n$. Since the bracketed expression becomes infinite with $\| P \|$, its infimum is achieved and is positive (by the uniqueness of P_X). From the definition of δ we have

$$\| P - P_X \| \geq \epsilon \Rightarrow \| f - P \| - \| f - P_X \| \geq \delta(\epsilon)$$

Thus in order to ensure that $\| P_Y - P_X \| < \epsilon$ it is sufficient to have $\| f - P_Y \| - \| f - P_X \| < \delta(\epsilon)$; by Theorem 1 this will occur when $| Y |$ is sufficiently small. ∎

We observe in passing that if our set of functions $\{g_1, \ldots, g_n\}$ satisfies the Haar condition, then from the strong unicity theorem of Sec. 5, we have the inequality

$$\| P_Y - P_X \| \leq \gamma^{-1}[\| P_Y - f \| - \| P_X - f \|]$$

From the proof of the theorem on page 86 we obtain then an estimate

$$\| P_Y - P_X \| \leq \gamma^{-1}[\omega(\delta) + 2\alpha\beta \| f \| \Omega(\delta)]$$

Here, the constant γ depends on f but α and β do not. (They come from Lemmas 1 and 2, respectively.)

Problems

1. Show that if $Y \subset X = [a,b]$ and $d(x,y) = | x - y |$, then

$$| Y | = \tfrac{1}{2} \max_{y \in Y \cup \{a\}} \min_{\substack{x \in Y \cup \{b\} \\ x > y}} (x - y)$$

2. For a subset Y of a metric space we define the *diameter* by the equation $\operatorname{diam}(Y) = \sup_{x,y \in Y} d(x,y)$. For an interval X with its usual metric, show that a subset of $n - 1$ points must have density at least $(1/2n) \operatorname{diam}(Y)$. Show that this is not true for general metric spaces. In particular, a finite subset can have density zero. If a metric d on an interval has the property that $d(x,y)$—as a function of x—is increasing when $y \leq x$, and decreasing when $x \leq y$, what is the minimum density of a set of $n - 1$ points?
3. If $d(x,y) = (2/\pi) | \cos^{-1} x - \cos^{-1} y |$ what is the density of a set Y of $n + 1$ equally spaced points on $[-1,1]$, assuming that $\pm 1 \in Y$? What is the minimum density of a set of $n + 1$ points?
4. For the functions $g_i(x) = x^i$ $(i = 0, \ldots, n)$ compute the joint modulus of continuity on $[-1,1]$, using the usual metric. *Ans.* $\Omega(\delta) = 1 - (1 - \delta)^n$.

*5. On $[-1,1]$, define metrics $d_1(x,y) = | x - y |$ and $d_2(x,y) = (2/\pi) | \cos^{-1} x - \cos^{-1} y |$. Show that $d_2 \leq \sqrt{2d_1}$. Thus if the density of a set Y is measured in both metrics, then $| Y |_2 \leq \sqrt{2 | Y |_1}$.

6. Solve Prob. 4, using the metric of Prob. 3.

7. Compute the moduli of continuity with the usual metric for some elementary functions. If f satisfies a Lipschitz condition, $|f(x) - f(y)| \leq \lambda |x - y|$, what can you say about ω? What is ω in the case of $f(x) = \sqrt{x}$ on $[0,1]$? If f is monotone, then is ω additive: $\omega(\delta_1 + \delta_2) = \omega(\delta_1) + \omega(\delta_2)$?
8. The modulus of continuity ω of a continuous function f on $[a,b]$ has the property $\omega(\delta_1 + \delta_2) \leq \omega(\delta_1) + \omega(\delta_2)$. It also has the property $\omega(\delta) \downarrow 0$ as $\delta \downarrow 0$. Show that ω is continuous. It therefore has a modulus of continuity of its own. Show that the modulus of continuity of ω is ω itself. (The usual metric is to be employed, and the interval should have 0 for its left end point.)
9. If X is a compact metric space and $\delta > 0$, then there exists a finite set $Y \subset X$ such that $|Y| < \delta$. *Hint*: If not, define a sequence $x_1, x_2, \ldots$ such that $d(x_{n+1},x_i) \geq \delta$ for $i = 1, \ldots, n$. Show that such a sequence has no convergent subsequence.
10. If two moduli of continuity of $f \in C[-1,1]$ are computed using two metrics

$$d_1(x,y) = |x - y| \quad \text{and} \quad d_2(x,y) = |\cos^{-1} x - \cos^{-1} y|$$

then $\omega_2 \leq \omega_1$.

11. If we compute moduli of continuity (with the usual metric) for two functions $f_1 \in C[-1,1]$ and $f_2 \in C[0,\pi]$, where $f_2(\theta) = f_1(\cos\theta)$, then $\omega_2 \leq \omega_1$.
12. An alternative proof of Theorem 1 may be given by applying Lemma 1 to the set of functions $\{f, g_1, \ldots, g_n\}$.
13. Lemma 1 can also be proved by use of the function $\Omega(\delta) = \max\limits_{\|P\|=1} \max\limits_{d(x,y)\leq\delta} |P(x) - P(y)|$

where $P = \sum c_i g_i$.

*14. What happens in Prob. 8 if we use a metric other than the usual one on $[a,b]$?

7. Discretization: Algebraic Polynomials. The Inequalities of Markoff and Bernstein

The results of the preceding section take on greater significance when applied to the compact metric space $X = [-1,1]$ (with its usual metric) and to the base functions $1, x, x^2, \ldots, x^n$. Thus the final inequality of page 88 ensures that the best approximations $P_Y = \sum_{k=0}^{n} c_k x^k$ for a function f computed on a subset $Y \subset X$ will converge to P_X as $|Y| \to 0$. We are now going to study this polynomial convergence problem *ab initio*, i.e., without appeal to the results of Sec. 6. The analysis will require the classical inequalities of Markoff and Bernstein, and we therefore will establish these first, using a method of proof due to Pólya and Szegö.

Lemma 1. *For any polynomial P of degree $\leq n - 1$,* $\max\limits_{-1\leq x\leq 1} |P(x)| \leq$

$$\max_{-1\leq x\leq 1} |n\sqrt{1 - x^2}P(x)|.$$

Proof. Let M denote the right member of this inequality. We must show that $|P(x)| \leq M$ for all $x \in [-1,1]$. Now, the zeros of the nth Tchebycheff polynomial T_n are the points $\xi_i = \cos[(2i-1)\pi/2n]$. If x is in the interval $[\xi_n,\xi_1]$, then

$$\sqrt{1-x^2} \geq \sqrt{1-\xi_1^2} = \sqrt{1-\cos^2\frac{\pi}{2n}} = \sin\frac{\pi}{2n} \geq \frac{1}{n}$$

Hence for such x, $|P(x)| \leq n\sqrt{1-x^2}\,|P(x)| \leq M$. For the remaining points of $[-1,1]$ we apply the Lagrange interpolation formula (with nodes ξ_i) to the polynomial P (cf. Prob. 5, Sec. 2):

$$P(x) = \frac{1}{n}\sum_{i=1}^{n} P(\xi_i)\,\frac{T_n(x)(-1)^{i-1}\sqrt{1-\xi_i^2}}{x-\xi_i}$$

Since either $x < \xi_n$ or $x > \xi_1$, the numbers $x - \xi_i$ have the same sign. Hence

$$|P(x)| \leq \frac{M}{n^2}\sum_{i=1}^{n}\left|\frac{T_n(x)}{x-\xi_i}\right| = \frac{M}{n^2}\left|\sum_{i=1}^{n}\frac{T_n(x)}{x-\xi_i}\right|$$

Since $T_n(x) = 2^{n-1}\prod(x-\xi_i)$, we have

$$T_n'(x) = \sum_{i=1}^{n}\left[2^{n-1}\prod_{\substack{j=1\\ j\neq i}}^{n}(x-\xi_j)\right] = \sum_{i=1}^{n}\frac{T_n(x)}{x-\xi_i}$$

Putting $x = \cos\theta$ and using the inequality $|\sin n\theta/\sin\theta| \leq n$, which may be proved by induction, we have $|T_n'(x)| = n\,|\sin n\theta/\sin\theta| \leq n^2$. Hence $|P(x)| \leq M$ as asserted. ∎

The next inequalities are concerned with *trigonometric polynomials*. These are expressions of the form

$$S(\theta) = \sum_{k=0}^{n}(a_k\cos k\theta + b_k\sin k\theta)$$

If $|a_n| + |b_n| \neq 0$, then S is said to be of degree n. If S is an even function, then $S(\theta) = \frac{1}{2}[S(\theta) + S(-\theta)]$, and S can therefore be written with cosine terms only. By appealing to the unicity of representation of a trigonometric polynomial (which we have not proved), we may conclude that an even trigonometric polynomial may contain *only* cosine terms. Similarly, an odd trigonometric polynomial contains only sine terms.

Lemma 2. *If S is an odd trigonometric polynomial of degree $\leq n$, then*
$$\max_{-\pi\leq\theta\leq\pi}|S(\theta)/\sin\theta| \leq n\max_{-\pi\leq\theta\leq\pi}|S(\theta)|.$$

Proof. By induction, using the identity

$$\sin(k+1)\theta = \sin k\theta\cos\theta + \sin\theta\cos k\theta$$

we see that $\sin k\theta/\sin\theta$ is a polynomial of degree $k-1$ in $\cos\theta$. Hence there is an algebraic polynomial P of degree $\leq n-1$ such that $P(\cos\theta) = S(\theta)/\sin\theta$. Applying Lemma 1 to P yields the desired inequality at once. ∎

Bernstein's Inequality. *For any trigonometric polynomial S of degree $\leq n$,*

$$\max_{-\pi\leq\theta\leq\pi} | S'(\theta) | \leq n \max_{-\pi\leq\theta\leq\pi} | S(\theta) |.$$

Proof. Define $f(\alpha,\theta) = \frac{1}{2}[S(\alpha + \theta) - S(\alpha - \theta)]$. Then for each fixed α, $f(\alpha,\theta)$ is an odd trigonometric polynomial in θ of degree $\leq n$. If $M = \max | S(\theta) |$, then $|f(\alpha,\theta) | \leq M$, and consequently by Lemma 2, $|f(\alpha,\theta)/\sin \theta | \leq nM$. But $S'(\alpha) = \lim_{\theta\to 0} \{[S(\alpha + \theta) - S(\alpha - \theta)]/2\theta\} = \lim_{\theta\to 0} [f(\alpha,\theta)/\sin \theta] \sin \theta/\theta$. Hence $| S'(\alpha) | \leq nM$, as required. ∎

Markoff's Inequality. *For any polynomial P of degree $\leq n$, $\max_{-1\leq x\leq 1} | P'(x) | \leq n^2 \max_{-1\leq x\leq 1} | P(x) |$.*

Proof. If $M = \max_{-1\leq x\leq 1} | P(x) |$, then $\max_{-\pi\leq\theta\leq\pi} | P(\cos \theta) | = M$. But $P(\cos \theta)$ is a trigonometric polynomial of degree $\leq n$ (cf. Prob. 2*b*, Sec. 2), and consequently by Bernstein's inequality $| P'(\cos \theta) \sin \theta | \leq nM$. In other words, $| P'(x) \sqrt{1 - x^2} | \leq nM$ when $-1 \leq x \leq 1$. Applying Lemma 1 to P', we have $| P'(x) | \leq n^2M$. ∎

We are now ready to prove some theorems analogous to those of Sec. 6, but specialized to ordinary polynomials on $[-1,1]$.

Lemma 3. *Let P be an algebraic polynomial of degree $\leq n$ defined on $X = [-1,1]$. The following are lower bounds for $\| P \|_Y/\| P \|$:*

(*i*) $1 - n^2 | Y |$ *when* $| Y | = \max_{x\in X} \min_{y\in Y} | x - y |$

(*ii*) $1 - n | Y |$ *when* $| Y | = \max_{x\in X} \min_{y\in Y} | \cos^{-1} x - \cos^{-1} y |$

(*iii*) $1 - \frac{1}{2}n^2 | Y |^2$ *when* $| Y | = \max_{x\in X} \min_{y\in Y} | \cos^{-1} x - \cos^{-1} y |$

Proof. (*i*) Select $x \in X$ where $| P(x) | = \| P \|$. If $| Y | < \delta$, then there is a $y \in Y$ such that $| x - y | < \delta$. Hence from Markoff's inequality above, $\| P \| = | P(x) | \leq | P(x) - P(y) | + | P(y) | = | P'(\xi) | | x - y | + | P(y) | \leq n^2 \| P \| \delta + \| P \|_Y$. Dividing by $\| P \|$ gives us $\| P \|_Y/\| P \| \geq 1 - n^2\delta$. Since this is true for all $\delta > | Y |$, it is true for $\delta = | Y |$.

(*ii*) Select $x \in X$ where $| P(x) | = \| P \|$. If $| Y | < \delta$, then there is a $y \in Y$ such that $| \cos^{-1} x - \cos^{-1} y | < \delta$. Define $S(\theta) = P(\cos \theta)$. Then S is a trigonometric polynomial of degree equal to that of P. (Indeed, if P is written in Tchebycheff polynomials, $P = \sum c_k T_k$, then $S(\theta) = \sum c_k \cos k\theta$). Now let $\alpha = \cos^{-1} x$, $\beta = \cos^{-1} y$. By Bernstein's inequality, $\| P \| = | S(\alpha) | \leq | S(\alpha) - S(\beta) | + | S(\beta) | \leq n \| S \| \delta + \| P \|_Y = n \| P \| \delta + \| P \|_Y$. Proceeding exactly as before, we obtain (*ii*).

(*iii*) Proceeding as in (*ii*), we observe that $|S(\alpha)| = \|S\|$, and consequently $S'(\alpha) = 0$. Thus $|S'(\theta)| = |S'(\theta) - S'(\alpha)| = |S''(\theta_1)|\,|\theta - \alpha| \leq n^2\|S\|\,|\theta - \alpha|$. We have used Bernstein's inequality twice here. Now write

$$\begin{aligned}\|P\| = |S(\alpha)| &= \left|S(\beta) + \int_\beta^\alpha S'(\theta)\,d\theta\right| \\ &\leq |S(\beta)| + n^2\|S\|\left|\int_\beta^\alpha |\theta - \alpha|\,d\theta\right| \\ &= |S(\beta)| + n^2\|S\|\tfrac{1}{2}|\alpha - \beta|^2 \\ &\leq \|P\|_Y + \tfrac{1}{2}n^2\|P\|\,\delta^2\end{aligned}$$

Dividing by $\|P\|$ gives (*iii*). ■

Theorem 1. *Given $X = [-1,1]$, $f \in C[X]$, and $Y \subset [-1,1]$, let P_Y denote the algebraic polynomial of degree $\leq n$ which best approximates f on Y. Then $P_Y \to P_X$ as $|Y| \to 0$, according to the following estimates:*

(*i*) $\|P_X - P_Y\| \leq \gamma^{-1}[\omega(\delta) + 2n^2\|f\|\,\delta(1 - n^2\delta)^{-1}]$, *valid when* $\delta = \max_{x\in X}\min_{y\in Y}|x - y| < n^{-2}$

(*ii*) $\|P_X - P_Y\| \leq \gamma^{-1}[\omega(\delta) + 2n\|f\|\,\delta(1 - n\delta)^{-1}]$, *valid when* $\delta = \max_{x\in X}\min_{y\in Y}|\cos^{-1}x - \cos^{-1}y| < n^{-1}$

Proof. We shall give the proof of (*i*) in detail, leaving part (*ii*) to the reader. From the strong unicity theorem (Sec. 5) we have

$$\|P_X - P_Y\| \leq \gamma^{-1}[\|f - P_Y\| - \|f - P_X\|]$$

Now let x be a point where $|f(x) - P_Y(x)| = \|f - P_Y\|$. If $|Y| < \delta$, then we may find $y \in Y$ such that $|x - y| < \delta$. Thus by the mean-value theorem and Markoff's inequality,

$$\begin{aligned}\|f - P_Y\| &= |f(x) - P_Y(x)| \\ &\leq |f(x) - f(y)| + |f(y) - P_Y(y)| + |P_Y(y) - P_Y(x)| \\ &\leq \omega(\delta) + \|f - P_Y\|_Y + n^2\|P_Y\|\,|x - y| \\ &\leq \omega(\delta) + \|f - P_X\|_X + n^2\|P_Y\|\,\delta\end{aligned}$$

The proof of (*i*) now requires only that we show $\|P_Y\| \leq 2\|f\|(1 - n^2\delta)^{-1}$. It is clear that $\|P_Y\|_Y \leq \|P_Y - f\|_Y + \|f\|_Y \leq \|0 - f\|_Y + \|f\|_Y \leq 2\|f\|$. Hence from the preceding lemma,

$$\|P_Y\| \leq \|P_Y\|_Y(1 - n^2\delta)^{-1} \leq 2\|f\|(1 - n^2\delta)^{-1}$$

Thus the inequality (*i*) is proved for all $\delta > |Y|$. Since the left member does not involve δ, and the right member is continuous in δ, it follows that (*i*) is true for $\delta = |Y|$. ■

Since inequality (*ii*) is more favorable than inequality (*i*), we see again the apparent advantage of distributing the discrete points for polynomial approximation in a nonuniform manner. Although this theorem does not disclose what the *optimum* distribution of points may be, it suggests that if we have m points at our disposal on $[-1,1]$, we should put $y_i = \cos[(2i-1)\pi/2m]$ for $i = 1, \dots, m$, thus achieving

$$\delta = \max_x \min_y |\cos^{-1} x - \cos^{-1} y| = \frac{\pi}{2m}$$

Inequality (*ii*) would then read, assuming $m > (\pi/2)n$,

$$\|P_X - P_Y\| \le \gamma^{-1}\left[\omega\left(\frac{\pi}{2m}\right) + \frac{2\pi n}{2m - \pi n}\|f\|\right]$$

In order to optimize the bound in (*i*) we would take $y_i = -1 + (2i-1)/m$, thus achieving $\delta = \max_x \min_y |x - y| = 1/m$. Inequality (*i*) would then read, assuming $m > n^2$,

$$\|P_X - P_Y\| \le \gamma^{-1}\left[\omega\left(\frac{1}{m}\right) + \frac{2n^2}{m - n^2}\|f\|\right]$$

The advantages of the former distribution of points will be more decisive if $n \to \infty$. We have the following theorem.

Theorem 2. *For each n let Y_n be a subset of $X = [-1,1]$ such that $\delta_n = \max_{x \in X} \min_{y \in Y_n} |\cos^{-1} x - \cos^{-1} y| \le \lambda n^{-1}$ for some $\lambda < \sqrt{2}$. Then the polynomials P_n of degree $\le n$ of best approximation on Y_n to $f \in C[-1,1]$ converge uniformly to f as $n \to \infty$.*

Proof. From part (*iii*) of Lemma 3, if $n\delta_n \le \lambda < \sqrt{2}$, then for any polynomial P of degree $\le n$, $\|P\| \le M\|P\|_{Y_n}$, with $M = (1 - \frac{1}{2}\lambda^2)^{-1}$. If Q_n is the best approximation to f of degree $\le n$, then

$$\begin{aligned}
\|P_n - f\| &\le \|P_n - Q_n\| + \|Q_n - f\| \\
&\le M\|P_n - Q_n\|_{Y_n} + \|Q_n - f\| \\
&\le M\|P_n - f\|_{Y_n} + M\|f - Q_n\|_{Y_n} + \|Q_n - f\| \\
&\le M\|Q_n - f\| + M\|f - Q_n\| + \|Q_n - f\| \\
&= (2M+1)\|Q_n - f\| \to 0
\end{aligned}$$

■

Problems

1. Complete the proof of Theorem 1.

2. Prove that for any polynomial P of degree $\leq n$, $|P'(x)| \leq n(1-x^2)^{-1/2} \|P\|$, where $\|P\| = \max_{-1\leq x\leq 1} |P(x)|$, and $-1 < x < 1$.

3. Prove that for any polynomial P of degree $\leq n$, $\|P^{(k+1)}\| \leq n^2(n-1)^2 \cdots (n-k)^2 \times \|P\|$ where $\|P\| = \max_{-1\leq x\leq 1} |P(x)|$.

4. Prove that for any polynomial P of degree $\leq n$,

$$\max_{a\leq x\leq b} |P'(x)| \leq \frac{2n^2}{b-a} \max_{a\leq x\leq b} |P(x)|$$

5. Prove that for any polynomial of the form $P(x) = \sum_{k=0}^{n} c_k x^k$, $\max_k |c_k| \leq n^{2n} \|P\|$.

What is the *best* inequality of this type that you can prove? Can you prove an inequality of the form $\|P\| \leq A_n \max_k |c_k|$?

6. Show that $\|P'\| \leq \frac{1}{2}n^2 [\max P(x) - \min P(x)]$. Thus for a polynomial which is positive on $[-1,1]$, $\|P'\| \leq (n^2/2) \|P\|$. *Hint*: P' is not altered if P is augmented by a constant.

***7.** In Markoff's inequality, the factor n^2 changes abruptly to $(n-1)^2$ as the leading coefficient c_n in the polynomial becomes zero. Find an inequality like Markoff's in which the factor depends continuously on c_n. *Hint*: The function

$$\phi(n,\lambda) = \max \left\{ \frac{\|P'\|}{\|P\|} : P(x) = \sum_{k=0}^{n} c_k x^k, |c_n| \leq \lambda \sum |c_k| \right\}$$

is continuous in λ. Also, $\phi(n,1) = n^2$ and $\phi(n,0) = (n-1)^2$.

8. If P is the nth Tchebycheff polynomial, then $\|P'\| / \|P\| = n^2$. Hence the constant n^2 in Markoff's inequality cannot be replaced by a smaller one.

9. Prove that $|\sin n\theta / \sin \theta| \leq n$, limiting values being used when $\sin \theta = 0$.

10. If the functions $f(\theta) = \sum_{k=0}^{n} (a_k \cos k\theta + b_k \sin k\theta)$ and $g(\theta) = \sum_{k=0}^{n} (\alpha_k \cos k\theta + \beta_k \sin k\theta)$ agree at $2n+1$ points of the interval $[0,2\pi)$, then $a_k = \alpha_k$ for $k = 0, \ldots, n$, and $b_k = \beta_k$ for $k = 1, \ldots, n$. *Hint*: Replace the sines and cosines by exponentials, and thus obtain $e^{in\theta} f(\theta)$ as an ordinary polynomial in $e^{i\theta}$.

11. As a consequence of Prob. 10, the set of functions $\{1, \cos\theta, \ldots, \cos n\theta, \sin\theta, \ldots, \sin n\theta\}$ satisfies the Haar condition on $[0,2\pi)$. By the use of the alternation theorem, show that 0 is the trigonometric polynomial of degree $<n$ which best approximates the function $f(\theta) = A \cos n\theta + B \sin n\theta$ in the uniform sense on $[0,2\pi)$. Similarly the best approximation of degree $<n$ to the function

$$f(\theta) = \sum_{k=0}^{n} (A_k \cos k\theta + B_k \sin k\theta)$$

is $\sum_{k=0}^{n-1} (A_k \cos k\theta + B_k \sin k\theta)$. [Bernstein, 1912b]

12. The fact that a constant c_n exists such that $\| P' \| \leq c_n \| P \|$ for all polynomials P of degree $\leq n$ follows easily from the following fact: *If E is a finite-dimensional normed linear space, and if L is a linear operator on E, then there exists c such that $\| Lf \| \leq c \| f \|$ for all $f \in E$.*

13. Let $\{g_1,\ldots,g_n\}$ be a system of functions satisfying the Haar condition on $[a,b]$. Let $\{x_1,\ldots,x_n\}$ be any n distinct points in $[a,b]$. Then there is a least constant $\lambda = \lambda(x_1,\ldots,x_n)$ such that for all generalized polynomials $P = \sum c_i g_i$,

$$\| P \| = \max_{a \leq x \leq b} | P(x) | \leq \lambda \max_i | P(x_i) |$$

For ordinary polynomials, show that $\lambda = \| F \|$, where $F(x) = \sum | l_i(x) |$, and l_i are as on page 59. Prove that there is a choice of nodes x_i which makes λ a minimum. Determine this minimum for ordinary polynomials of degree ≤ 1. (Cf. Prob. 21, Sec. 2.)

8. Algorithms

One consequence of the results in Sec. 6 is that the computation of a best approximation on an interval (or any other compact metric space) can be effected by computing a succession of best approximations on finer and finer discrete sets. For such discrete problems the algorithms of Chap. 2 may be applied. That is to say, if we seek an approximation $\sum_{j=1}^{n} c_j g_j$ to a function f on set X, we may select a finite set $Y = \{x_1, \ldots, x_m\}$ and minimize the function

$$\Delta(c_1,\ldots,c_n) = \max_{1 \leq i \leq m} \left| \sum_{j=1}^{n} c_j g_j(x_i) - f(x_i) \right|$$

This is the same as solving an overdetermined system of linear equations

$$\sum_{j=1}^{n} c_j g_j(x_i) = f(x_i) \qquad (i = 1, \ldots, m)$$

in the minimax sense.

We shall be interested too in algorithms which solve the original problem rather than discrete analogues of it. The algorithm now to be described requires only that the functions $f, g_1, \ldots, g_n$ be continuous on the compact metric space X. The problem is to determine a coefficient vector $c = (c_1,\ldots,c_n)$ for which the deviation

$$\Delta(c) = \left\| \sum_{i=1}^{n} c_i g_i - f \right\| = \max_{x \in X} \left| \sum_{i=1}^{n} c_i g_i(x) - f(x) \right|$$

is a minimum. If the set $\{g_1,\ldots,g_n\}$ is not independent, we may replace it by a smaller set that *is* independent without raising the minimum deviation,

$p = \inf \Delta(c)$. We assume that this has been done. We define *residual* functions $r(c,x) = \sum c_i g_i(x) - f(x)$.

First Algorithm of Remes. At the kth step we are given a finite subset X^k of X. Select a coefficient vector c^k to minimize the function $\Delta^k(c) = \max \{|r(c,x)| : x \in X^k\}$. Select $x^k \in X$ to maximize the expression $|r(c^k,x)|$. Thus $|r(c^k,x^k)| = \Delta(c^k)$. Now start anew with the finite set $X^{k+1} = X^k \cup \{x^k\}$. At the beginning, X^1 may be arbitrary except that the set of n-tuples $\hat{x} = [g_1(x), \ldots, g_n(x)]$ corresponding to $x \in X^1$ should be of rank n.

Theorem. $\Delta^k(c^k) \uparrow p = \inf \Delta(c)$. *Furthermore, the sequence* $\{c^k\}$ *is bounded, and its cluster points minimize* Δ.

Proof. Define $|c| = \sum_{i=1}^{n} |c_i|$. By our assumption on X^1 it follows that the number $\theta = \min_{|c|=1} \max_{x \in X^1} |\sum c_i g_i(x)|$ is positive. Consequently, $\Delta^1(c) = \max_{x \in X^1} |\sum c_i g_i(x) - f(x)| \geq \max_{x \in X^1} |\sum c_i g_i(x)| - \|f\| \geq \theta |c| - \|f\|$.
If $|c| > 2\|f\|/\theta$, then $\Delta^k(c) \geq \Delta^1(c) > \|f\| \geq \Delta^k(0)$, so that the vector c does not enter the competition to minimize any of the functions Δ^k. Thus the sequence $\{c^k\}$ generated by the algorithm is bounded. Now it follows from the inclusions $X^k \subset X^{k+1} \subset X$ that $\Delta^k(c) \leq \Delta^{k+1}(c) \leq \Delta(c)$. Hence $\Delta^k(c^k) \leq \Delta^{k+1}(c^{k+1}) \leq p$. Thus for some $\epsilon \geq 0$, $\Delta^k(c^k) \uparrow p - \epsilon$. We must prove that $\epsilon = 0$. Since

$$|r(b,x) - r(c,x)| = |\sum (b_i - c_i) g_i(x)| \leq M |b - c|$$

with $M = \max_i \max_x |g_i(x)|$, it follows that $|r(b,x)| \leq |r(c,x)| + M|b - c|$ and $\Delta(b) \leq \Delta(c) + M|b - c|$. Suppose now that $\epsilon > 0$. Let b denote any cluster point of the sequence $\{c^k\}$. For any $\delta > 0$, we may find an index k such that $|b - c^k| < \delta$ and an index $i > k$ such that $|b - c^i| < \delta$. Then $|c^i - c^k| \leq 2\delta$, and

$$\begin{aligned} p \leq \Delta(b) &\leq \Delta(c^k) + M\delta \\ &= |r(c^k,x^k)| + M\delta \\ &\leq |r(c^i,x^k)| + 3M\delta \\ &\leq \Delta^i(c^i) + 3M\delta \\ &\leq p - \epsilon + 3M\delta \end{aligned}$$

If $3M\delta < \epsilon$, this is a contradiction. The same inequalities show that $\Delta(b) = p$. ∎

It should be especially noted that this theorem does not guarantee that the coefficient vectors c^k converge. In order to establish such convergence we must add the *Haar condition* (page 74) as an hypothesis.

Theorem. *In Algorithm 1, if the set of base functions $\{g_1, \ldots, g_n\}$ satisfies the Haar condition, then the coefficient vectors c^k converge to the (unique) solution vector.*

Proof. Under the Haar conditions, there is a unique c^* for which $\Delta(c^*) = \inf_c \Delta(c)$ (strong unicity theorem, Sec. 5). But by the preceding theorem, every cluster point of the sequence $\{c^k\}$ minimizes Δ. Thus the bounded sequence $\{c^k\}$ possesses exactly one cluster point c, and must therefore *converge* to it. (See Prob. 10 of Chap. 1, Sec. 2.) ■

The next algorithm to be described depends on our having the Haar conditions fulfilled by the basis $\{g_1, \ldots, g_n\}$.

Second Algorithm of Remes. We seek a coefficient vector c^* which renders the uniform norm of the function $r(x) = f(x) - \sum_{j=1}^{n} c_j g_j(x)$ a minimum on the interval $[a,b]$. The set of functions $\{g_1, \ldots, g_n\}$ is assumed to satisfy the Haar condition (page 74). In each cycle of this algorithm we are given an ordered set of $n + 1$ points from the preceding cycle: $a \leq x_0 < \cdots < x_n \leq b$. (In the beginning this set may be arbitrary.) We now compute a coefficient vector for which $\max_i |r(x_i)|$ is a minimum. From the alternation theorem (Sec. 4) it follows that the numbers $r(x_i)$ are of equal magnitude but of alternating sign. Hence $r(x)$ possesses a root z_i in each interval (x_{i-1}, x_i). In addition, let $z_0 = a$ and $z_{n+1} = b$. Let $\sigma_i = \operatorname{sgn} r(x_i)$. For each $i = 0, \ldots, n$ select a point y_i in $[z_i, z_{i+1}]$ where $\sigma_i r(y)$ is a maximum. This determines a *trial* set $\{y_0, \ldots, y_n\}$. If $\|r\| > \max_i |r(y_i)|$, then the definition of $\{y_0, \ldots, y_n\}$ must be altered as follows. Let y be a point where $|r(y)|$ is a maximum. Now insert y in its correct position in the set $\{y_0, \ldots, y_n\}$, and then remove a y_i in such a way that on the resulting ordered set the values of r still alternate in sign. (The precise rules for accomplishing this are given in Prob. 2.) This completes the definition of $\{y_0, \ldots, y_n\}$. Now the next cycle begins with $\{y_0, \ldots, y_n\}$ in place of $\{x_0, \ldots, x_n\}$.

A typical situation is pictured in the following graph, where we show the function r for $n = 4$.

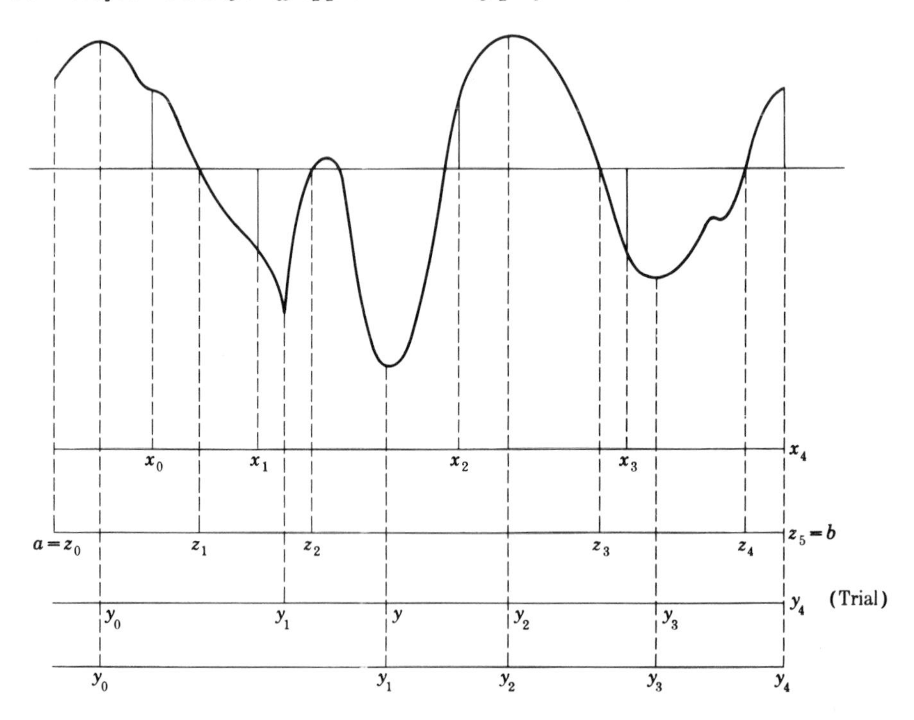

The complexity of the algorithm is justified by the fact that for many functions f the convergence is *quadratic*. This means that the error in the coefficient vector at the kth step is majorized by a quantity of the form $A\theta^{2k}$, where A and θ are positive numbers independent of k, and $\theta < 1$. We shall not prove this fact here but content ourselves with a proof of the linear convergence. For the theorem on quadratic convergence, see [Veidinger, 1960].

Theorem. *The successive generalized polynomials P^k generated in the second algorithm converge uniformly to the best approximation P^* according to an inequality of the form $\| P^k - P^* \| \leq A\theta^k$ where $0 < \theta < 1$.*

Proof. Referring to the description of a single cycle in the algorithm, let us define $\alpha = \max_i | r(x_i) | = \min_i | r(x_i) |$, $\beta = \max_i | r(y_i) | = \| r \|$, and $\gamma = \min_i | r(y_i) |$. In the next cycle of the algorithm let the corresponding quantities be distinguished by primes: α', β', γ'. Define $\beta^* = \| f - P^* \|$. From the theorem of de La Vallée Poussin (Sec. 4) we have

$$(1) \qquad \alpha \leq \gamma \leq \beta^* \leq \beta$$

In the next cycle of computation, the coefficient vector c' is selected to minimize the expression $\max_i |f(y_i) - \sum_j c_j' g_j(y_i)|$. The solution is obtained by solving the linear system of equations $(-1)^i\lambda + \sum_j c_j' g_j(y_i) = f(y_i)$ for the unknown vector c' and the unknown number λ. It will then follow that $\alpha' = |\lambda|$. The solution for λ via Cramer's rule is

$$\lambda = \begin{vmatrix} f(y_0) & g_1(y_0) & \cdots & g_n(y_0) \\ f(y_1) & g_1(y_1) & \cdots & g_n(y_1) \\ \cdots & \cdots & \cdots & \cdots \\ f(y_n) & g_1(y_n) & \cdots & g_n(y_n) \end{vmatrix} \div \begin{vmatrix} 1 & g_1(y_0) & \cdots & g_n(y_0) \\ -1 & g_1(y_1) & \cdots & g_n(y_1) \\ \cdots & \cdots & \cdots & \cdots \\ (-1)^n & g_1(y_n) & \cdots & g_n(y_n) \end{vmatrix}$$

Let the minors belonging to the column of ± 1s in the denominator be denoted by M_i. Then $\lambda = \sum f(y_i)(-1)^i M_i / \sum M_i$. Now if f were any generalized polynomial, $f = \sum a_j g_j$, then λ would vanish, because in that case the approximation would be exact. Hence $\sum P(y_i)(-1)^i M_i = 0$. We may therefore replace $f(y_i)$ by $r(y_i)$ in the expression for λ. Taking advantage of the alternation property of $r(y_i)$ and of the fact that the minors M_i all have the same sign, we obtain $\alpha' = |\lambda| = \sum |M_i|\,|r(y_i)|/\sum |M_i|$. (This fact about the signs of the minors was observed in the proof of the lemma on page 74.) Now let $\theta_i = |M_i|/\sum |M_j|$ so that

$$\alpha' = \sum \theta_i |r(y_i)| \geq \gamma \tag{2}$$

Note that $\theta_i \in (0,1)$ and that $\sum \theta_i = 1$. Assume for the moment that throughout the application of the algorithm the numbers θ_i remain larger than a fixed positive number $1 - \theta$. Then the proof may be completed by combining (1) and (2) as follows:

$$\begin{aligned} \gamma' - \gamma \geq \alpha' - \gamma = \sum \theta_i[|r(y_i)| - \gamma] &\geq (1 - \theta)(\beta - \gamma) \\ &\geq (1 - \theta)(\beta^* - \gamma) \end{aligned} \tag{3}$$

$$\begin{aligned} \beta^* - \gamma' = (\beta^* - \gamma) - (\gamma' - \gamma) &\leq (\beta^* - \gamma) - (1 - \theta)(\beta^* - \gamma) \\ &= \theta(\beta^* - \gamma) \end{aligned} \tag{4}$$

Inequality (4) shows that if we label the values of γ in successive cycles as $\gamma^{(0)}, \gamma^{(1)}, \ldots,$ then $\beta^* - \gamma^{(k)} \leq \theta^k(\beta^* - \gamma^{(0)}) = B\theta^k$. Inequality (3) now yields

$$\begin{aligned} \beta^{(k)} - \beta^* \leq \beta^{(k)} - \gamma^{(k)} &\leq (1 - \theta)^{-1}(\gamma^{(k+1)} - \gamma^{(k)}) \\ &\leq (1 - \theta)^{-1}(\beta^* - \gamma^{(k)}) \leq C\theta^k. \end{aligned}$$

Finally, by the strong unicity theorem (Sec. 5) we obtain $\| P^k - P^* \| \leq A\theta^k$.

There now remains the problem of showing that the numbers θ_i are bounded away from zero. This will certainly be true if, at the kth stage of our algorithm, an inequality $y_{i+1}^{(k)} - y_i^{(k)} \geq \epsilon > 0$ is valid, with ϵ independent of k. If such an inequality is not valid, then the $(n+1)$-tuples $[y_0^{(k)},\ldots,y_n^{(k)}]$ will possess a cluster point $[\bar{y}_0,\ldots,\bar{y}_n]$ in which two or more points $\bar{y}_i$ are identical. The generalized polynomial P of best approximation to f on $\{\bar{y}_i\}$ then actually *interpolates* to f at $\bar{y}_i$. Let ω denote the modulus of continuity of $P - f$ and select δ so that $\omega(\delta) < \alpha^{(1)}$. Then take k such that $| y_i^{(k)} - \bar{y}_i | < \delta$. Since the best approximation of f on $\{y_i^{(k)}\}$ is at least as good as P, we have $\alpha^{(k+1)} \leq \max_i | (P-f)(y_i^{(k)}) | \leq \max_i | (P-f)(\bar{y}_i) | + \omega(\delta) < \alpha^{(1)}$. This contradicts the fact that the numbers $\alpha^{(k)}$ are monotonically increasing. ∎

Problems

1. In the second algorithm, show that $\beta^* - \alpha' \leq \theta(\beta^* - \alpha)$.
2. In the second algorithm, complete the following rules for the introduction of the point y into the set $\{y_i\}$. If $y \in [z_i, y_i)$ for some $i \geq 1$, then replace y_{i-1} by y. If $y \in [a, y_0)$, then replace y_n by y_{n-1}, y_{n-1} by y_{n-2}, etc., y_1 by y_0, and y_0 by y.
3. Give the details for the application of the Remes second algorithm to ordinary polynomial approximation. For example, show that
$$\theta_i = m_i / \sum_{j=0}^{n} m_j \qquad \text{with} \qquad m_i = \prod \{ | y_i - y_j |^{-1} : 0 \leq j \leq n, j \neq i \}$$
4. In the second algorithm, show that the coefficient vectors $c^{(k)}$ converge according to an inequality $\| c^{(k)} - c^* \| \leq D\theta^k$. *Hint*: Prob. 15, of Chap. 1, Sec. 3. Why is it not necessary to be explicit about the norm in this statement?

*5. An approximation problem $f \approx \sum_{i=1}^{n} c_i g_i$ involving discontinuous functions can be transformed into an equivalent one involving continuous functions. *Hint*: For each x in the domain of the functions, define a point in $(n+1)$ space by $\hat{x} = [g_1(x),\ldots,g_n(x), f(x)]$. The coordinates of $\hat{x}$ are continuous functions of $\hat{x}$. [Krein]

Chapter 4

Least-squares approximation and related topics

1. Introduction

In the preceding chapters we have been concerned almost exclusively with approximations in the sense of Tchebycheff. That is to say, the measurement of the discrepancy between the approximated and the approximating functions has been effected with the norm

$$\|f\| = \max_{a \leq x \leq b} |f(x)|$$

or with the analogous seminorm

$$\|f\| = \max_{1 \leq i \leq m} |f(x_i)|$$

In the present chapter attention will be directed to approximation problems involving the norm

$$\|f\| = \left[\int_a^b |f(x)|^2 w(x)\, dx\right]^{1/2}$$

or an analogous seminorm,

$$\|f\| = \left[\sum_{i=1}^{m} |f(x_i)|^2 w(x_i)\right]^{1/2}$$

In each of these equations, w is a positive function called a *weight function.* For obvious reasons the minimum problems involving these norms are called *least-squares* problems, and the norms are called *quadratic* norms.

To the subject of least-squares approximation belong also the theories of orthogonal polynomials, of Fourier series, and of harmonic analysis in general. Obviously, only an introduction to this vast subject can be given here.

We shall continue to work with the linear space of all functions which are continuous on an interval $[a,b]$, but now normed by a quadratic norm. The quadratic norm defined above arises in the natural way (see page 13) from an inner product

$$\langle f,g \rangle = \int_a^b f(x)g(x)w(x)\,dx$$

We shall assume about $w(x)$ that its integral over any subinterval of $[a,b]$ is positive and that $\int_a^b f(x)w(x)\,dx$ exists for all continuous f. It is necessary to verify that the postulates of page 12 are fulfilled by this alleged inner product. Only postulate (i) is questionable:

$$f \neq 0 \Rightarrow \langle f,f \rangle > 0$$

This may be proved as follows. If $f \neq 0$, then there is a point ξ such that the number $\epsilon = |f(\xi)|$ is positive. By the continuity of f, there is a subinterval J contained in $[a,b]$ and containing ξ in which the inequality $|f(x)| > \frac{1}{2}\epsilon$ remains valid. Then

$$\langle f,f \rangle = \int_a^b f^2(x)w(x)\,dx \geq \int_J f^2(x)w(x)\,dx \geq \frac{\epsilon^2}{4}\int_J w(x)\,dx$$

The discrete analogue of the inner product just considered would be

$$(1) \qquad \langle f,g \rangle = \sum_{i=1}^{m} f(x_i)g(x_i)w(x_i)$$

This is an inner product *except* for satisfying postulate (i). Thus we can have $\langle f,f \rangle = 0$ although f is not the zero function; it is only necessary that $f(x_i) = 0$ for all i. Nevertheless we shall use the expression (1) as an inner product, exercising a little care to avoid the postulate (i).

In Chap. 1, Sec. 4, the advantages of orthonormal systems were discussed from the viewpoint of approximation theory. Indeed, we were able to give an explicit and elegant solution to the (least-squares) approximation problem of minimizing

$$\left\| f - \sum_{j=1}^{n} c_j g_j \right\|$$

when the set of base functions $\{g_1, \ldots, g_n\}$ was orthonormal. In the present context, a set of vectors $\{f_1, f_2, \ldots\}$ would be *orthonormal* if for all n and m

$$\int_a^b f_n(x)f_m(x)w(x)\,dx = \delta_{nm}$$

If on the right we have $a_n{}^2\delta_{nm}$ instead of δ_{nm}, then the set of vectors would be termed simply *orthogonal*. In that case it is clear that $a_n = \|f_n\|$. If each $a_n \neq 0$, then the set $\{f_n/a_n\}$ is orthonormal.

It should not be surmised that an orthonormal base is *necessary* for the solution of a least-squares approximation problem. Indeed, suppose that we have any linearly independent set $\{g_1, \ldots, g_n\}$ in $C[a,b]$ and wish to approximate a given function f in the least-squares sense by a generalized polynomial, $\sum c_j g_j$. Then we must determine the coefficient vector $c = [c_1, \ldots, c_n]$ in such a way as to minimize the function

$$\Delta(c) = \int_a^b [\sum c_j g_j(x) - f(x)]^2 w(x)\, dx$$

Now this function is differentiable in each component c_i, and thus the conditions $\partial\Delta/\partial c_i = 0$ must be met at any extremum. Performing the differentiations, we obtain

$$\frac{\partial\Delta}{\partial c_i} = \int_a^b 2[\sum_{j=1}^n c_j g_j(x) - f(x)] g_i(x) w(x)\, dx = 0 \qquad (i = 1, \ldots, n)$$

This can be written in the alternative form

$$\sum_j \left[\int_a^b g_j(x) g_i(x) w(x)\, dx\right] c_j = \int_a^b f(x) g_i(x) w(x)\, dx \qquad (i = 1, \ldots, n)$$

This is a system of n linear equations in $c_1, \ldots, c_n$ of the form

$$\sum_{j=1}^n A_{ij} c_j = b_i \tag{2}$$

in which $A_{ij} = \langle g_i, g_j \rangle$ and $b_i = \langle f, g_i \rangle$. In order for (2) to be most easily solved, we would like the matrix (A_{ij}) to be the *identity* matrix, $A_{ij} = \delta_{ij}$. This will occur if and only if the system $\{g_1, \ldots, g_n\}$ is orthonormal. However, the system (2) can be solved by standard procedures such as Gaussian elimination in any case, once the elements A_{ij} and b_i have been computed. This statement depends upon the following lemma.

Lemma. *In an inner-product space, if a set $\{g_1, \ldots, g_n\}$ is independent then the (Gram) matrix having elements $A_{ij} = \langle g_i, g_j \rangle$ is nonsingular.*

Proof. If the Gram-Schmidt process (page 15) is applied to $\{g_1, \ldots, g_n\}$, we obtain an orthonormal set $\{f_1, \ldots, f_n\}$ in which each f_i is of the form $f_i = \sum_{j=1}^n B_{ij} g_j$. (Actually $B_{ij} = 0$ if $j > i$.) Thus

$$\delta_{ij} = \langle f_i, f_j \rangle = \langle \sum_\nu B_{i\nu} g_\nu, \sum_\mu B_{j\mu} g_\mu \rangle = \sum_\nu \sum_\mu B_{i\nu} \langle g_\nu, g_\mu \rangle B_{j\mu}$$

This last equation may be written in the matrix form $I = BAB^T$, from which it follows that A is nonsingular. ■

We shall give some examples of orthonormal systems of functions.

Example 1. The sequence $\{1/\sqrt{2}, \cos x, \sin x, \cos 2x, \sin 2x, \ldots\}$ is orthonormal with respect to the inner product $\frac{1}{\pi}\int_{-\pi}^{\pi} f(x)g(x)\,dx$. Part of the verification is as follows. From the identities $\cos(A \pm B) = \cos A \cos B \mp \sin A \sin B$ we obtain $\cos(A + B) + \cos(A - B) = 2\cos A \cos B$. Hence

$$\frac{1}{\pi}\int_{-\pi}^{\pi} \cos nx \cos mx\,dx = \frac{1}{2\pi}\int_{-\pi}^{\pi} [\cos(n+m)x + \cos(n-m)x]\,dx$$

If $n \neq m$, then this becomes

$$\frac{1}{2\pi}\left[\frac{\sin(n+m)x}{n+m} + \frac{\sin(n-m)x}{n-m}\right]_{-\pi}^{\pi} = 0$$

If $n = m \geq 1$, this becomes

$$\frac{1}{2\pi}\left[\frac{\sin 2nx}{2n} + x\right]_{-\pi}^{\pi} = 1$$

Example 2. The finite system

$$\left\{\frac{1}{\sqrt{2}}, \cos x, \sin x, \ldots, \cos(N-1)x, \sin(N-1)x\right\}$$

is orthonormal with respect to the inner product

$$\langle f,g\rangle = \frac{1}{N}\sum_{j=1}^{2N} f(x_j)g(x_j) \qquad \left(x_j = \frac{j\pi}{N}\right)$$

Some results of this type are developed in Sec. 5.

Example 3. The sequence of Tchebycheff polynomials

$$\left\{\frac{T_0}{\sqrt{2}}, T_1, T_2, \ldots\right\}$$

is orthonormal with respect to the inner product

$$\langle f,g\rangle = \frac{2}{\pi}\int_{-1}^{1} f(x)g(x)\,\frac{dx}{\sqrt{1-x^2}}$$

The verification consists in making a change of variable $x = \cos\theta$ in the integral and using Example 1:

$$\frac{2}{\pi}\int_{-1}^{1} T_n(x)T_m(x)\frac{dx}{\sqrt{1-x^2}} = \frac{2}{\pi}\int_0^{\pi}\cos n\theta\cos m\theta\,d\theta = \begin{cases} 0 & (n \neq m) \\ 1 & (n = m \neq 0) \\ 2 & (n = m = 0) \end{cases}$$

Example 4. Let the positive zeros of the Bessel function

$$J_0(x) = \sum_{k=0}^{\infty}(-1)^k\left(\frac{x^k}{2^k k!}\right)^2$$

be denoted by $r_1, r_2, \ldots$. Then the sequence

$$\left\{\sqrt{2}, \sqrt{2}\frac{J_0(r_1x)}{J_0'(r_1)}, \sqrt{2}\frac{J_0(r_2x)}{J_0'(r_2)}, \ldots\right\}$$

is orthonormal on [0,1] with respect to the inner product

$$\int_0^1 f(x)g(x)x\,dx$$

The proof may be found, for example, in [Churchill, 1941].

Problems

1. *Tchebycheff polynomials of the second kind.* These are defined on $[-1,1]$ by $U_n(x) = \sin(n+1)\theta/\sin\theta$ where $x = \cos\theta$. Show that these functions are polynomials and are orthogonal with respect to the inner product $\int_{-1}^{1} f(x)g(x)\sqrt{1-x^2}\,dx$. What is the *orthonormal* set? *Ans.* $\{\sqrt{2/\pi}\,U_n\}$.
2. If the sequence $\{g_n\}$ is orthonormal on $[a,b]$ with weight function $w(x)$, then the sequence $\{\sqrt{w(x)}\,g_n(x)\}$ is orthonormal with unit weight function.
3. The orthogonal system of sines is generated by an equation of the form $f_n(x) = \phi(nx)$. The Tchebycheff polynomials arise from an equation of the form $f_n(x) = \phi[n\,\phi^{-1}(x)]$. The Rademacher functions of Prob. 5 come from an equation $f_n(x) = \phi(2^n x)$. Identify ϕ in each case, and look for generalizations.
4. The formula

$$\langle f,g\rangle = \int_a^b f(\alpha(x))g(\alpha(x))w(x)\,dx$$

defines an inner product which is apparently more general than the ones considered in the text. Show that the additional generality is spurious.

5. An example of an orthonormal system of *discontinuous* functions is provided by the *Rademacher functions* R_n. Given n, divide $[0,1)$ into 2^n subintervals $[0,1/2^n)$, $[1/2^n,2/2^n)$, $[2/2^n,3/2^n), \ldots$, and define R_n to be alternately $+1$ and -1 in these

successive intervals. Prove that these form an orthonormal system with inner product $\int_0^1 f(x)g(x)\,dx$.

6. Any two finite products of Rademacher functions are orthogonal to each other under suitable hypotheses.
7. For continuous functions on [0,1],

$$\int_0^1 |f(x)|\,dx \le \left(\int_0^1 |f(x)|^2\,dx\right)^{1/2}$$

Hint: Recall the Cauchy-Schwarz inequality of page 13.

8. Consider a symmetric interval $[-a,a]$ and an *even* weight function, $w(-x) = w(x)$. If $\{f_1,f_2,\ldots\}$ is an orthonormal system of *even* functions, and if $\{g_1,g_2,\ldots\}$ is an orthonormal system of *odd* functions, then $\{f_1,g_1,f_2,g_2,\ldots\}$ is orthonormal.
9. Show that $\{1, \cos x, \cos 2x,\ldots\}$ is an orthogonal system on $[0,\pi]$.
10. The *Hermite polynomials* H_n afford an example of an orthogonal system on an *infinite* interval. Let $f(x) = e^{-x^2}$. Then H_n is defined by $H_n(x) = e^{x^2}f^{(n)}(x)$. Show that H_n is a polynomial of degree n. If P is any polynomial, show that $\lim_{|x|\to\infty} f^{(n)}(x)P(x) = 0$. If $m < n$, verify that integration by parts gives us

$$\int_{-\infty}^{\infty} H_m(x)H_n(x)e^{-x^2}\,dx = \int_{-\infty}^{\infty} H_m(x)\,f^{(n)}(x)\,dx$$

$$= [f^{(n-1)}(x)H_m(x) - f^{(n-2)}H_m'(x) + \cdots]_{-\infty}^{\infty}$$

$$\pm \int_{-\infty}^{\infty} f(x)H_m{}^{(n)}(x)\,dx = 0$$

Hence $\{H_n\}$ is an orthogonal system on $(-\infty,\infty)$ with weight e^{-x^2}.

11. *More about Tchebycheff polynomials of the second kind.* (See Prob. 1.) Prove the following:
    (*a*) $U_{n+1} = 2xU_n - U_{n-1}$
    (*b*) $|U_n(x)| \le n+1$ when $-1 \le x \le 1$
    (*c*) $U_n - U_{n-2} = 2T_n$
    (*d*) $nU_{n-1} = T_n'$
    (*e*) $(1-x^2)U_n'' - 3xU_n' + n(n+2)U_n = 0$
    (*f*) $U_n = xU_{n-1} + T_n$
    (*g*) $T_n = xU_{n-1} - U_{n-2}$
    (*h*) $(1 - 2xy + y^2)^{-1} = 1 + U_1(x)y + U_2(x)y^2 + \cdots$
    $(1-xy)(1-2xy+y^2)^{-1} = 1 + T_1(x)y + T_2(x)y^2 + \cdots$

    Hint: Equate real and imaginary parts in the formula $\sum y^n e^{in\theta} = (1 - ye^{i\theta})^{-1}$.

2. Orthogonal Systems of Polynomials

Given a linearly independent sequence of vectors $\{f_1, f_2,\ldots\}$ in any inner-product space, the Gram-Schmidt process (page 15) may be applied to generate an orthonormal system $\{g_1,g_2,\ldots\}$ with the property that for each n the subspaces spanned by $\{f_1,\ldots,f_n\}$ and by $\{g_1,\ldots,g_n\}$ are identical. In this section we investigate the special cases that arise when the functions f_n are polynomials, in particular $f_n(x) = x^n$. We begin with a simple criterion for the linear independence of a set of polynomials.

Theorem 1. *Any sequence of polynomials $\{Q_0,Q_1,\ldots\}$ in which (for each n) Q_n is a polynomial of exact degree n is linearly independent. An arbitrary polynomial of degree $\le n$ is uniquely expressible as a linear combination of $Q_0, \ldots, Q_n$.*

Proof. The second assertion would imply the first, for if the polynomial 0 can be written *only* in the form $0 = 0Q_0 + 0Q_1 + \cdots$, then the sequence $\{Q_0,Q_1,\ldots\}$ is independent. If the second assertion is false, let n be the first index for which it fails. Certainly $n > 0$, since every constant is a unique multiple of the nonzero constant Q_0. Now let P be an arbitrary polynomial of degree $\le n$, say $P(x) = \sum_{i=0}^{n} c_i x^i$. This is to be expressed in the form $P = \sum_{i=0}^{n} \lambda_i Q_i$. The term x^n occurs on the right only in Q_n, and has, say, the coefficient a_n. Then λ_n is uniquely determined by the requirement that $c_n x^n = \lambda_n a_n x^n$. By the minimality of n and by the fact that $P - \lambda_n Q_n$ is of degree $<n$ it follows that $P - \lambda_n Q_n$ is uniquely expressible in the form $\sum_{i=0}^{n-1} \lambda_i Q_i$. ∎

In the application of the Gram-Schmidt process to a sequence $\{f_1,f_2,\ldots\}$, each member g_n of the orthonormal set is defined as a linear combination of f_n and all preceding g_k. A simplification occurs in the case of the sequence $\{1,x,x^2,\ldots\}$ in accordance with the following important result. In this theorem, the interval $[a,b]$ and the weight function w have been prescribed, and the inner product is defined as $\langle f,g\rangle = \int_a^b f(x)g(x)w(x)\,dx$.

Theorem 2. *The sequence of polynomials defined inductively in the following way is* orthogonal:

$$Q_n = (x - a_n)Q_{n-1} - b_nQ_{n-2}$$

with $Q_0 = 1$, $Q_1 = x - a_1$, $a_n = \langle xQ_{n-1},Q_{n-1}\rangle/\langle Q_{n-1},Q_{n-1}\rangle$, *and* $b_n = \langle xQ_{n-1},Q_{n-2}\rangle/\langle Q_{n-2},Q_{n-2}\rangle$.

Proof. We see from the formulas that for each n, Q_n is a *monic* polynomial (i.e., its leading coefficient is unity) and is therefore not zero. Hence the denominators in the formulas for a_n and b_n are not zero. We now show by induction on n that $\langle Q_n,Q_i\rangle = 0$ for $i < n$. For $n = 0$ there is nothing to prove. For $n = 1$ we have $\langle Q_1,Q_0\rangle = \langle xQ_0 - a_1Q_0,\ Q_0\rangle = \langle xQ_0,Q_0\rangle - \langle a_1Q_0,Q_0\rangle = \langle xQ_0,Q_0\rangle - \langle xQ_0,Q_0\rangle = 0$. Now assume that our assertion is true for $n - 1$. Then we have $\langle Q_n,Q_{n-1}\rangle = \langle xQ_{n-1} - a_nQ_{n-1} - b_nQ_{n-2},\ Q_{n-1}\rangle = \langle xQ_{n-1},Q_{n-1}\rangle - a_n\langle Q_{n-1},Q_{n-1}\rangle = 0$. Similarly $\langle Q_n,Q_{n-2}\rangle = \langle xQ_{n-1},Q_{n-2}\rangle - a_n\langle Q_{n-1},Q_{n-2}\rangle - b_n\langle Q_{n-2},Q_{n-2}\rangle = 0$. Now if $i < n - 2$, we have $\langle Q_n,Q_i\rangle =$

$\langle xQ_{n-1},Q_i\rangle - a_n\langle Q_{n-1},Q_i\rangle - b_n\langle Q_{n-2},Q_i\rangle = \langle Q_{n-1},xQ_i\rangle = \langle Q_{n-1}, Q_{i+1} + a_{i+1}Q_i + b_{i+1}Q_{i-1}\rangle = 0$. Here we have employed the recurrence formula to get an expression for xQ_i. ■

Example. If the above process is employed on the interval $[-1,1]$ with $w(x) = 1$, the resulting polynomials are termed *Legendre* polynomials and are denoted by $X_0, X_1, \ldots$. Let us compute the first few of these, to illustrate the use of the recurrence relation.

(*i*) $X_0 = 1$

(*ii*) $a_1 = \langle xX_0,X_0\rangle/\langle X_0,X_0\rangle = 0$

(*iii*) $X_1 = x$

(*iv*) $a_2 = \langle xX_1,X_1\rangle/\langle X_1,X_1\rangle = 0$

(*v*) $b_2 = \langle xX_1,X_0\rangle/\langle X_0,X_0\rangle = \frac{1}{3}$

(*vi*) $X_2 = x^2 - \frac{1}{3}$

The next two Legendre polynomials are $X_3 = x^3 - \frac{3}{5}x$ and $X_4 = x^4 - \frac{6}{7}x^2 + \frac{3}{35}$.

Corollary. *If* $f = \sum_{k=0}^{n} c_kQ_k$, *then* $f(x)$ *may be evaluated with* $2n - 1$ *multiplications by means of the recurrence formula* $d_{n+2} = d_{n+1} = 0$, $d_k = c_k + (x - a_{k+1})\, d_{k+1} - b_{k+2}\, d_{k+2}$, $f(x) = d_0$.

Proof. $f(x) = \sum_{k=0}^{n} c_kQ_k(x) = \sum_{k=0}^{n} [d_k - (x - a_{k+1})\, d_{k+1} + b_{k+2}\, d_{k+2}]Q_k(x) = d_0Q_0(x) + d_1[Q_1(x) - (x - a_1)Q_0(x)] + \sum_{k=2}^{n} d_k[Q_k(x) - (x - a_k)Q_{k-1}(x) + b_kQ_{k-2}(x)] = d_0$. In order to count the number of multiplications required, note that $d_n = c_n$ and $d_{n-1} = c_{n-1} + (x - a_n)\, d_n$, while the computation of $d_{n-2}, \ldots, d_0$ each requires two multiplications. Hence the total is $1 + 2(n - 1)$ or $2n - 1$. ■

At an earlier stage when considering the errors in Lagrange interpolation, we proved that the Tchebycheff polynomial $2^{-n+1}T_n$ had the minimum Tchebycheff norm on $[-1,1]$ among all monic polynomials of degree n. We are now going to see that the same polynomial minimizes also the quadratic norm

$$\left[\int_{-1}^{1} |f(x)|^2(1 - x^2)^{-1/2}\, dx\right]^{1/2}$$

Moreover, this fact will emerge as a special case of a more general assertion as follows.

Theorem 3. *The polynomials Q_n given in Theorem 2 are the monic polynomials which make the expression $\langle \cdot,\cdot \rangle$ a minimum.*

Proof. Since each Q_n is monic, an arbitrary monic polynomial f is expressible in the form $f = Q_n - a_{n-1}Q_{n-1} - \cdots - a_0Q_0$. In accordance with the theorem on page 14, the quantity $\|f\|^2 = \langle f,f \rangle$ will be a minimum if and only if the coefficients a_k are the Fourier coefficients of Q_n with respect to the orthogonal system $\{Q_0,\ldots,Q_{n-1}\}$. Thus $a_k = \langle Q_n,Q_k \rangle / \langle Q_k,Q_k \rangle = 0$, and $f = Q_n$ as asserted. ■

One of the important applications of orthogonal polynomials is to the problem of numerical integration. We shall depart briefly from our main course in order to give an account of this application.

Numerical integration is the process of guessing the value of a definite integral $\int_a^b f(x)\,dx$ from a finite number of values or "samples" of the function f. The definition of the integral as a limit of Riemann sums, $\sum f(\xi_i)(x_{i+1} - x_i)$, suggests that this should be possible. We shall look at *linear* processes of this type. They must take the form

$$\int_a^b f(x)\,dx \approx \sum_{k=1}^{n} A_k f(x_k) \tag{1}$$

Actually we can handle the problem of integration with a weight function $w(x)$ with no difficulty. If the sample points x_k are fixed, the coefficients A_k may be determined according to various error criteria, the usual one being that the formula shall be exact for all polynomials of degree $<n$. It is possible to arrange this by integrating the Lagrange interpolating formula (page 59) with nodes x_k:

$$P(x) = \sum_{k=1}^{n} f(x_k) l_k(x)$$

$$\begin{aligned} \int_a^b f(x)w(x)\,dx &\approx \int_a^b P(x)w(x)\,dx \\ &= \sum_{k=1}^{n} f(x_k) \int_a^b l_k(x)w(x)\,dx \\ &= \sum_{k=1}^{n} A_k f(x_k) \end{aligned} \tag{2}$$

If f is a polynomial of degree $<n$, then $P = f$, and the integration formula is exact.

We owe to Gauss the discovery that by an adroit placement of the nodes

x_k the formula (2) can be made exact for all polynomials of degree $<2n$: The x_k must be the zeros of the orthogonal polynomial Q_n of degree n. The resulting formulas are called *Gaussian integration* (or *quadrature*) *formulas.*

Theorem 4. *Let the integration formula* $\int_a^b f(x)w(x)\,dx \approx \sum_{k=1}^{n} A_k f(x_k)$ *be exact for all polynomials of degree* $<n$. *It will be exact for all polynomials of degree* $<2n$ *if and only if the nodes* x_k *are the zeros of the polynomial* Q_n *defined in Theorem* 2.

Proof. First let $x_1, \ldots, x_n$ be the zeros of Q_n. If f is a polynomial of degree $<2n$, then by the division algorithm we may find polynomials P and R such that $f = Q_nP + R$, the degrees of R and P being $<n$. The quadrature formula is exact for R, and Q_n is orthogonal to P. Hence $\int fw = \int Q_nPw + \int Rw = \int Rw = \sum A_kR(x_k) = \sum A_k f(x_k)$. For the converse, let the formula be exact for all polynomials of degree $<2n$. Then it is exact for $f(x) = Q_k(x) \prod_{i=1}^{n} (x - x_i)$ if $k < n$. Consequently $\int fw = \sum A_j f(x_j) = 0$, and $\prod (x - x_i)$ is orthogonal to each Q_k for $k = 0, \ldots, n-1$. Thus $\prod (x - x_i)$ is a multiple of Q_n. ■

We have left a logical gap in the development of the Gaussian quadrature theory. If the domain of the function f is the interval $[a,b]$, the integration formula can be applied to f only if the sample points x_k lie in $[a,b]$. The following theorems fill this gap.

Theorem 5. *Let* $\{Q_0,Q_1,\ldots\}$ *be a sequence of polynomials (subscripts denoting degrees) which is orthogonal with respect to an inner product* $\langle f,g\rangle = \int_a^b f(x)g(x)w(x)\,dx$. *If* f *is any continuous function on* $[a,b]$ *orthogonal to* $Q_0, \ldots, Q_{n-1}$, *then* f *must change sign at least* n *times in* (a,b) *or vanish identically.*

Proof. Since $f \perp Q_0$ and $Q_0 = 1$, $\int_a^b f(x)w(x)\,dx = 0$. Thus if $f \neq 0$, then f must change sign at least once in (a,b). If f changes sign fewer than n times, let $r_1 < r_2 < \cdots < r_k$ be the points of (a,b) where f changes sign. Then in each interval (a,r_1), (r_1,r_2), $\ldots$, (r_k,b), f does not change sign, but has opposite signs in adjacent intervals. Since this property is shared by the polynomial $P(x) = \prod_{i=1}^{k} (x - r_i)$, it follows that $\int_a^b f(x)P(x)w(x)\,dx \neq 0$. But this is a contradiction since P (being a polynomial of degree $k < n$) is a linear combination of $Q_0, \ldots, Q_{n-1}$ and is therefore orthogonal to f. ∎

Corollary 1. *Let $\{Q_0, \ldots\}$ be an orthogonal system of polynomials (subscripts denoting degrees) on the interval $[a,b]$ with weight function w. Then the roots of Q_n are simple and lie in (a,b).*

Proof. This follows from Theorem 5 because Q_n is orthogonal to $Q_0, \ldots, Q_{n-1}$. ∎

Corollary 2. *Let $\{Q_0, \ldots\}$ be an orthogonal system as in Corollary 1. Let P be the best least-squares approximation of the form $P = \sum_{i=0}^{n} c_i Q_i$ to some continuous f. Then P interpolates to f in at least $n + 1$ points of (a,b).*

Proof. This follows from Theorem 5 because $P - f$ is orthogonal to $Q_0, \ldots, Q_n$ and must therefore change sign at least $n + 1$ times. ∎

In the foregoing discussion, the value of n has been static, and the notation has not shown the dependence of the coefficients A_k and the nodes on n. In order to discuss the behavior of errors as $n \to \infty$, let us write the nth Gaussian formula in the form

$$\int_a^b f(x)w(x)\,dx \approx \sum_{k=1}^{n} A_{nk} f(x_{nk}) \tag{3}$$

Stieltjes' Theorem. *The errors in the Gaussian quadrature formulas (3) converge to zero (as $n \to \infty$) for every continuous function f on $[a,b]$.*

Proof. We begin by showing that $A_{nk} > 0$. Let P be the polynomial which results when the factor $x - x_{nk}$ is removed from Q_n. Since P^2 is of degree $<2n$, formula (3) would be exact for it. Thus $0 < \int P^2 w = \sum_{i=1}^{n} A_{ni} P^2(x_{ni}) = A_{nk} P^2(x_{nk})$. Since x_{nk} is a simple root of Q_n, $P(x_{nk}) \neq 0$. Hence $A_{nk} > 0$.

We shall also require the fact that $\sum_{k=1}^{n} A_{nk}$ is independent of n. This follows from the observation that (3) is exact for the function $f(x) = 1$, so that

$$\int_a^b w(x)\,dx = \sum_{k=1}^{n} A_{nk}.$$

For the proof proper, let f be an arbitrary continuous function on $[a,b]$, and let a positive number ϵ be given. By the Weierstrass theorem (Chap. 3, Sec. 3), we may determine a polynomial P such that $|f(x) - P(x)| < \epsilon/c$ on $[a,b]$, where c denotes the constant $2\int_a^b w(x)\,dx$. If the degree of P is less than $2n$, formula (3) is exact for P. Consequently by the triangle inequality

$$\left|\int fw - \sum A_{nk} f(x_{nk})\right| \le \left|\int fw - \int Pw\right| + \left|\sum A_{nk}P(x_{nk}) - \sum A_{nk} f(x_{nk})\right|$$
$$\le \int |f - P|\, w + \sum A_{nk} |P(x_{nk}) - f(x_{nk})|$$
$$\le \frac{\epsilon}{c}\left(\int w + \sum A_{nk}\right) = \epsilon \qquad \blacksquare$$

The calculation of the numbers A_i in the Gaussian quadrature formula can be effected directly from the definition

$$A_i = \int_a^b l_i(x) w(x) = \int_a^b \frac{Q_n(x)}{(x - x_i)Q_n'(x_i)} w(x)\,dx$$

However, there is a more convenient alternative which we now discuss. Let us define a new sequence of polynomials $\phi_0, \phi_1, \ldots,$ using the same recurrence relation as for Q_n,

$$\phi_n(x) = (x - a_n)\phi_{n-1}(x) - b_n\phi_{n-2}(x)$$

but with different starting values: $\phi_0(x) = 0$ and $\phi_1(x) = b_1 = \int_a^b w(x)\,dx$. It is clear that ϕ_n will be a polynomial of degree $n - 1$. The alternative formula for A_i is then

$$A_i = \frac{\phi_n(x_i)}{Q_n'(x_i)}$$

and this is a direct consequence of the following theorem, since $Q_n(x_i) = 0$.

Theorem. *With ϕ_n and Q_n as above, we have*

$$\phi_n(x) = \int_a^b \frac{Q_n(t) - Q_n(x)}{t - x} w(t)\,dt$$

Proof. The proof will be by induction. We start with the inductive step. Suppose that our equation is true for the integers $0, 1, \ldots, n - 1$. If $n \ge 2$, then we have by the recurrence relations

$$\int_a^b \frac{Q_n(t) - Q_n(x)}{t - x} w(t)\, dt$$

$$= \int_a^b \frac{(t - a_n)Q_{n-1}(t) - b_nQ_{n-2}(t) - (x - a_n)Q_{n-1}(x) + b_nQ_{n-2}(x)}{t - x} w(t)\, dt$$

$$= (x - a_n) \int_a^b \frac{Q_{n-1}(t) - Q_{n-1}(x)}{t - x} w(t)\, dt - b_n \int_a^b \frac{Q_{n-2}(t) - Q_{n-2}(x)}{t - x} w(t)\, dt$$

$$+ \int_a^b Q_{n-1}(t) w(t)\, dt$$

$$= (x - a_n)\phi_{n-1}(x) - b_n\phi_{n-2}(x)$$

$$= \phi_n(x)$$

Since the above argument requires that $n \geq 2$, a separate verification is necessary for $n = 0$ and $n = 1$. For $n = 0$ we observe that both sides of the asserted equation reduce to zero. For $n = 1$, the right side reduces to $\int_a^b w(t)\, dt$, and this is $\phi_1(x)$ by definition. ∎

Problems

1. Prove Theorem 1 by setting up an isomorphism between $(n + 1)$-tuples and polynomials of degree $\leq n$: $(a_0, \ldots, a_n) \leftrightarrow \sum_{i=0}^{n} a_i x^i$.
2. If the sequence of functions $\{g_0, g_1, \ldots\}$ is orthogonal on $[a,b]$, and if g_0 does not change sign, then every function of the form $f = \sum_{i=1}^{n} c_i g_i$ has a root in $[a,b]$. Generalize.
3. If $\tilde{T}_n$ is defined as the *monic* multiple of T_n, then $\tilde{T}_n = x\tilde{T}_{n-1} - \frac{1}{4}\tilde{T}_{n-2}$ for $n \geq 3$ and $\tilde{T}_2 = x\tilde{T}_1 - \frac{1}{2}\tilde{T}_0$.
4. Prove that with a symmetric interval $[-a,a]$ and an *even* weight function $w(-x) = w(x)$, we always have $a_n = 0$. Prove that Q_n is even or odd in accord with its subscript.
5. Show that b_n is always positive, by establishing the equation

$$b_n = \frac{\langle Q_{n-1}, Q_{n-1} \rangle}{\langle Q_{n-2}, Q_{n-2} \rangle}$$

6. If $T_n^*(x) = T_n(2x - 1)$, then $T_n^* = (4x - 2)T_{n-1}^* - T_{n-2}^*$. What orthogonality property have these polynomials?
7. If $C_n(x) = 2T_n(x/2)$ then $C_n = xC_{n-1} - C_{n-2}$. These polynomials are orthogonal with respect to the inner product $\int_{-2}^{2} f(x)g(x)(4 - x^2)^{-1/2}\, dx$.

8. Work out a Gaussian quadrature formula of the form $\int_{-1}^{+1} f(x)\,dx = A_1 f(x_1) + A_2 f(x_2)$. *Ans.* $A_i = 1$, $x_i = \pm\sqrt{\tfrac{1}{3}}$.

9. Prove the following theorem. Let a sequence of quadrature formulas $\int_a^b f(x)w(x)\,dx \approx \sum_{k=1}^{n} B_{nk} f(x_{nk})$ be given, not restricted in any way except that $B_{nk} \geq 0$ and that the error goes to zero (as $n \to \infty$) for all *polynomials* f. Then the error goes to zero for arbitrary *continuous* f.

***10.** Let $\{g_0, g_1, \ldots\}$ be a *Markoff* system on an interval $[a,b]$. This means that every initial segment $\{g_0, \ldots, g_n\}$ satisfies the Haar condition. Let w be a weight function on $[a,b]$. Let $\{G_0, G_1, \ldots\}$ be the orthonormal system obtained by the Gram-Schmidt process. Generalize to this situation as many of the theorems of this section as you can.

11. If $f = \sum_{k=0}^{n} a_k T_k$ (the coefficients a_k being given), then $f(x)$ may be evaluated with $n + 1$ multiplications. If multiplication by 2 or by $\frac{1}{2}$ is not counted, then n multiplications suffice. [Clenshaw, 1955]

12. In Prob. 9 the theorem remains true if the family of all polynomials is replaced by the sequence $\{1, x, x^2, \ldots\}$ or indeed by any *fundamental* set. (This term is defined on page 87.)

13. The Gaussian quadrature formula for nodes at the roots of the Tchebycheff polynomial T_n is, with $x_i = \cos[(2i - 1)\pi/2n]$,

$$\int_{-1}^{1} f(x)(1 - x^2)^{-1/2}\,dx \approx \frac{\pi}{n} \sum_{i=1}^{n} f(x_i)$$

In order to verify that $A_i = \pi/n$, we use the formula $A_i = \phi_n(x_i)/Q_n'(x_i)$. The Q_n are $Q_0 = 1$ and $Q_n = 2^{1-n}T_n$ for $n \geq 1$. The constants in the recurrence relation are $a_n = 0$, $b_1 = \pi$, $b_2 = \frac{1}{2}$, and $b_n = \frac{1}{4}$ for $n \geq 3$. Hence the polynomials ϕ_n are $\phi_0 = 0$, $\phi_1 = \pi$, $\phi_2 = \pi x$, $\phi_3 = \pi(x^2 - \frac{1}{4})$, and generally, $\phi_{n+1} = \pi 2^{-n} U_n$, where U_n is the Tchebycheff polynomial of the second kind. (See Probs. 1 and 11, Sec. 1.) Thus

$$A_i = \frac{\pi 2^{-n+1} U_{n-1}(x_i)}{2^{1-n} T_n'(x_i)}$$

But $T_n' = nU_{n-1}$, and so $A_i = \pi/n$. The above quadrature formula is called *Hermite's formula*.

14. If the Gram-Schmidt process is applied to $\{x^n, x^{n-1}, \ldots, 1\}$ in that order, will the resulting polynomials be connected by three-term recurrence relations?

***15.** For each n let P_n be a polynomial of degree n. Suppose that $P_n(x) = (Ax - B)P_{n-1}(x) - CP_{n-2}(x)$ with A, B, and C independent of n, and $C > 0$. Then for appropriate constants, $\alpha, \beta, \ldots$, $P_n(x) = \gamma\lambda^n T_n(\alpha x + \beta) + \delta\lambda^n U_n(\alpha x + \beta)$. In other words, the only polynomials which satisfy three-term recurrence relations with scalars *independent* of n are related in a simple manner to the Tchebycheff polynomials.

16. If the quadrature formula $\int f \approx \sum_{i=1}^{n} A_i f(x_i)$ is exact for all polynomials of degree $<n$, then $A_i = \int l_i$ where $l_i(x) = \prod_{\substack{j=1 \\ j \neq i}}^{n} \frac{x - x_j}{x_i - x_j}$.

17. Starting with the formula $f \approx B_n f$, where B_n is the Bernstein polynomial operator, we obtain the quadrature formula $\int_0^1 f \approx \frac{1}{n+1} \sum_{k=0}^{n} f\left(\frac{k}{n}\right)$. Prove this, and discuss the convergence.

18. Given points $x_i^{(n)} \in [-1,1]$, form the operator L_n such that $L_n f$ is the polynomial of degree $\leq n$ which interpolates to f at $x_0^{(n)},\ldots,x_n^{(n)}$. If $L_n f$ converges to f in the sense that $\lim \int | L_n f - f | = 0$, then show that $\int f = \lim \int L_n f$. Is the converse true? What conclusion can be drawn if $\lim \int | L_n f - f |^p = 0$?

19. If a quadrature formula $\int_a^b f \approx \sum_{i=1}^{n} A_i f(x_i)$ is obtained by integrating the Lagrange interpolating polynomial with nodes at $x_1,\ldots,x_n$, and if $f^{(n)} \in C[a,b]$, then the quadrature error does not exceed

$$\frac{1}{n!} \, \| f^{(n)} \| \int_a^b \prod_{i=1}^{n} | x - x_i | \, dx$$

The problem of determining the nodes optimally is solved in Chap. 6, Sec. 6.

20. Show that Theorem 2 is true for any inner product that has the property $\langle fg,h \rangle = \langle f,gh \rangle$.

21. The Gaussian quadrature formula for nodes at the zeros of the Tchebycheff polynomial $U_n(x) = \sin (n+1)\theta / \sin \theta$, $x = \cos \theta$, is

$$\int_{-1}^{1} f(x) \sqrt{1 - x^2} \, dx = \frac{\pi}{n+1} \sum_{i=1}^{n} (1 - x_i^2) \, f(x_i)$$

with $x_i = \cos [i\pi/(n+1)]$. This may be verified as in Prob. 13. Some of the intermediate steps are $Q_n = 2^{-n} U_n$, $\phi_n = (\pi/2) U_{n-1}$, $U_{n-1}(x_i) = (-1)^i$, and $U_n'(x_i) = (n+1)(-1)^i (1 - x_i^2)^{-1}$.

3. Convergence of Orthogonal Expansions

In the linear space of continuous functions on $[a,b]$ let an inner product be defined by the equation

$$\langle f,g \rangle = \int_a^b f(x) g(x) w(x) \, dx$$

and let $\{\bar{Q}_0, \bar{Q}_1, \ldots\}$ be the orthonormal system of polynomials obtained by applying the Gram-Schmidt process to $\{1, x, x^2, \ldots\}$. The theorem on page 14 tells us that if a given function f is to be approximated by a linear combination $\sum_{i=0}^{n} c_i \bar{Q}_i$ in the least-squares norm, then the coefficients c_i must be the Fourier coefficients of f: $c_i = \langle f, \bar{Q}_i \rangle$. A remarkable feature of this approximation is that the optimum coefficients c_i are independent of n. This contrasts sharply with optimum *Tchebycheff* approximations, in which the coefficients generally depend on n. (Examples to the contrary will occur in Sec. 4.)

Thus to each continuous function f there corresponds a *formal* orthogonal expansion, $\sum_{i=0}^{\infty} \langle f, \bar{Q}_i \rangle \bar{Q}_i$, which has the property that its partial sums are best approximations to f in the least-squares sense. It is not clear at the moment whether the series converges at any point, or whether it represents $f(x)$ at points where it does converge. If f is a polynomial of degree n, then the series converges uniformly to f by virtue of the fact that $f = \sum_{i=0}^{n} \langle f,\bar{Q}_i\rangle \bar{Q}_i$ and $\langle f,\bar{Q}_i\rangle = 0$ for $i > n$.

In order to discuss the convergence questions, let us define $S_n f = \sum_{i=0}^{n} \langle f,\bar{Q}_i\rangle \bar{Q}_i$. Further, let $\mathfrak{J}_n f$ denote the polynomial of degree $\leq n$ which best approximates f in the Tchebycheff sense. The least-squares norm with weight w and the Tchebycheff norm will be distinguished by subscripts w and T. Now for continuous functions f generally we have

$$\| f - S_n f \|_T \nrightarrow 0$$

This result is a corollary of a deeper theorem to which Sec. 5 of Chap. 6 is devoted. For the other types of convergence a more favorable situation prevails, however.

Theorem. *For all $f \in C[a,b]$ we have*

$$(i) \quad \| f - \mathfrak{J}_n f \|_T \to 0$$

$$(ii) \quad \| f - \mathfrak{J}_n f \|_w \to 0$$

$$(iii) \quad \| f - S_n f \|_w \to 0$$

Proof. The Weierstrass theorem (Chap. 3, Sec. 3) implies (i). Statement (ii) follows from (i) and from the following inequality

$$\int_a^b | f(x) - (\mathfrak{J}_n f)(x) |^2 w(x)\, dx \leq \| f - \mathfrak{J}_n f \|_T^2 \int_a^b w(x)\, dx$$

Finally, statement (iii) follows from (ii) and from the fact that $S_n f$ is the best approximation to f in the least-squares norm, so that $\| f - S_n f \|_w \leq \| f - \mathfrak{J}_n f \|_w$. ■

The partial sums $S_n f$ of the orthogonal expansion $\sum \langle f,\bar{Q}_k\rangle \bar{Q}_k$ may not only fail to converge *uniformly* for a particular continuous function f, but they may fail to converge point by point. That is, there may exist points ξ such that $\lim_{n\to\infty} (S_n f)(\xi)$ does not exist. There is here a certain analogy with the Lagrange interpolating polynomials: Usually by imposing certain *smoothness* conditions on f we can prove that $\| f - S_n f \|_T \to 0$. The following theorem—although it is far from the best possible—illustrates this remark.

Theorem. *If f possesses a continuous second derivative in $[-1,1]$, then the expansion of f in Tchebycheff polynomials converges uniformly to f.*

Proof. The expansion referred to is $\frac{1}{2}A_0 + \sum_{k=1}^{\infty} A_k T_k$, where

$$A_k = \frac{2}{\pi} \int_{-1}^{1} f(x)\, T_k(x) \frac{dx}{\sqrt{1-x^2}}$$

By making the change of variable $x \to \cos\theta$, we obtain

$$A_k = \frac{2}{\pi} \int_{0}^{\pi} g(\theta) \cos k\theta \, d\theta$$

where $g(\theta) = f(\cos\theta)$. Integrating by parts twice in succession gives

$$A_k = \frac{-2}{\pi k^2} \int_{0}^{\pi} \cos k\theta \, g''(\theta)\, d\theta$$

From the hypothesis on f it follows then that A_k satisfies an inequality of the form $|A_k| \leq Mk^{-2}$. Thus the series $\sum |A_k|$ converges, and the Tchebycheff series for f converges uniformly by the Weierstrass M test (Prob. 19). By another theorem of Weierstrass, or by the theorem on page 9, the function F to which the series converges is continuous. It remains to prove that $F = f$. Let $S_n f$ denote the nth partial sum of the expansion of f. Then

$$\| f - F \|_w \leq \| f - S_n f \|_w + \| S_n f - F \|_w$$

The first term on the right converges to zero, by the preceding theorem. The second term also goes to zero, because $S_n f$ converges uniformly to F. Hence $\| f - F \|_w = 0$ and $f = F$. ∎

As an illustration of this theorem, consider the function $f(x) = e^{\lambda x}$. The coefficients of the expansion $f = \frac{1}{2}A_0 + \sum A_n T_n$ are given by

$$A_n = \frac{2}{\pi} \int_{0}^{\pi} e^{\lambda \cos\theta} \cos n\theta \, d\theta$$

In terms of Bessel functions, this integral turns out to be $A_n = 2i^{-1} J_n(i\lambda) = 2I_n(\lambda)$. The first few of these coefficients are as follows when $\lambda = 1$.

$\frac{1}{2}A_0 = 1.2660658778$	$A_5 = 0.0005429263$
$A_1 = 1.1303182080$	$A_6 = 0.0000449773$
$A_2 = 0.2714953395$	$A_7 = 0.0000031984$
$A_3 = 0.0443368498$	$A_8 = 0.0000001992$
$A_4 = 0.0054742404$	$A_9 = 0.0000000110$

In order to give a general discussion of expansions in orthogonal polynomials, it is convenient to interpret the operator S_n, defined by the equation $S_n f = \sum_{i=0}^{n} \langle f,\bar{Q}_i\rangle \bar{Q}_i$, as an *integral* operator. Indeed, we have

$$(S_n f)(x) = \sum_{i=0}^{n} \int_a^b f(t)\bar{Q}_i(t)w(t)\,dt\,\bar{Q}_i(x) \tag{1}$$

$$= \int_a^b f(t) \sum_{i=0}^{n} \bar{Q}_i(t)\bar{Q}_i(x)w(t)\,dt$$

$$= \int_a^b f(t)K_n(t,x)w(t)\,dt$$

where we have set $K_n(t,x) = \sum_{i=0}^{n} \bar{Q}_i(t)\bar{Q}_i(x)$. This function is called the *kernel* of the orthonormal system. What we have just written is not at all limited to orthonormal systems of polynomials, but for the next several theorems we shall suppose that the $\bar{Q}_n$ are obtained by applying the Gram-Schmidt process to $\{1,x,x^2,\ldots\}$. The polynomials $Q_n = \lambda_n\bar{Q}_n$ are the *monic* polynomials discussed in the previous section. In other words, λ_n^{-1} is the coefficient of x^n in $\bar{Q}_n$.

Lemma. (Christoffel-Darboux Identity). *The kernel takes the form*

$$\sum_{i=0}^{n} \bar{Q}_i(x)\bar{Q}_i(t) = \lambda_{n+1}\lambda_n^{-1}\,\frac{\bar{Q}_{n+1}(x)\bar{Q}_n(t) - \bar{Q}_n(x)\bar{Q}_{n+1}(t)}{x - t}$$

Proof. From the three-term recurrence relation (Theorem 2, Sec. 2) we have the equations

$$Q_{n+1}(x)Q_n(t) = (x - a_{n+1})Q_n(x)Q_n(t) - b_{n+1}Q_{n-1}(x)Q_n(t)$$

$$Q_{n+1}(t)Q_n(x) = (t - a_{n+1})Q_n(t)Q_n(x) - b_{n+1}Q_{n-1}(t)Q_n(x)$$

If we subtract the second of these equations from the first, we obtain

$$\begin{aligned} &Q_{n+1}(x)Q_n(t) - Q_{n+1}(t)Q_n(x) \\ &\quad = (x - t)Q_n(t)Q_n(x) + b_{n+1}[Q_n(x)Q_{n-1}(t) - Q_n(t)Q_{n-1}(x)] \end{aligned} \tag{2}$$

Note that $\lambda_n = \sqrt{\langle Q_n,Q_n\rangle}$ since $Q_n = \lambda_n\bar{Q}_n$. From the recurrence formula written in the form $b_{n+1}Q_{n-1} = xQ_n - a_{n+1}Q_n - Q_{n+1}$ we obtain, on taking inner products with Q_{n-1},

$$\begin{aligned} b_{n+1}\lambda_{n-1}^2 &= \langle xQ_n,Q_{n-1}\rangle = \langle Q_n,xQ_{n-1}\rangle \\ &= \langle Q_n, Q_n + a_nQ_{n-1} + b_nQ_{n-2}\rangle = \lambda_n^2 \qquad (Q_{-1} = 0) \end{aligned}$$

Thus if equation (2) is divided by λ_n^2 the result is

$$\begin{aligned}(3)\qquad &\lambda_n^{-2}[Q_{n+1}(x)Q_n(t) - Q_{n+1}(t)Q_n(x)]\\ &= (x-t)\bar{Q}_n(x)\bar{Q}_n(t) + \lambda_{n-1}^{-2}[Q_n(x)Q_{n-1}(t) - Q_n(t)Q_{n-1}(x)]\end{aligned}$$

Now reapply the recurrence formula (3) in order to simplify the last term. We obtain eventually

$$\begin{aligned}&\lambda_n^{-2}[Q_{n+1}(x)Q_n(t) - Q_{n+1}(t)Q_n(x)]\\ &= (x-t)\sum_{i=1}^{n}\bar{Q}_i(x)\bar{Q}_i(t) + \lambda_0^{-2}[Q_1(x)Q_0(t) - Q_1(t)Q_0(x)]\\ &= (x-t)\sum_{i=0}^{n}\bar{Q}_i(x)\bar{Q}_i(t)\end{aligned}$$

The proof is completed by writing $\lambda_n\bar{Q}_n$ for Q_n, etc. ∎

Bessel's Inequality. *In an inner-product space, let* $\{g_n\}$ *be any orthonormal sequence, and let f be any element. Then* $\sum \langle f,g_n\rangle^2 \le \langle f,f\rangle$. *In particular, the Fourier coefficients of f converge to zero.*

Proof. Put $F_n = \sum_{i=0}^{n} \langle f,g_i\rangle g_i$. By an argument familiar from page 14, $f - F_n \perp F_n$. Indeed, since F_n is a linear combination of $g_0, \ldots, g_n$, we need only observe (for $j \le n$) that $\langle f - F_n, g_j\rangle = \langle f,g_j\rangle - \sum_{i=0}^{n}\langle f,g_i\rangle\langle g_i,g_j\rangle = 0$. Consequently by the Pythagorean law (page 13) we have $\|f\|^2 = \|F_n\|^2 + \|f - F_n\|^2 \ge \|F_n\|^2 = \sum_{i=0}^{n}\langle f,g_i\rangle^2$. Since this is true for all n, it remains true in the limit. ∎

Theorem. *If the values of the polynomials* $\bar{Q}_n$ *remain bounded at a certain point* $x_0 \in [a,b]$, *if f is continuous, and if f satisfies at* x_0 *the Lipschitz condition* $|f(x_0) - f(x)| \le \alpha|x_0 - x|$, *then at that point* $f(x_0) = \sum \langle f,\bar{Q}_n\rangle \bar{Q}_n(x_0)$.

Proof. We observe first that the numbers $\lambda_n\lambda_{n-1}^{-1}$ are bounded. Indeed,

$$\begin{aligned}\lambda_n^2 &= \langle Q_n,Q_n\rangle = \langle Q_n, xQ_{n-1} - a_nQ_{n-1} - b_nQ_{n-2}\rangle = \langle Q_n,xQ_{n-1}\rangle\\ &\le \int_a^b |x|\,|Q_n(x)|\,|Q_{n-1}(x)|\,w(x)\,dx \le c\langle|Q_n|,|Q_{n-1}|\rangle\\ &\le c\,\|Q_n\|\,\|Q_{n-1}\| = c\lambda_n\lambda_{n-1}\end{aligned}$$

Thus $\lambda_n\lambda_{n-1}^{-1} \leq c \equiv \max_{a\leq x\leq b} |x|$. Now from the observation that $(S_n1)(t) = 1$, it follows that the error at x_0 is

$$\epsilon_n \equiv f(x_0) - (S_nf)(x_0) = f(x_0)(S_n1)(x_0) - (S_nf)(x_0)$$

From the integral form of S_n [equation (1)] we have

$$\epsilon_n = \int_a^b [f(x_0) - f(x)] \sum_{i=0}^n \bar{Q}_i(x_0)\bar{Q}_i(x)w(x)\,dx$$

By the Christoffel-Darboux identity this becomes

$$\epsilon_n = \lambda_{n+1}\lambda_n^{-1} \int_a^b \frac{f(x_0) - f(x)}{x_0 - x} [\bar{Q}_{n+1}(x_0)\bar{Q}_n(x) - \bar{Q}_n(x_0)\bar{Q}_{n+1}(x)]w(x)\,dx$$

$$= \lambda_{n+1}\lambda_n^{-1}[\langle h,\bar{Q}_n\rangle\bar{Q}_{n+1}(x_0) - \langle h,\bar{Q}_{n+1}\rangle\bar{Q}_n(x_0)]$$

where $h(x) = [f(x_0) - f(x)]/(x_0 - x)$. By the Lipschitz condition on f, $|h(x)| \leq \alpha$. Also we have $\lambda_{n+1}\lambda_n^{-1} \leq c$. Finally we observe that from Bessel's inequality the Fourier coefficients of h, $\langle h,\bar{Q}_n\rangle$, approach zero. Since the numbers $\bar{Q}_n(x_0)$ remain bounded, $\epsilon_n \to 0$. ∎

We shall conclude this section with two celebrated theorems concerning the convergence of ordinary Fourier series. Again certain integral operators play a role. Let us denote the nth partial sum of the Fourier series of f by the symbol S_nf:

$$(S_nf)(x) = \frac{a_0}{2} + \sum_{k=1}^n (a_k \cos kx + b_k \sin kx)$$

where $$a_k = \frac{1}{\pi}\int_{-\pi}^{\pi} f(t) \cos kt\,dt$$

and $$b_k = \frac{1}{\pi}\int_{-\pi}^{\pi} f(t) \sin kt\,dt$$

In accordance with our earlier observation about general orthonormal systems, this operator S_n can be put into integral form. The kernel that occurs here is known as the *Dirichlet* kernel.

Lemma. *The Fourier-series operator S_n has the following integral form for 2π-periodic continuous functions f:*

$$(S_nf)(x) = \frac{1}{\pi}\int_{-\pi}^{\pi} f(t + x) \frac{\sin (n + \frac{1}{2})t}{2 \sin (t/2)}\,dt \tag{4}$$

Proof. In the definition of S_n, insert the integrals by which a_k and b_k are defined. The result is

$$(S_n f)(x) = \frac{1}{\pi}\int_{-\pi}^{\pi} f(t)\left[\frac{1}{2} + \sum_{k=1}^{n} (\cos kx \cos kt + \sin kx \sin kt)\right] dt$$

Application of the trigonometric identity $\cos(A - B) = \cos A \cos B + \sin A \sin B$ produces

$$(S_n f)(x) = \frac{1}{\pi}\int_{-\pi}^{\pi} f(t)\left[\frac{1}{2} + \sum_{k=1}^{n} \cos k(x - t)\right] dt$$

Since the integrand is periodic we may integrate instead over the interval $[-\pi + x, \pi + x]$. Then the change of variable $t \to x + t$ gives us

$$(S_n f)(x) = \frac{1}{\pi}\int_{-\pi}^{\pi} f(t + x)\left(\frac{1}{2} + \sum_{k=1}^{n} \cos kt\right) dt$$

The proof now continues with a verification of the equation

$$\left(\frac{1}{2} + \sum_{k=1}^{n} \cos kt\right) 2 \sin \frac{t}{2} = \sin (n + \tfrac{1}{2})t \tag{5}$$

This is true because with the aid of the identity $2 \cos A \sin B = \sin(A + B) - \sin(A - B)$, the left member of (5) becomes

$$\sin \frac{t}{2} + \sum_{k=1}^{n} [\sin (k + \tfrac{1}{2})t - \sin (k - \tfrac{1}{2})t] = \sin (n + \tfrac{1}{2})t$$

Equation (5) establishes the correct form of the kernel in (4) at all points where $\sin(t/2) \neq 0$. At the points where $\sin(t/2) = 0$, a limiting value is to be taken, the existence of the limit being a consequence of the continuity of the functions in (5). ∎

Theorem. *If f is 2π-periodic, continuous everywhere, and differentiable at x, then its Fourier series converges at x to $f(x)$.*

Proof. Since $S_n 1 = 1$, we have from the lemma,

$$\epsilon_n \equiv (S_n f)(x) - f(x) = \frac{1}{\pi}\int_{-\pi}^{\pi} [f(t + x) - f(x)] \frac{\sin (n + \frac{1}{2})t}{2 \sin \frac{1}{2}t} dt$$

Let x be fixed, and set $g(t) = [f(t+x) - f(x)]/(2 \sin \frac{1}{2}t)$. This function is continuous except at $t = 0$, where it is undefined. But

$$\lim_{t\to 0} g(t) = \lim_{t\to 0} \frac{f(x+t) - f(x)}{t} \frac{t}{2 \sin \frac{1}{2}t} = f'(x)$$

so that simply by defining $g(0) = f'(x)$ we make g continuous. Since

$$\frac{1}{\pi}\int_{-\pi}^{\pi} g(t) \sin (n + \tfrac{1}{2})t \, dt$$

$$= \frac{1}{\pi}\int_{-\pi}^{\pi} g(t) \cos \tfrac{1}{2}t \sin nt \, dt + \frac{1}{\pi}\int_{-\pi}^{\pi} g(t) \sin \tfrac{1}{2}t \cos nt \, dt$$

we have here the Fourier coefficients of the functions $g(t) \cos \frac{1}{2}t$ and $g(t) \sin \frac{1}{2}t$. These coefficients go to zero as $n \to \infty$ by the Bessel inequality. Hence $\epsilon_n \to 0$. ∎

A theorem similar to the one above was proved as long ago as 1829 by Dirichlet. His result was that the Fourier series of a function which is continuous and monotone in sections converges at each point x to $f(x)$ provided that at a point of discontinuity, $f(x)$ was defined as $\frac{1}{2}[f(x+0) + f(x-0)]$. For almost 50 years after this it was conjectured that every continuous function was represented by its Fourier series. However, in 1876 Du Bois–Reymond succeeded in constructing an example of a continuous function whose Fourier series diverged at a point. In 1965, Carleson established that the Fourier series of a continuous function converges everywhere, except possibly on a set of measure 0.

In spite of these negative results, a very simple transformation of the Fourier series can be employed to provide *uniform* approximations of arbitrary precision to all continuous functions. This discovery by Fejér in 1900 may be very easily stated: *The first Cesàro means of the Fourier series of a continuous function converge uniformly to the function.* The Cesàro means of a sequence $\{g_n\}$ are the averages $G_n = \frac{1}{n}\sum_{k=1}^{n} g_k$. The Cesàro means of a series are then the averages of its partial sums. In the case of the Fourier series, if we retain the symbol $S_n f$ to denote the partial sums, then the Cesàro means are the functions

$$G_n f = \frac{1}{n}[S_0 f + S_1 f + \cdots + S_{n-1} f]$$

We can put this operator into the form of an integral operator as follows, the kernel which occurs being known as the *Fejér kernel.*

Lemma. *The Fejér operator G_n has the alternative form*

$$(G_n f)(x) = \frac{1}{2n\pi} \int_{-\pi}^{\pi} f(t + x) \left(\frac{\sin \frac{1}{2}nt}{\sin \frac{1}{2}t} \right)^2 dt$$

Proof. Using the integral form of the operators S_n (page 120), we have

$$\begin{aligned} (G_n f)(x) &= \frac{1}{n} \sum_{k=0}^{n-1} \frac{1}{\pi} \int_{-\pi}^{\pi} f(t + x) \frac{\sin (k + \frac{1}{2})t}{2 \sin \frac{1}{2}t} dt \\ &= \frac{1}{2n\pi} \int_{-\pi}^{\pi} f(t + x) \sum_{k=0}^{n-1} \frac{\sin (k + \frac{1}{2})t}{\sin \frac{1}{2}t} dt \end{aligned}$$

In order to complete the proof we must show that

$$\sum_{k=0}^{n-1} \frac{\sin (k + \frac{1}{2})t}{\sin \frac{1}{2}t} = \left(\frac{\sin \frac{1}{2}nt}{\sin \frac{1}{2}t} \right)^2$$

It will be sufficient to establish that

$$\sum_{k=0}^{n-1} \sin (k + \tfrac{1}{2})t \sin \tfrac{1}{2}t = (\sin \tfrac{1}{2}nt)^2$$

Using the identity $2 \sin A \sin B = \cos (A - B) - \cos (A + B)$, we may write the left member of this equation as

$$\frac{1}{2} \sum_{k=0}^{n-1} [\cos kt - \cos (k + 1)t] = \tfrac{1}{2}(1 - \cos nt)$$

From the "half-angle" formula, the latter is $(\sin \frac{1}{2}nt)^2$. ∎

In order to prove Fejér's theorem we require the trigonometric analogue of the monotone-operator theorem (Chap. 3, Sec. 3). We state it without proof.

Korovkin's Theorem. *Let $\{L_n\}$ denote a sequence of monotone linear operators on $C_{2\pi}$. In order that $L_n f \to f$ (uniformly) for all $f \in C_{2\pi}$, it is necessary and sufficient that such convergence occur for $f = 1$,* cos, *and* sin.

Fejér's Theorem. *The first Cesàro means of the Fourier series of a continuous 2π-periodic function converge uniformly to the function.*

Proof. We see from the preceding lemma that the Fejér operators G_n are monotone operators; i.e., if $f \geq g$, then $G_n f \geq G_n g$. By Korovkin's theorem, we may complete the proof by verifying that $G_n f \to f$ when $f = 1$, cos, or sin. We calculate, then,

$$G_n 1 = \frac{1}{n}(1 + \cdots + 1) = 1 \to 1$$

$$(G_n \cos)(x) = \frac{1}{n}(0 + \cos x + \cdots + \cos x) = \frac{n-1}{n}\cos x \to \cos x$$

$$(G_n \sin)(x) = \frac{1}{n}(0 + \sin x + \cdots + \sin x) = \frac{n-1}{n}\sin x \to \sin x \quad \blacksquare$$

Problems

1. Show that the kernel defined on page 118 has the *reproducing* property: $P(x) = \int_a^b P(t)K_n(t,x)w(t)\,dt$ whenever P is a polynomial of degree $\leq n$. The Dirichlet kernel (page 120) also has a reproducing property. What is it? Has the Fejér kernel (page 123) any reproducing property?

2. If ξ is fixed in $(-\infty,a)$, then the polynomials $K_n(\xi,x)$ form an orthogonal system on $[a,b]$ with weight function $w(x)(x-\xi)$. *Hint*: Use Prob. 1.

***3.** Prove the Christoffel-Darboux formula by computing the Fourier coefficients of the function $f(x) = (x-y)\sum_{i=0}^{n} \bar{Q}_i(x)\bar{Q}_i(y)$.

4. If $g_n \to g$, then the first Cesáro means G_n of $\{g_n\}$ also converge to g. *Hint*: Select m so that $|g - g_i| < \epsilon$ whenever $i > m$. Then let $n > m$, and write

$$|g - G_n| = \left|g - \frac{1}{n}\sum_{i\leq n} g_i\right| = \left|\frac{1}{n}\sum_{i\leq n}(g - g_i)\right| \leq \frac{1}{n}\sum_{i\leq m}|g - g_i| + \frac{1}{n}\sum_{i>m}|g - g_i|$$

5. At a point x_0, the Fourier series of a continuous 2π-periodic function f either converges to $f(x_0)$ or diverges. *Hint*: Use Fejér's theorem and Prob. 4.

6. If all the Fourier coefficients of a continuous 2π-periodic function f vanish, then $f = 0$. *Hint*: Use Fejér's theorem.

7. The set of functions $\{\cos nx, \sin nx : n = 0, 1, \ldots\}$ is *fundamental* in the space $C_{2\pi}$ of continuous 2π-periodic functions with the uniform norm.

8. One might hope that the smoother the functions f, the more rapid would be the convergence of $G_n f$ to f. (G_n = Fejér operator.) Give a counterexample.

9. If a_n, b_n are the Fourier coefficients of f, then Fejér's operator takes the form $(G_n f)(x) = \frac{1}{2}a_0 + \sum_{k=1}^{n-1}\left(1 - \frac{k}{n}\right)(a_k \cos kx + b_k \sin kx)$.

10. $\frac{1}{\pi}\int_{-\pi}^{\pi}(G_n f)^2 = \frac{1}{2}a_0^2 + \sum_{k=1}^{n}\left(1 - \frac{k}{n}\right)^2(a_k^2 + b_k^2)$. *Hint*: Use Prob. 9.

11. Prove Parseval's theorem: If f is a continuous 2π-periodic function having Fourier coefficients a_n, b_n, then $\frac{1}{\pi}\int_{-\pi}^{\pi} f^2 = \frac{1}{2}a_0^2 + \sum_{n=1}^{\infty}(a_n^2 + b_n^2)$. *Hint*: Use Fejér's theorem and Prob. 10.

12. By observing that $\frac{1}{2} + \cos t + \cos 2t + \cdots$ is the real part of the geometric series $\frac{1}{2} + \sum e^{int}$, provide an alternative proof for equation (5).

13. Two different functions can have the same Fourier series. For example, if f and g differ only on a finite set of points, then the integrals defining the Fourier coefficients are the same for f and g. Can this happen for continuous functions? *Hint*: Use Prob. 6.

14. Let f and g be two continuous 2π-periodic functions having Fourier coefficients $\{a_n,b_n\}$ and $\{\alpha_n,\beta_n\}$, respectively. Then $\frac{1}{\pi}\int_{-\pi}^{\pi} fg = \frac{1}{2}\alpha_0 a_0 + \sum (\alpha_n a_n + \beta_n b_n)$. *Hint*: Apply Parseval's theorem in Prob. 11 to the function $f - g$.

15. Deduce the Weierstrass approximation theorem from Fejér's theorem. *Hint*: If f is continuous on $[-1,1]$, then $g(x) = f(\cos x)$ is continuous and 2π-periodic.

16. A number L is said to be the $(C,1)$ sum of a series $\sum u_n$ if the Cesàro means of the partial sums converge to L. Show that the $(C,1)$ sum of $\frac{1}{2} + \cos x + \cos 2x + \cdots$ is 0 if x is not a multiple of 2π. Is this series the Fourier series of some continuous function?

17. If f is a continuous 2π-periodic function whose Fourier coefficients have the property $\sum (|a_n| + |b_n|) < \infty$, then the Fourier series converges uniformly to f.

18. If f is 2π-periodic and has modulus of continuity ω, then its Fourier coefficients satisfy $|a_n| \le \omega(\pi/n)$ and $|b_n| \le \omega(\pi/n)$. *Hint*: In the usual formula for a_n make the change of variable $x \to x + \pi/n$. [de La Vallée Poussin, 1919, p. 16]

19. Weierstrass M test. If $|f_n(x)| \le M_n$ and $\sum M_n < \infty$, then $\sum f_n$ converges uniformly.

20. Let $\{g_n\}$ be an orthonormal system of continuous functions on $[a,b]$ for which $\|g_n\|_T$ remains bounded. If the series $\sum |\langle f,g_n\rangle|$ converges for a given $f \in C[a,b]$, does the generalized Fourier series $\sum \langle f,g_n\rangle g_n$ converge uniformly to f?

21. If f is a continuous 2π-periodic function with Fourier coefficients a_k and b_k then $\sum_{k\le n} (|a_k| + |b_k|) \le \pi\sqrt{2n-1}\,\|f\|_T$. More generally, if $\{g_1,\ldots,g_n\}$ is an orthonormal system on $[a,b]$ with weight $w(x)$, then

$$\sum_{k=1}^{n} |a_k| \le \sqrt{n}\left(\int_a^b w\right)^{1/2} \max_x \left|\sum_{k=1}^{n} a_k g_k(x)\right|$$

Hint: Use Bessel's inequality and the Cauchy-Schwarz inequality.

22. For each n let P_n be a trigonometric polynomial of degree n such that $P_n(x) \ge 0$ and $\int_{-\pi}^{\pi} P_n(x)\,dx = 1$. Assume further that to each pair (ϵ,δ) there corresponds an N such that $\int_{-\delta}^{\delta} P_n(x)\,dx > 1 - \epsilon$ for $n \ge N$. Then the operators

$$(L_n f)(x) = \int_{-\pi}^{\pi} f(t)P_n(t-x)\,dt$$

have the property that for $f \in C_{2\pi}$, $L_n f \to f$ and $L_n f$ is a trigonometric polynomial of degree $\le n$. [Perron, 1941]

23. Show that the Fejér operator G_n has the property $\|G_n f\| \le \|f\|$. [de La Vallée Poussin, 1919, p. 32]

24. Show that the Fejér operator G_n has the property $P = 2G_{2n}P - G_nP$ for any trigonometric polynomial P of degree $\le n$.

25. Establish the inequality $E_n(f) \geq \|f - 2G_{2n}f + G_nf\|_T$ by use of the preceding two problems. Here $f \in C[-\pi,\pi]$ and $E_n(f) = \|f - P\|_T$ if P is the trigonometric polynomial of degree $\leq n$ which best approximates f. [de La Vallée Poussin, 1918]

26. Among the polynomials P of degree $\leq n$ which take a prescribed value at a certain point x_0, the one of minimum norm $(\int P^2w)^{1/2}$ is a multiple of $K_n(x_0,x) \equiv \sum_{i=0}^{n} \bar{Q}_i(x_0)\bar{Q}_i(x)$.

In [Stiefel, 1958] this fact becomes the cornerstone of an analysis of matrix iteration.

27. If f is expanded formally in a series of orthogonal polynomials $f \sim \sum c_nQ_n$ and if $\sum |c_n| \|Q_n\|_T$ converges, then the orthogonal expansion converges uniformly to f. Where does the fact that the Q_n are *polynomials* enter?

28. Is every series $\sum (A_n \cos nx + B_n \sin nx)$ the Fourier series of some $f \in C_{2\pi}$? Answer the same question when we assume $\sum |A_n| + |B_n| < \infty$, or only that $|A_n| + |B_n| \to 0$.

29. Prove the Korovkin theorem. *Hint*: Employ the change of variable $x \to \cos\theta$ and the monotone operator theorem of Chap. 3, Sec. 3.

4. Approximation by Series of Tchebycheff Polynomials

The possibility of representing functions in the form $f = \sum_{k=0}^{\infty} a_kT_k$ was considered briefly in the preceding section, along with examples of expansions in other orthogonal systems. It will be seen presently that the Tchebycheff polynomials have decisive advantages over other orthogonal systems if it is our purpose to provide good approximations in the *uniform* norm. Indeed, for many functions f, a truncated Tchebycheff expansion $S_nf = \sum_{k=0}^{n} a_kT_k$ is very close to a best polynomial approximation in the Tchebycheff norm. Of course, it follows from the theorems of Sec. 1 that S_nf is the best approximation of f in the least-squares norm:

$$\|f - S_nf\|_w^2 = \int_{-1}^{1} |f(x) - (S_nf)(x)|^2 (1 - x^2)^{-1/2}\, dx$$

For functions which are highly "regular" (possess many derivatives) the Tchebycheff norm,

$$\|f - S_nf\|_T = \max_{-1\leq x\leq 1} |f(x) - (S_nf)(x)|$$

is generally within a few percent of its absolute minimum. When we take into account the fact that a polynomial of best approximation is often difficult to obtain, while S_nf is relatively easy to obtain, we recognize that the Tchebycheff expansions are an important device in uniform approximations. Actually, even if f is assumed to be only continuous (and if $n < 400$), we can never get more than one extra decimal place of accuracy in passing from $S_n f$ to the polynomial of *best* approximation (Prob. 13, Sec. 6).

In order to make comparisons of the type just mentioned, we employ again the notation $E_n(f)$ for the distance (in uniform norm) from f to the subspace of polynomials having degree $\leq n$. Thus

$$E_n(f) = \min_{c_0,\ldots,c_n} \max_{-1\leq x\leq 1} \left| f(x) - \sum_{i=0}^{n} c_i x^i \right|$$

With this notation, the Weierstrass theorem states simply that $E_n(f) \to 0$ for all $f \in C[-1,1]$. Later on we shall establish some theorems of Jackson which state that $E_n(f)$ converges to zero all the faster when f is smooth. At the moment we shall employ Tchebycheff expansions to establish a "lethargy" theorem: $E_n(f)$ can converge to zero very slowly for some continuous functions.

Theorem 1. *If $\{\epsilon_n\}$ is any sequence converging downward to zero, then there exists an element $f \in C[-1,1]$ such that $E_n(f) \geq \epsilon_n$ for all n.*

Proof. Define $\alpha_k = \epsilon_{k-1} - \epsilon_k$ and $f = \sum_{k=1}^{\infty} \alpha_k T_{3^k}$. By hypothesis, $\alpha_k \geq 0$. Since $\| T_n \| = 1$, the series for f is majorized by the series $\sum \alpha_k$ and consequently converges uniformly by the Weierstrass M test. (Prob. 19, Sec. 3.) It follows that $f \in C[-1,1]$. Now we show that the best approximation of degree $\leq 3^n$ to f is simply the partial sum $P = \sum_{k=1}^{n} \alpha_k T_{3^k}$. It will suffice to show that the error function $r = f - P = \sum_{k=n+1}^{\infty} \alpha_k T_{3^k}$ achieves its maximum deviation from zero with alternating signs in at least $3^n + 2$ points of the interval $[-1,1]$. Consider the points $x_i = \cos(i\pi/3^{n+1})$ with $i = 0, \ldots, 3^{n+1}$. If $k \geq n + 1$, then $T_{3^k}(x_i) = \cos(3^k i\pi/3^{n+1}) = (-1)^i$. Thus $r(x_i) = (-1)^i \sum_{k=n+1}^{\infty} \alpha_k = (-1)^i \epsilon_n$. Since $\epsilon_n = | r(x_i) | \leq \| r \| \leq \sum_{k=n+1}^{\infty} | \alpha_k | = \epsilon_n$, we have proved actually that r alternates at least $3^{n+1} + 1$ times between the values $\pm\epsilon_n$. Consequently, $E_n(f) \geq E_{3^n}(f) = \epsilon_n$. ∎

It should be observed that the proof remains valid if the sequence $\{3^n\}$ is replaced by any sequence of odd integers, each dividing the next.

Since a more general theorem of the same type is known, we shall state it here but refer the reader to [Timan, 1960] or [Golomb, 1960] for its proof.

Theorem 2. *Let $\{g_1, g_2, \ldots\}$ be a linearly independent sequence in a Banach space B. If $\{\epsilon_n\}$ is any sequence of numbers converging downward to zero, then there exists an $f \in B$ such that, for all n, $\epsilon_n =$*

$$\inf_c \| f - \sum_{i=1}^{n} c_i g_i \|.$$

In the proof of Theorem 1 there was exhibited a class of functions $f \in C[-1,1]$ having the property that $P_n f = S_n f$ for all n, $P_n f$ and $S_n f$ being respectively the best approximations in the uniform norm and in the least-squares norm with weight $(1 - x^2)^{-1/2}$. It would be a very desirable state of affairs if many common functions requiring approximation possessed this property. Unfortunately this is not the case. Moreover, the functions considered in Theorem 1 can be highly irregular. The classic example of this (in terms of trigonometric series) is the *Weierstrass function*,

$$f(x) = \sum_{k=0}^{\infty} a^k \cos b^k x \tag{1}$$

in which $0 < a < 1$, and b is an odd integer greater than a^{-1}. It is clear that this function is continuous everywhere. But it is not differentiable at any point whatsoever. A proof of this (in slightly weakened form) is included here because of its general interest.

Theorem 3. *If $0 < a < 1$ and if b is an odd integer greater than $6a^{-1}$, then the Weierstrass function (1) is nowhere differentiable.*

Proof. (Titchmarsh) Let x be an arbitrary point. The theorem will be proved by showing how to make $h \to 0$ in such a way that $h^{-1} | f(x+h) - f(x) | \to \infty$. First write, with obvious abbreviations,

$$\begin{aligned} \left| \frac{f(x+h) - f(x)}{h} \right| &= \left| \sum_{k=0}^{\infty} h^{-1} a^k [\cos b^k (x+h) - \cos b^k x] \right| \\ &= \left| \sum_{k<n} A_k + \sum_{k \geq n} A_k \right| \\ &\geq \left| \sum_{k \geq n} A_k \right| - \sum_{k<n} | A_k | \end{aligned}$$

With n fixed, select an integer ν such that the number $h = -x + \nu\pi b^{-n}$ lies in the interval $[\frac{1}{2}\pi b^{-n}, \frac{3}{2}\pi b^{-n})$. For $k \geq n$ we have $\cos b^k(x+h) = \cos b^{k-n} b^n (x+h) = \cos b^{k-n} \nu\pi = (-1)^\nu$. Thus the terms A_k, for $k \geq n$, all have the same sign, viz., $(-1)^\nu$. Consequently

$$\left| \sum_{k \geq n} A_k \right| = \sum_{k \geq n} | A_k | \geq | A_n |$$

Since $-\cos b^n x = -\cos(\nu\pi - b^n h) = -(-1)^\nu \cos b^n h$ and $b^n h \in [\pi/2, 3\pi/2)$

it follows that $-\cos b^n x$ has the sign $(-1)^\nu$. Consequently

$$
\begin{aligned}
|A_n| &= h^{-1}a^n \,|\cos b^n(x+h) - \cos b^n x| \\
&\geq h^{-1}a^n \geq \frac{2}{3\pi}\, a^n b^n
\end{aligned}
$$

On the other hand, for $k < n$ we may use the mean-value theorem to write

$$\sum_{k<n} |A_k| = \sum_{k<n} h^{-1}a^k \,|hb^k \sin b^k\xi_k| \leq \sum_{k<n} a^k b^k \leq \frac{a^n b^n}{ab - 1}$$

Combining the inequalities above yields

$$\left|\sum_{k=0}^{\infty} A_k\right| \geq \left(\frac{2}{3\pi} - \frac{1}{ab-1}\right) a^n b^n$$

It is clear that for large ab the term in parentheses is positive, and the entire expression will become infinite as $n \to \infty$. Specifically, one may check that $b > 6a^{-1}$ is a sufficient condition. ■

The Tchebycheff series that we have considered up to now in this section have had a very simple convergence behavior: namely, the series $f = \sum a_k T_k$ has converged because $\sum |a_k|$ converged, and the Weierstrass M test was applicable. More sophisticated convergence criteria, involving the function f, are available. It is appropriate to state one of these here, although for reasons of economy the proof is deferred to Sec. 6. As before, ω denotes the modulus of continuity of the function f: $\omega(\delta) = \max\limits_{|x-y|\leq\delta} |f(x) - f(y)|$.

Dini-Lipschitz Theorem. *If the function f satisfies the Dini-Lipschitz condition on $[-1,1]$, $\lim\limits_{\delta\to 0} \omega(\delta) \log \delta = 0$, then its development in Tchebycheff polynomials converges uniformly to it.*

We conclude this section with some theorems which enable us to estimate $E_n(f)$ from a formal expansion of f in orthogonal polynomials. We suppose therefore that we have a sequence of polynomials $\{Q_0, Q_1, \ldots\}$ (the subscripts indicate the degrees) which is orthogonal with respect to the inner product

$$\langle f,g\rangle = \int_a^b f(x)g(x)w(x)\,dx$$

Any function $f \in C[a,b]$ then possesses a formal expansion

$$f \sim \sum_{k=0}^{\infty} c_k Q_k$$

where the coefficients are given by $c_k = \langle f,Q_k\rangle / \langle Q_k,Q_k\rangle$. The series may or may not converge to f.

Theorem 4. *Let f be formally expanded in a series of orthogonal polynomials, $f \sim \sum c_k Q_k$. Then*

(*i*) $E_n(f) \geq \max \{\alpha_{n+1} | c_{n+1} |, \alpha_{n+2} | c_{n+2} |, \ldots \}$

(*ii*) $E_n(f) \geq \sqrt{\beta_{n+1} c_{n+1}^2 + \beta_{n+2} c_{n+2}^2 + \cdots}$

(*iii*) $E_n(f) \leq \gamma_{n+1} | c_{n+1} | + \gamma_{n+2} | c_{n+2} | + \cdots$

$$\text{where } \alpha_k = \int_a^b Q_k^2 w \Big/ \int_a^b | Q_k | w,$$

$$\beta_k = \int_a^b Q_k^2 w \Big/ \int_a^b w,$$

$$\text{and } \quad \gamma_k = \max_{a \leq x \leq b} | Q_k(x) |$$

Proof. Let P be the polynomial of degree $\leq n$ which best approximates f in the uniform norm. If $k > n$, then the orthogonality relations imply

$$\begin{aligned} | c_k | \int_a^b Q_k^2 w &= \left| \int_a^b f Q_k w \right| \\ &= \left| \int_a^b (f - P) Q_k w \right| \\ &\leq \int_a^b | f - P | \, | Q_k | w \\ &\leq E_n(f) \int_a^b | Q_k | w \end{aligned}$$

This proves (*i*). In order to prove (*ii*) let $S_n = \sum_{k=0}^{n} c_k Q_k$. Also put $\langle f, g \rangle = \int_a^b fgw$, and let $\{\bar{Q}_k\}$ be the orthonormal set of polynomials. Now apply Bessel's inequality to $f - S_n$, noting that $f - S_n \perp \{Q_0, \ldots, Q_n\}$ and $S_n \perp \{Q_{n+1}, \ldots \}$. Thus

$$\begin{aligned} \sum_{k=n+1}^{\infty} \langle f, \bar{Q}_k \rangle^2 &= \sum_{k=0}^{\infty} \langle f - S_n, \bar{Q}_k \rangle^2 \\ &\leq \langle f - S_n, f - S_n \rangle \\ &\leq \langle f - P, f - P \rangle \\ &\leq E_n^2(f) \int_a^b w \end{aligned}$$

Since $\bar{Q}_k = Q_k/\langle Q_k,Q_k\rangle^{1/2}$, we have $\langle f,\bar{Q}_k\rangle^2 = \langle f,Q_k\rangle^2/\langle Q_k,Q_k\rangle = c_k^2\langle Q_k,Q_k\rangle$. This proves (*ii*). For the proof of (*iii*), observe that it is trivial unless $\sum \gamma_k |c_k|$ is convergent. In the latter case, the series $\sum c_kQ_k$ converges uniformly and represents f (Prob. 27 of the preceding section). Thus

$$E_n(f) \le \|f - S_n\| = \Big\|\sum_{k=n+1}^{\infty} c_kQ_k\Big\|$$

$$\le \sum_{k=n+1}^{\infty} |c_k| \, \|Q_k\| = \sum_{k=n+1}^{\infty} |c_k| \gamma_k \qquad \blacksquare$$

The chief application of the preceding theorem is to expansions of functions in Tchebycheff series, $f \sim \sum a_kT_k$, where we mean by the symbol $\sim$ only that $a_k = \dfrac{2}{\pi}\displaystyle\int_0^\pi f(\cos\theta)\cos k\theta\, d\theta$ for $k \ge 1$, and $a_0 = \dfrac{1}{\pi}\displaystyle\int_0^\pi f(\cos\theta)\, d\theta$. It is not at all necessary in these estimates that the series *represent* f.

Theorem 5. *If* $f \in C[-1,1]$ *and if* $f \sim \sum a_kT_k$, *then*

(*i*) $E_n(f) \ge \dfrac{\pi}{4} \max\{|a_{n+1}|, |a_{n+2}|, \ldots\}$

(*ii*) $E_n(f) \ge \sqrt{\tfrac{1}{2}(a_{n+1}^2 + a_{n+2}^2 + \cdots)}$

(*iii*) $E_n(f) \le |a_{n+1}| + |a_{n+2}| + \cdots$

Proof. We must compute the constants α_k, β_k, γ_k that occur in the preceding theorem:

$$\int_{-1}^{1} T_k^2(x)(1 - x^2)^{-1/2}\,dx = \int_0^\pi \cos^2 k\theta\, d\theta = \frac{\pi}{2}$$

$$\int_{-1}^{1} |T_k(x)|(1 - x^2)^{-1/2}\,dx = \int_0^\pi |\cos k\theta|\, d\theta$$

$$= 2k\int_0^{\pi/2k} \cos k\theta\, d\theta = 2$$

$$\int_{-1}^{1} (1 - x^2)^{-1/2}\,dx = \int_0^\pi d\theta = \pi$$

Hence $\alpha_k = \pi/4$, $\beta_k = \frac{1}{2}$. Of course, $\gamma_k = 1$. $\blacksquare$

Theorem 6. [Rivlin, 1962a] *If* $f \sim \sum a_kT_k$, *then*

$$\big|E_{n-1}(f) - |a_n|\big| \le \sum_{k>n} |a_k|$$

Proof. Half of the asserted inequality comes from the preceding theorem:

$$E_{n-1}(f) - |a_n| \le \sum_{k\ge n} |a_k| - |a_n| = \sum_{k>n} |a_k|$$

For the other half, we must show that $E_{n-1}(f) \ge |a_n| - \sum_{k>n} |a_k| \equiv \epsilon$. This is trivial if $\epsilon \le 0$, and we may therefore assume the contrary. Then it follows that

$$\|\sum_{k>n} a_k T_k\| \le \sum_{k>n} |a_k| < |a_n| = \|a_n T_n\|$$

Consequently the function $f - \sum_{k<n} a_k T_k = a_n T_n + \sum_{k>n} a_k T_k$ possesses $n + 1$ points at which it assumes alternately positive and negative values, these points being the extrema of T_n. These values are at least ϵ in magnitude, and thus by de La Vallée Poussin's theorem (Chap. 3, Sec. 4), $E_{n-1}(f) \ge \epsilon$. ∎

Problems

1. If f satisfies a Lipschitz condition on $[-1,1]$, then its development in a series of Tchebycheff polynomials converges uniformly. Show that this follows from the Dini-Lipschitz theorem. Also give a proof based on the proof of the theorem on page 119.
2. The estimate $E_n(f) \le \sum_{k>n} |a_k|$ cannot be improved because the equality sign can occur for some f and n. Give an example of this. Does there exist a function f such that $E_n(f) = \sum_{k>n} |a_k|$ for *all* n?
3. Compute $E_n(T_k)$ and $E_n(x^k)$.
4. What can you say about the differentiability of the function $f(x) = \sum_{k=0}^{\infty} c_k \cos b^k x$ when c_k is not of the form a^k?
5. Starting with $E_n(f) \le \sum_{k>n} |a_k|$ (page 131), and proceeding as in the proof on page 117, show that $E_n(f) \le (4/\pi n)(\|f'\| + \|f''\|)$. This elementary inequality is much worse than that given by Jackson's theorem in Sec. 6.
6. If b is an odd integer >1 and if $0 < \alpha < 1$, then the function $f = \sum_{k=0}^{\infty} \alpha^k T_{b^k}$ is continuous on $[-1,1]$ and has the property that its best polynomial approximation of degree $\le N$ is $\sum_{k=0}^{n} \alpha^k T_{b^k}$, with n selected so that $b^n \le N < b^{n+1}$.
7. For the function f of Prob. 6 show that $E_n(f) \ge (1 - \alpha)^{-1} \exp\left(\log \alpha + \dfrac{\log \alpha \log n}{\log b}\right)$.
8. Under what conditions does the equation $f = \sum_{n=0}^{\infty} \phi_n$ imply the equation $f' = \sum_{n=0}^{\infty} \phi_n'$? Prove that if $f = \sum c_n T_n$ and $\sum |c_n| n^2 < \infty$, then $f' = \sum c_n T_n'$.
9. Establish the formal expansion, and investigate its convergence:

$$|x| \sim \frac{2}{\pi} - \frac{4}{\pi} \sum_{n=1}^{\infty} \frac{(-1)^n}{4n^2 - 1} T_{2n}(x)$$

10. Consider any real-valued function Φ defined on $C[-1,1]$ which has the properties
(*a*) $\Phi(f-g) = \Phi(f) - \Phi(g)$
(*b*) $|\Phi(f)| \le \lambda \|f\|$
(*c*) $\Phi(P) = 0$ when P is a polynomial of degree $\le n$
Then $E_n(f) \ge \lambda^{-1}|\Phi(f)|$.

11. Show that the proof of Theorem 4 illustrates the reasoning in the preceding problem. Prove that the following affords another illustration. Let $x_0,\ldots,x_n$ be any $n+1$ nodes, and let $\sum f(x_i)l_i$ be the Lagrange interpolating operator for these nodes. Define $\Phi(f) = f(\xi) - \sum f(x_i)l_i(\xi)$. Show that in this case λ may be taken to be $2\sum |l_i(\xi)|$.

12. Let $f \in C[-1,1]$ and $f \sim \frac{a_0}{2} + \sum_{n=1}^{\infty} a_nT_n$. Establish the implication

$$\frac{|a_{n+1}| + |a_{n+2}| + \cdots}{|a_n|} \to 0 \Rightarrow \frac{E_n(f)}{|a_{n+1}|} \to 1$$

[Bernstein, 1912b]

13. Theorem 2 may be easily proved in a Hilbert space. Indeed, let $\{h_1,h_2,\ldots\}$ be the sequence which results from applying the Gram-Schmidt process to $\{g_1,g_2,\ldots\}$. Define

$$f = \sum_{k=0}^{\infty} (\epsilon_k^2 - \epsilon_{k+1}^2)^{1/2} h_{k+1}.$$

5. Discrete Least-squares Approximation

We consider now a "pseudo inner product"

$$\langle f,g\rangle = \sum_{i=1}^{m} f(x_i)g(x_i)w(x_i)$$

in which the points x_i and the weights $w(x_i)$ are prescribed and held fixed. We assume that f and g belong to $C[a,b]$ and that $x_i \in [a,b]$. We term this a *pseudo* inner product because there exist nonzero functions f satisfying $\langle f,f\rangle = 0$. Corresponding to this pseudo inner product there is a pseudo norm or seminorm

$$\|f\| = \langle f,f\rangle^{1/2}$$

and it is reasonable to ask about approximation problems relative to it. It turns out that some systems of functions which were orthogonal with an *integral* inner product are also orthogonal with respect to a *discrete* inner product.

Theorem 1. *Let $\{\bar{Q}_0,\bar{Q}_1,\ldots\}$ be the system of polynomials (the subscripts denoting their degrees) which is orthonormal with respect to the inner product*

$$\langle f,g\rangle = \int_a^b f(x)g(x)w(x)\,dx$$

Then the system $\{\bar{Q}_0, \ldots, \bar{Q}_n\}$ *is also orthonormal with respect to the pseudo inner product*

$$\langle f,g \rangle = \sum_{i=1}^{N} A_i f(x_i) g(x_i)$$

where $N > n$, x_i *are roots of* $\bar{Q}_N$, *and* A_i *are the Gaussian quadrature coefficients.*

Proof. For each N there is a Gaussian integration formula (page 110),

$$\int_a^b f(x) w(x)\, dx \approx \sum_{i=1}^{N} A_i f(x_i)$$

which is exact when f is a polynomial of degree $<2N$. Here x_i are the roots of Q_N, and the coefficients are given by the formulas of page 112:

$$A_i = \frac{\phi_N(x_i)}{Q'_N(x_i)} = \frac{1}{Q'_N(x_i)} \int_a^b \frac{Q_N(x)}{x - x_i} w(x)\, dx$$

The positivity of the coefficients A_i was observed in the course of proving Stieltjes' theorem (Sec. 2). Thus the pseudo inner product has the property $\langle f,f \rangle \geq 0$. Taking $k + m < 2N$ in the Gaussian formula, we have

$$\delta_{km} = \int_a^b \bar{Q}_k(x) \bar{Q}_m(x) w(x)\, dx = \sum_{i=1}^{N} A_i \bar{Q}_k(x_i) \bar{Q}_m(x_i)$$ ∎

Corollary. *The Tchebycheff polynomials have the following property, when* $x_i = \cos[(2i - 1)\pi/2N]$ *and* $n + m < 2N$,

$$\frac{2}{N} \sum_{i=1}^{N} T_n(x_i) T_m(x_i) = \begin{cases} 0 & (n \neq m) \\ 1 & (n = m > 0) \\ 2 & (n = m = 0) \end{cases}$$

Proof. From Prob. 13, Sec. 2, the Gaussian quadrature formula for this case has coefficients $A_i = \pi/N$. As noted in Example 3, Sec. 1, the Tchebycheff polynomials have the property

$$\frac{2}{\pi} \int_{-1}^{1} T_n(x) T_m(x) (1 - x^2)^{-1/2}\, dx = \begin{cases} 0 & (n \neq m) \\ 1 & (n = m > 0) \\ 2 & (n = m = 0) \end{cases}$$

from which the desired equation follows, upon our substituting the Gaussian sum for the integral. ∎

Theorem 2. *The polynomial P of degree $\le n - 1$ which interpolates to f at the zeros $x_1, \ldots, x_n$ of $\bar{Q}_n$ is given by*

$$P(x) = \sum_{k=0}^{n-1} a_k \bar{Q}_k \qquad a_k = \sum_{i=1}^{n} A_i \bar{Q}_k(x_i) f(x_i)$$

Proof. The equations by which P is determined are

$$\sum_{k=0}^{n-1} a_k \bar{Q}_k(x_i) = f(x_i) \qquad (i = 1, \ldots, n)$$

Now multiply both sides of this equation by $A_i \bar{Q}_j(x_i)$ and sum for $i = 1, \ldots, n$. By the preceding theorem, we obtain

$$a_j = \sum_{i=1}^{n} A_i \bar{Q}_j(x_i) f(x_i) \qquad \blacksquare$$

Corollary. *The polynomial P of degree $\le n - 1$ which interpolates to f at the roots x_i of T_n is given by the formulas*

$$P = \tfrac{1}{2} a_0 T_0 + \sum_{k=1}^{n-1} a_k T_k \qquad a_k = \frac{2}{n} \sum_{i=1}^{n} f(x_i) T_k(x_i)$$

Actually, the Tchebycheff polynomials have several discrete orthogonality properties, derived from analogous properties of the sine, cosine, and exponential functions. One of the most important of these properties is contained in the next theorem.

Theorem 3. *The Tchebycheff polynomials have the orthogonality property*

$$\sum_{j=0}^{N}{}'' T_n(x_j) T_m(x_j) = \Delta(N,n,m) \frac{N}{2}$$

where $x_i = \cos(i\pi/N)$, the double prime signifies that the first and last terms are to be halved, and $\Delta(N,n,m)$ denotes the number of integers in the set $\{(n + m)/2N, (n - m)/2N\}$.

Proof. Since $x_j = \cos(j\pi/N) = \cos[(2N - j)\pi/N] = x_{2N-j}$, we have

$$\sum_{j=0}^{N}{}'' T_n(x_j) T_m(x_j) = \frac{1}{2} \sum_{j=0}^{2N-1} T_n(x_j) T_m(x_j)$$

$$= \frac{1}{2} \sum_{j=0}^{2N-1} \cos \frac{nj\pi}{N} \cos \frac{mj\pi}{N}$$

An application of the identity $\cos A \cos B = \tfrac{1}{2} \cos(A + B) + \tfrac{1}{2} \cos(A - B)$ yields

$$\frac{1}{4} \sum_{j=0}^{2N-1} \left(\cos \frac{n + m}{N} j\pi + \cos \frac{n - m}{N} j\pi \right)$$

which (because of the formula $e^{i\theta} = \cos\theta + i\sin\theta$) may be recognized as the real part of

$$\frac{1}{4}\sum_{j=0}^{2N-1}\{[e^{(n+m)i\pi/N}]^j + [e^{(n-m)i\pi/N}]^j\}$$

These geometric series are easily summed using the formula

$$\sum_{j=0}^{k-1}\lambda^j = \begin{cases}(1-\lambda^k)(1-\lambda)^{-1} & (\lambda \neq 1)\\ k & (\lambda = 1)\end{cases}$$

In fact, if neither $n - m$ nor $n + m$ is a multiple of $2N$, the sum is

$$\frac{1 - e^{2(n+m)i\pi}}{1 - e^{(n+m)i\pi/N}} + \frac{1 - e^{2(n-m)i\pi}}{1 - e^{(n-m)i\pi/N}} = 0$$

If one of $n - m$ or $n + m$ is a multiple of $2N$, the sum is $2N$, and if both are multiples of $2N$, the sum is $4N$. ∎

Theorem 4. *If* $f \in C[-1,1]$, *then, with* $x_i = \cos(i\pi/n)$,

$$E_{n-1}(f) \geq \frac{1}{n}\left|\sum_{i=0}^{n}{}'' (-1)^i f(x_i)\right|$$

(The primes indicate that the first and last terms in the sum are to be halved.)

Proof. Let P denote the polynomial of degree $<n$ which best approximates f on the $n + 1$ points $x_0, \ldots, x_n$. By the alternation theorem (Chap. 3, Sec. 4), P satisfies the following equations

$$(1) \qquad (-1)^i\lambda + P(x_i) = f(x_i)$$

Since P is a best approximation on $\{x_i\}$, we will have

$$E_{n-1}(f) \geq \max_i |f(x_i) - P(x_i)| = |\lambda|$$

It remains then to prove that

$$(2) \qquad n\lambda = \sum_{i=0}^{n}{}'' (-1)^i f(x_i)$$

In order to do this, multiply equation (1) by $T_n(x_i)$, and then apply the operator $\sum''$ to both sides, thus:

$$(3) \qquad \lambda\sum_{i=0}^{n}{}'' (-1)^i T_n(x_i) + \sum_{i=0}^{n}{}'' P(x_i)T_n(x_i) = \sum_{i=0}^{n}{}'' T_n(x_i)f(x_i)$$

Now P can be expressed as a linear combination of the Tchebycheff polynomials $T_0, \ldots, T_{n-1}$. Hence by the preceding theorem, the term $\sum'' P(x_i)T_n(x_i)$ vanishes. For the other terms it is only necessary to observe that $T_n(x_i) = \cos i\pi = (-1)^i$. ∎

It is now an easy matter to establish a theorem from [Bernstein, 1912] which provides a lower bound for $E_{n-1}(f)$ in those cases when the Tchebycheff series for f converges *absolutely*.

Theorem 5. *If* $\sum |a_n| < \infty$ *and* $f = \sum a_n T_n$, *then on* $[-1,1]$,

$$E_{n-1}(f) \geq |a_n + a_{3n} + a_{5n} + \cdots|$$

Proof. By the preceding theorem $E_{n-1}(f) \geq |\lambda|$ where

$$\lambda = \frac{1}{n} \sum_{i=0}^{n}{}'' (-1)^i f(x_i)$$

$$= \frac{1}{n} \sum_{i=0}^{n}{}'' (-1)^i \sum_{k=0}^{\infty} a_k T_k(x_i)$$

$$= \frac{1}{n} \sum_{k=0}^{\infty} a_k \sum_{i=0}^{n}{}'' T_n(x_i) T_k(x_i) \qquad \left(x_i = \cos \frac{i\pi}{n}\right)$$

We have used here the fact that $T_n(x_i) = (-1)^i$ and a theorem on changing the order of summations (Prob. 4). Now by Theorem 3, $\sum'' T_n(x_i) T_k(x_i) = n$ when k is an odd multiple of n, and vanishes otherwise. Hence $\lambda = a_n + a_{3n} + \cdots$. ■

We have remarked earlier (but the proof was postponed to Chap. 6, Sec. 5) that the Lagrange interpolating polynomials with nodes fixed in advance do not provide *uniform* approximations of arbitrary precision to all continuous functions. What is the situation for norms other than the uniform norm? One positive result is the following.

Erdös-Turán Theorem. *Let* $Q_0, Q_1, \ldots$ *be the system of polynomials which is orthogonal on* $[a,b]$ *with weight function* w. *For each* $f \in C[a,b]$ *let* $L_n f$ *denote the polynomial of degree* $\leq n$ *which interpolates to* f *at the zeros of* Q_{n+1}. *Then* $\| L_n f - f \|_w \to 0$. *That is,*

$$\int_a^b |(L_n f - f)(x)|^2 w(x)\, dx \to 0$$

Proof. The Lagrange interpolation formula may be written in the form

$$(L_n f)(x) = \sum_{i=0}^{n} f(x_i) l_i(x) \qquad l_i(x) = \frac{Q_{n+1}(x)}{(x - x_i) Q'_{n+1}(x_i)}$$

From this it will follow that $l_i \perp l_j$ for $i \neq j$; we simply write

$$\langle l_i, l_j \rangle = \frac{1}{Q'_{n+1}(x_i) Q'_{n+1}(x_j)} \int_a^b Q_{n+1}(x) \frac{Q_{n+1}(x)}{(x - x_i)(x - x_j)} w(x)\, dx$$

and observe that the fraction under the integral sign is a polynomial of degree

$n - 1$. We shall need also the identity $\sum \int l_i^2(x)w(x)\,dx = \int w(x)\,dx$. To prove this, start from the equation $[\sum l_i(x)]^2 = 1$ (Prob. 6 of Chap. 3, Sec. 2). If we multiply by $w(x)$ and integrate, the left member simplifies (through the use of the orthogonality property $l_i \perp l_j$) to $\sum \int l_i^2(x)w(x)\,dx$.

Now to prove the theorem, let P_n denote the polynomial of degree $\leq n$ which best approximates f in the uniform norm, $\|\cdot\|_T$. According to a theorem on page 116, $\| P_n - f \|_w \to 0$. Thus it will be sufficient for the present purpose to prove that $\| L_n f - P_n \|_w \to 0$. Since $L_nP_n = P_n$, we have, with the help of the above facts,

$$\| L_n f - P_n \|_w^2 = \| L_n(f - P_n) \|_w^2$$

$$= \int \{\sum [f(x_i) - P_n(x_i)]l_i(x)\}^2 w(x)\,dx$$

$$= \sum [f(x_i) - P_n(x_i)]^2 \int l_i^2(x)w(x)\,dx$$

$$\leq \| f - P_n \|_T^2 \int w(x)\,dx \to 0 \qquad \blacksquare$$

Problems

1. The system $\{T_0/\sqrt{2}, T_1, \ldots, T_{N-1}\}$ is orthonormal with respect to the inner product.
$$\langle f,g \rangle = \frac{2}{N} \sum_{i=1}^{N} f(x_i)g(x_i) \qquad x_i = \cos \frac{2i - 1}{2N} \pi$$
2. The system $\{T_0/\sqrt{2}, T_1, \ldots T_{N-1}, T_N/\sqrt{2}\}$ is orthonormal with respect to the inner product
$$\langle f,g \rangle = \frac{2}{N} \sum_{i=0}^{N}{}'' f(x_i)g(x_i) \qquad \left(x_i = \cos \frac{i\pi}{N}\right)$$
3. Prove that $E_{n-1}(f) \geq | \sum_{i=0}^{n} f(x_i)T_n(x_i) / \sum_{i=0}^{n} (-1)^i T_n(x_i) |$ where x_i are the zeros of T_{n+1}. *Hint*: Obtain the polynomial of degree $\leq n - 1$ which best approximates f at $x_0, \ldots, x_n$. Use the corollaries of pages 76 and 135.
4. If $\sum_{k=0}^{\infty} | A_{ki} | < \infty$ for each i, then $\sum_{i=0}^{n} \sum_{k=0}^{\infty} A_{ki} = \sum_{k=0}^{\infty} \sum_{i=0}^{n} A_{ki}$.
5. Let $f, w \in C[a,b]$ and $w(x) \geq 0$. Prove that
$$\int_a^b | f(x) | w(x)\,dx \leq \sqrt{\int_a^b w(x)\,dx} \sqrt{\int_a^b [f(x)]^2 w(x)\,dx}$$
Hint: Cauchy-Schwarz.
6. With the help of the preceding problem, show that in the Erdös-Turán theorem we may draw the further conclusion that $\int | L_n f - f | w \to 0$. Then give a brief proof of Stieltjes' theorem (Sec. 4).
7. If $w(x) \geq \epsilon > 0$ in $[a,b]$, then in the Erdös-Turán theorem we may draw the further conclusions that $\int | L_n f - f | \to 0$ and $\int (L_n f - f)^2 \to 0$.

8. The polynomial of degree $<n$ which best approximates f at the extrema of $T_n\ [x_i = \cos\ (i\pi/n)]$ is given by $P = a_0/2 + \sum_{k=1}^{n-1} a_k T_k$ where $a_k = \frac{2}{n} \sum_{i=0}^{n}{}'' f(x_i)\, T_k(x_i)$.
9. Given the function $f \in C[-1,1]$, let $\mathfrak{L}_{n-1} f$ denote the polynomial P defined in Prob. 8. Prove that $\mathfrak{L}_{n-1}$ is a linear operator. Show that $\| \mathfrak{L}_{n-1} f \| \le 2(n+1) \| f \|$. (A much better inequality can be established.)
10. Establish the formula, valid on $[-1,1]$,

$$\cos^{-1} x = \frac{\pi}{2} T_0(x) - \frac{4}{\pi} T_1(x) - \frac{4}{9\pi} T_3(x) - \frac{4}{25\pi} T_5(x) - \cdots$$

Then show that $E_n \ge (\pi/2)(n+2)^{-2}$. Thus in order to approximate $\cos^{-1} x$ on $[-1,1]$ with an accuracy of 10^{-8}, a polynomial of degree 10^4 would be required.

6. The Jackson Theorems

In Sec. 4, we derived some estimates of the quantity

$$E_n(f) = \inf_{c_0,\ldots,c_n} \sup_{-1 \le x \le 1} \left| f(x) - \sum_{i=0}^{n} c_i x^i \right|$$

under the assumption that f could be represented by a Tchebycheff series, $\sum_{k=0}^{\infty} a_k T_k$. The only upper bound on $E_n(f)$ obtained there was given by the elementary inequality

$$E_n(f) \le |a_{n+1}| + |a_{n+2}| + \cdots$$

This could be used indirectly in the case of a twice-differentiable function f to bound $E_n(f)$ in terms of $\| f'' \|$. Theorems of such a nature, relating $E_n(f)$ to the smoothness properties of f, were first given by Jackson in 1911. In the intervening years the quantitative (but not the qualitative) aspects of these theorems have been improved by other workers. In this section we shall derive a number of these theorems, pointing out (but not always proving) the best possible results currently known.

Our plan is to obtain the estimates of $E_n(f)$ first for approximation by *trigonometric* polynomials. Let $C_{2\pi}$ stand for the space of continuous 2π-periodic functions, with supremum norm. For $f \in C_{2\pi}$ we write

$$E_n(f) = \inf_{a_k, b_k} \max_{\theta} \left| f(\theta) - \sum_{k=0}^{n} (a_k \cos k\theta + b_k \sin k\theta) \right|$$

The first Jackson theorem, in the improved form due to Favard and Achieser-Krein, states that $E_n(f) \le (\pi/2)(n+1)^{-1} \| f' \|$, the constant $(\pi/2)(n+1)^{-1}$ being best possible. The proof requires three lemmas.

Lemma 1. *If $k < n$, then* $\int_0^{\pi} (\sin kx)\ \mathrm{sgn} \sin nx\, dx = 0$.

Proof. Since the integrand is an even function, it will suffice to prove

$$\int_{-\pi}^{\pi} (\sin kx) \operatorname{sgn} \sin nx \, dx = 0$$

Since $\sin kx$ is a linear combination of e^{ikx} and e^{-ikx}, it will suffice to prove that when $|m| < n$,

$$\int_{-\pi}^{\pi} e^{imx} \operatorname{sgn} \sin nx \, dx = 0$$

Now denote the integral by I, and make the change of variable $x = y + \pi/n$. Thus

$$I = \int_{-\pi-\pi/n}^{\pi-\pi/n} e^{im(y+\pi/n)} \operatorname{sgn} \sin (ny + \pi) \, dy$$

Since the integrand has period 2π, the interval of integration may be replaced by $[-\pi,\pi]$. Hence

$$I = -e^{im\pi/n} \int_{-\pi}^{\pi} e^{imy} \operatorname{sgn} \sin ny \, dy = -e^{im\pi/n} I$$

Since $|m| < n$, $m\pi/n$ is not an odd multiple of π, and $e^{im\pi/n} \neq -1$. Hence $I = 0$. ∎

In connection with this, see Prob. 6 of Chap. 6, Sec. 6.

Lemma 2. *The minimum value of*

$$\int_0^{\pi} \left| x - \sum_{k=1}^{n-1} \alpha_k \sin kx \right| dx$$

(for all possible choices of α_k*) is* $\pi^2/2n$.

Proof. No matter how we choose $\alpha_1, \ldots, \alpha_{n-1}$, we can write, using the orthogonality property of Lemma 1,

$$\int_0^{\pi} \left| x - \sum_{k=1}^{n-1} \alpha_k \sin kx \right| dx \geq \left| \int_0^{\pi} \left(x - \sum_{k=1}^{n-1} \alpha_k \sin kx\right) \operatorname{sgn} \sin nx \, dx \right|$$

$$= \left| \int_0^{\pi} x \operatorname{sgn} \sin nx \, dx \right|$$

$$= \left| \sum_{k=0}^{n-1} (-1)^k \int_{k\pi/n}^{(k+1)\pi/n} x \, dx \right|$$

$$= \left| \sum_{k=0}^{n-1} (-1)^k \frac{1}{2} \left[\left(\frac{k+1}{n} \pi \right)^2 - \left(\frac{k}{n} \pi \right)^2 \right] \right|$$

$$= \frac{\pi^2}{2n^2} \left| \sum_{k=0}^{n-1} (-1)^k (2k+1) \right| = \frac{\pi^2}{2n}$$

In the last step we have employed a simple formula which may be proved by induction (see Prob. 1). It now remains to be seen whether the lower bound $\pi^2/2n$ can be achieved by a particular choice of $\alpha_1, \ldots, \alpha_{n-1}$. Noticing how the *inequalities* enter into the above calculation, we see that we should make the function $\phi(x) = x - \sum \alpha_k \sin kx$ change sign precisely at the points of $(0,\pi)$ where $\sin nx$ changes sign. Hence the following equations are to be satisfied for $i = 1, \ldots, n - 1$.

$$\sum_{k=1}^{n-1} \alpha_k \sin kx_i = x_i \qquad \left(x_i = \frac{i\pi}{n}\right)$$

That this may be done is a consequence of the fact that $\{\sin x, \ldots, \sin (n-1)x\}$ satisfies the Haar condition on $(0,\pi)$. That the resulting function ϕ actually *changes sign* at the points x_i will be proved by assuming the contrary. Then ϕ' vanishes once in each interval (x_i,x_{i+1}), once in $(0,x_1)$, and at one or more of the points x_i, for a total of at least n times. But ϕ' is of the form $1 + \sum_{k=1}^{n-1} \beta_k \cos kx$ and can vanish in no more than $n - 1$ points of $(0,\pi)$. ∎

Fixing n, let us define an operator L as follows. Given a continuous and 2π-periodic function f, set

$$(Lf)(x) = \frac{a_0}{2} + \sum_{k=1}^{n} A_k(a_k \cos kx + b_k \sin kx)$$

in which the coefficients A_k remain at our disposal, and the a_k and b_k are the ordinary Fourier coefficients of f. The proof of Jackson's theorem will depend upon a certain choice of A_k which makes Lf a good approximation to f. We require an integral formula for Lf.

Lemma 3. *If f is 2π-periodic and if f' is continuous, then*

$$(Lf - f)(x) = \frac{1}{\pi}\int_{-\pi}^{\pi} \left[\tfrac{1}{2}t + \sum_{k=1}^{n} \frac{(-1)^k}{k} A_k \sin kt\right] f'(x + \pi - t)\, dt$$

Proof. If we denote the expression in brackets by $\phi(t)$ and then integrate the right member of the equation by parts, the result is

$$\frac{-1}{\pi}\, \phi(t)f(x + \pi - t) \Big|_{-\pi}^{\pi} + \frac{1}{\pi}\int_{-\pi}^{\pi} \phi'(t)f(x + \pi - t)\, dt$$

Using the equations $\phi(\pm\pi) = \pm\frac{1}{2}\pi$ and $f(x) = f(x + 2\pi)$, we obtain

$$-f(x) + \frac{1}{\pi}\int_{-\pi}^{\pi} [\tfrac{1}{2} + \sum_{k=1}^{n} (-1)^k A_k \cos kt] f(x + \pi - t)\, dt$$

Changing the variable $t \to x + \pi - s$, and using $\cos k(x + \pi - s) = \cos k(x + \pi) \cos ks + \sin k(x + \pi) \sin ks = (-1)^k(\cos kx \cos ks + \sin kx \sin ks)$, we arrive at

$$-f(x) + \frac{1}{\pi}\int_{-\pi}^{\pi} [\tfrac{1}{2} + \sum_{k=1}^{n} A_k(\cos kx \cos ks + \sin kx \sin ks)]f(s)\, ds$$

Since $a_k = \frac{1}{\pi}\int_{-\pi}^{\pi} f(s) \cos ks\, ds$ and $b_k = \frac{1}{\pi}\int_{-\pi}^{\pi} f(s) \sin ks\, ds$, this last expression becomes immediately $-f(x) + (Lf)(x)$. ■

Jackson's Theorem I. *For all 2π-periodic and continuously differentiable functions f,*

$$E_n(f) \le \frac{\pi}{2(n+1)} \|f'\|$$

and the constant $\pi/2(n + 1)$ is best possible.

Proof. From Lemma 3, no matter how we select $A_1, \ldots, A_n$,

$$\begin{aligned} E_n(f) &\le \| Lf - f \| \\ &\le \|f'\| \frac{1}{\pi}\int_{-\pi}^{\pi} \left| \frac{t}{2} + \sum_{k=1}^{n} \frac{(-1)^k}{k} A_k \sin kt \right| dt \\ &= \|f'\| \frac{1}{\pi}\int_{0}^{\pi} \left| t + \sum_{k=1}^{n} \frac{2(-1)^k}{k} A_k \sin kt \right| dt \end{aligned}$$

By Lemma 2, a choice of A_k's exists for which this last upper bound becomes the one in the theorem.

In order to prove that the constant is best possible, we will have to exhibit functions for which the upper bound is nearly reached. In order to see what properties such functions must have, look at the formula in Lemma 3. The coefficients A_k are so chosen that the expression in brackets changes sign with $\sin (n + 1)t$. Thus the integral reaches its maximum magnitude when f' changes sign with $\sin (n + 1)t$. Hence we shall consider continuously differentiable functions which are close to the nondifferentiable function

$$f_0(x) = \int_0^x \operatorname{sgn} \sin (n + 1)t\, dt$$

In the following sketch we have shown the integrand (with dashes) and its integral f_0 (solid line).

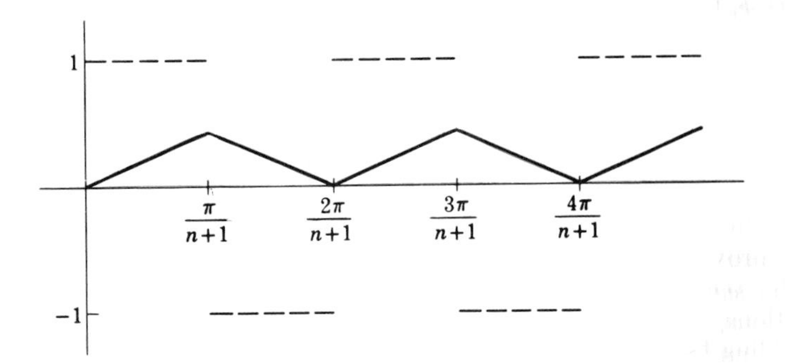

The norm of f_0 is clearly $\int_0^{\pi/(n+1)} 1\,dx = \pi/(n+1)$. The trigonometric polynomial of degree $\leq n$ which best approximates f_0 on $[0,2\pi)$ is the constant $\frac{1}{2}\,\|f_0\|$, since the error then has $2n+2$ points of alternation, viz., the points $k\pi/(n+1)$ for $k = 0, \ldots, 2n+1$. Then $E_n(f_0) = \frac{1}{2}\,\|f_0\| = \pi/2(n+1) = [\pi/2(n+1)]\,\|f_0'\|$. Since this function f_0 is the limit of other functions which have continuous derivatives of norm 1, we conclude that the constant $\pi/2(n+1)$ is best possible. ∎

Jackson's Theorem II. *For all $f \in C_{2\pi}$ which satisfy $|f(x) - f(y)| \leq \lambda\,|x - y|$,*

$$E_n(f) \leq \frac{\pi\lambda}{2(n+1)}$$

and the constant $\pi/2$ is best possible.

Proof. Fixing $\delta > 0$, define $\phi(x) = \frac{1}{2\delta}\int_{x-\delta}^{x+\delta} f(t)\,dt$. Then

$$|\phi'(x)| = \frac{1}{2\delta}\,|f(x+\delta) - f(x-\delta)| \leq \lambda$$

Consequently, by the first Jackson theorem, $E_n(\phi) \leq \pi\lambda/2(n+1)$. Furthermore,

$$|\phi(x) - f(x)| \leq \frac{1}{2\delta}\int_{x-\delta}^{x+\delta} |f(t) - f(x)|\,dt$$

$$\leq \frac{\lambda}{2\delta}\int_{x-\delta}^{x+\delta} |t - x|\,dt = \frac{\lambda}{2}\,\delta$$

If P denotes the trigonometric polynomial of degree $\leq n$ which best approximates ϕ, then

$$\begin{aligned} E_n(f) &\leq \|f - P\| \\ &\leq \|f - \phi\| + \|\phi - P\| \\ &\leq \frac{\lambda}{2}\,\delta + \frac{\pi\lambda}{2(n+1)} \end{aligned}$$

Since this is true for all $\delta > 0$, it is true for $\delta = 0$, and this is the inequality to be proved. The constant $\pi/2$ is best possible here because it is best possible in the *smaller* class of continuously differentiable functions, and for such functions, $\lambda \leq \|f'\|$. [Actually for the special function f_0 considered in the preceding theorem, the bound $\pi\lambda/2(n + 1)$ is attained.] ■

A strengthened form of this theorem will be given in Sec. 3 of Chap. 6 (page 202).

Jackson's Theorem III. *For all* $f \in C_{2\pi}$,

$$E_n(f) \leq \omega\left(\frac{\pi}{n+1}\right)$$

where ω *is the modulus of continuity of* f. *The coefficient* 1 *of* $\omega[\pi/(n + 1)]$ *is the best possible one independent of* f *and* n.

This precise form of Jackson's theorem is due to Korneicuk. Since the proof is rather technical, we shall prove instead an easy but weaker result, viz.,

$$E_n(f) \leq \tfrac{3}{2}\omega\left(\frac{\pi}{n+1}\right)$$

Proof. Employing the function ϕ defined in the proof of the preceding theorem, we have

$$|\phi'(x)| = \frac{1}{2\delta}\,|f(x + \delta) - f(x - \delta)| \leq \frac{1}{2\delta}\,\omega(2\delta)$$

Proceeding just as before, we obtain $|\phi(x) - f(x)| \leq \omega(\delta)$ and then

$$\begin{aligned} E_n(f) &\leq \omega(\delta) + \frac{\pi}{2(n+1)}\,\frac{1}{2\delta}\,\omega(2\delta) \\ &\leq \omega(2\delta)\left[1 + \frac{\pi}{\delta 4(n+1)}\right] \end{aligned}$$

If 2δ is taken to be $\pi/(n + 1)$, this last becomes $\frac{3}{2}\omega[\pi/(n + 1)]$. ■

Example. By exhibiting a special function, we can show that the inequality $E_n(f) \leq \omega[\pi/(n + 1)]$ is best possible. Let $0 < \epsilon < \frac{1}{2}$, and set $h = \pi/(n + 1)$.

Select $\beta \in (0, 2\epsilon/(n+1)^2)$, and put $x_i = ih - (n - i + 1)\beta$ for $i = 1, \ldots, n + 1$. Thus $x_{i+1} - x_i = h + \beta > 0$ and $x_{n+1} = (n+1)h = \pi$. Now define f by the following graph. Set $f(-x) = f(x)$. From the graph we see that

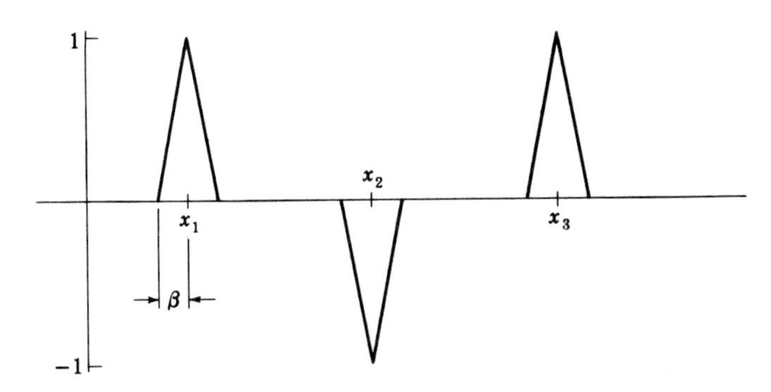

$\omega(h) = 1$. The proof will therefore be complete if we can show that $E_n(f) \geq (2n+1)/(2n+2) - \epsilon$. Let $P(x) = [1/(n+1)](\frac{1}{2} + \cos x + \cdots + \cos nx)$. This function (similar to the Dirichlet kernel) was shown on page 121 to be the same as

$$P(x) = \frac{\sin (n + \frac{1}{2})x}{2(n+1) \sin \frac{1}{2}x}$$

We note that $P(0) = (n + \frac{1}{2})/(n+1)$ and $P(ih) = (-1)^{i+1}/(2n+2)$. Further, $\| P' \| \leq [1/(n+1)](1 + 2 + \cdots + n) = n/2$. Hence $| P(x_i) - P(ih) | = | P'(\xi) | | x_i - ih | \leq (n/2)(n - i + 1)\beta \leq \epsilon$. Finally, $f(x_i) - P(x_i) = [f(x_i) - P(ih)] + [P(ih) - P(x_i)] = (-1)^{i+1} - (-1)^{i+1}/(2n+2) + \delta_i$ with $| \delta_i | \leq \epsilon$. Thus $f - P$ takes on at $2n + 2$ points of $[-\pi,\pi)$ values which alternate in sign and are at least $(2n+1)/(2n+2) - \epsilon$ in magnitude, the point 0 being one of these. By de La Vallée Poussin's theorem, $E_n(f) \geq (2n+1)/(2n+2) - \epsilon$. ■

Jackson's Theorem IV. *If $f \in C_{2\pi}$ and if f possesses a continuous kth derivative, then*

$$E_n(f) \leq \frac{\pi}{2}\left(\frac{1}{n+1}\right)^k \| f^{(k)} \|$$

and the coefficient $\pi/2$ is the best possible one independent of f, k, and n.

We shall not give the proof of this theorem, which depends upon an analysis similar to that leading up to Theorem I, but content ourselves with a proof of the weaker inequality

$$E_n(f) \leq \left(\frac{\pi}{2n+2}\right)^k \| f^{(k)} \| \tag{1}$$

Proof. Let us denote by $e_n(f)$ the minimum of $\| f - P \|$ as P ranges over all trigonometric polynomials of degree $\leq n$ with zero constant term. The proof of (1) consists in establishing the succession of inequalities

$$E_n(f) \leq \frac{\pi}{2n+2} e_n(f') \leq \left(\frac{\pi}{2n+2}\right)^2 e_n(f'') \tag{2}$$

$$\leq \cdots \leq \left(\frac{\pi}{2n+2}\right)^{k-1} e_n(f^{(k-1)})$$

$$\leq \left(\frac{\pi}{2n+2}\right)^{k} \| f^{(k)} \|$$

To verify the first of these, let p be the best approximation to f' free of a constant term. Let P be an indefinite integral of p. Then $\| (f - P)' \| = \| f' - p \| = e_n(f')$. Hence by Jackson's Theorem I,

$$E_n(f) = E_n(f - P) \leq \frac{\pi}{2n+2} \| (f - P)' \| = \frac{\pi}{2n+2} e_n(f')$$

Now in this argument, we would actually obtain

$$e_n(f) \leq \frac{\pi}{2n+2} e_n(f')$$

if the Fourier series of f were free of a constant term, for the operator L used in the proof of the Jackson Theorem I produces a trigonometric polynomial with the same constant as in the Fourier series. In the case of f' and all higher derivatives the constant in the Fourier series is zero, due to periodicity:

$$a_0 = \frac{2}{\pi} \int_{-\pi}^{\pi} f'(x)\,dx = \frac{2}{\pi} [f(\pi) - f(-\pi)] = 0$$

Thus we have always $e_n(f^{(\nu)}) \leq [\pi/(2n+2)] e_n(f^{(\nu+1)})$, for $\nu = 1, 2, \ldots$. The final inequality in (2) follows from Jackson's Theorem I, together with the remark just made about the operator L. ∎

Among the corollaries to be reaped from Jackson's theorems is the Dini-Lipschitz theorem cited in Sec. 4. We state it here in terms of Fourier series.

Dini-Lipschitz Theorem. *If $f \in C_{2\pi}$ and if $\omega(\delta) \log \delta \to 0$ as $\delta \to 0$, then the Fourier series of f converges uniformly to f.*

Proof. The $(n + 1)$st partial sum of the Fourier series of f was shown on page 120 to be of the form

$$(S_n f)(x) = \frac{1}{\pi} \int_{-\pi}^{\pi} f(t + x) \frac{\sin (n + \frac{1}{2})t}{2 \sin \frac{1}{2}t}\,dt$$

From this we obtain immediately

$$\| S_n f \| \le \| f \| \int_0^{\pi} \left| \frac{\sin (n + \frac{1}{2})t}{\pi \sin \frac{1}{2}t} \right| dt$$

The integration on the right yields a number known as the nth *Lebesgue constant.* It is bounded above by $3 + \log n$, as we shall see by integrating separately over $[0,1/n]$ and $[1/n,\pi]$ as follows:

$$\frac{2}{\pi} \int_0^{1/n} \left| \frac{\sin (n + \frac{1}{2})t}{2 \sin \frac{1}{2}t} \right| dt = \frac{2}{\pi} \int_0^{1/n} | \tfrac{1}{2} + \cos t + \cdots + \cos nt \,| \, dt$$

$$\le \frac{2}{\pi} \frac{1}{n} (\tfrac{1}{2} + n) < 1$$

Here we have used a trigonometric identity from page 121. On the other interval we use the fact that $\sin (t/2) \ge t/\pi$ [which is evident from the graph of $\sin (t/2)$] to obtain

$$\frac{1}{\pi} \int_{1/n}^{\pi} \left| \frac{\sin (n + \frac{1}{2})t}{\sin \frac{1}{2}t} \right| dt \le \frac{1}{\pi} \int_{1/n}^{\pi} \frac{1}{t/\pi} \, dt = \log \pi - \log \frac{1}{n}$$

$$< 2 + \log n$$

Now let P be the trigonometric polynomial of degree $\le n$ which best approximates f. Then by the above remarks and by Jackson's Theorem III,

$$\begin{aligned} \| S_n f - f \| &= \| S_n(f - P) - (f - P) \| \\ &\le \| S_n(f - P) \| + \| f - P \| \\ &\le (3 + \log n) \| f - P \| + \| f - P \| \\ &= (4 + \log n) E_n(f) \\ &\le (4 + \log n) \omega \left(\frac{\pi}{n + 1} \right) \to 0 \qquad \text{as } n \to \infty \end{aligned}$$

■

Jackson's Theorem V. *Let $E_n(f)$ now denote the minimax error in approximating $f \in C[-1,1]$ by algebraic polynomials of degree $\le n$. Then*

(*i*) $E_n(f) \le \omega(\pi/(n + 1))$

(*ii*) $E_n(f) \le [\pi\lambda/(2n + 2)]$ *if* $|f(x) - f(y)| \le \lambda |x - y|$

(*iii*) $E_n(f) \le (\pi/2)^k \| f^{(k)} \| / [(n + 1)(n) \cdots (n - k + 2)]$ *if* $f^{(k)} \in C[-1,1]$ *and* $n \ge k$

Proof. The function $g(\theta) = f(\cos \theta)$ is an even 2π-periodic continuous function. Its best approximations by trigonometric polynomials must therefore be even. To see this, let P be a best approximation and put $Q(\theta) = P(-\theta)$. Then

$$\|Q - g\| = \max_{-\pi\leq\theta\leq\pi} |Q(\theta) - g(\theta)| = \max_{-\pi\leq\theta\leq\pi} |Q(-\theta) - g(-\theta)| = \|P - g\|$$

whence it follows (by the unicity of best approximations) that $P = Q$.

We recall that every even trigonometric polynomial can be expressed as an algebraic polynomial in the variable $\cos\theta$, and conversely. Hence the error in the best approximation of g by *trigonometric* polynomials is the same as the error in the best approximation of f by *algebraic* polynomials:

$$\max_{-1\leq x\leq 1} |f(x) - P(x)| = \max_{-\pi\leq\theta\leq\pi} |f(\cos\theta) - P(\cos\theta)|$$

Assertion (*i*) will now follow directly from Jackson's Theorem III if we can establish that $\omega_g \leq \omega_f$. That this is the case may be seen by using the mean-value theorem to obtain

$$|\cos\theta_1 - \cos\theta_2| = |-\sin\theta_3|\,|\theta_1 - \theta_2| \leq |\theta_1 - \theta_2|$$

and by writing

$$\begin{aligned}\omega_g(\delta) &= \max_{|\theta_1-\theta_2|\leq\delta} |g(\theta_1) - g(\theta_2)| \\ &\leq \max_{|\cos\theta_1-\cos\theta_2|\leq\delta} |f(\cos\theta_1) - f(\cos\theta_2)| = \omega_f(\delta)\end{aligned}$$

Assertion (*ii*) is proved in a similar manner from Jackson's Theorem II. It is only necessary to observe that if $|f(x) - f(y)| \leq \lambda|x - y|$, then $|g(\theta_1) - g(\theta_2)| \leq \lambda|\cos\theta_1 - \cos\theta_2| \leq \lambda|\theta_1 - \theta_2|$.

The proof of assertion (*iii*) starts from the general inequality

$$(1) \qquad E_n(f) \leq \frac{\pi}{2(n+1)} E_{n-1}(f')$$

In order to verify this, take P_{n-1} to be the polynomial of degree $\leq n - 1$ which best approximates f', and let $P_n = \int P_{n-1}$. Then $\|(f - P_n)'\| = E_{n-1}(f')$. Consequently, $f - P_n$ satisfies a Lipschitz condition with constant $\lambda = E_{n-1}(f')$. By assertion (*ii*), it follows that $E_n(f - P_n) \leq \pi\lambda/(2n + 2)$, and this is equivalent to inequality (1).

Now apply inequality (1) k times and then use the obvious fact $E_n(f) \leq \|f\|$. The result is

$$\begin{aligned}E_n(f) &\leq \frac{\pi}{2(n+1)} E_{n-1}(f') \leq \frac{\pi^2}{4(n+1)n} E_{n-2}(f'') \leq \cdots \\ &\leq \left(\frac{\pi}{2}\right)^k \frac{1}{(n+1)(n)\cdots(n-k+2)} E_{n-k}(f^{(k)}) \\ &\leq \left(\frac{\pi}{2}\right)^k \frac{\|f^{(k)}\|}{(n+1)(n)\cdots(n-k+2)} \qquad \blacksquare\end{aligned}$$

Problems

1. Prove that for $n = 0, 1, 2, \ldots, \sum_{k=0}^{n} (-1)^k(2k+1) = (-1)^n(n+1)$. (This equation was required in Lemma 2.)
2. If f is not constant, then the inequality in Jackson's Theorem I is strict.
3. The modulus of continuity of a function has the property

$$\omega((n+\theta)\delta) \leq n\omega(\delta) + \omega(\theta\delta)$$

where $\delta > 0$, $1 > \theta \geq 0$, and n is any nonnegative integer. *Hint*: If $0 \leq y - x \leq (n+\theta)\delta$, then the interval $[x,y]$ may be divided into n subintervals of length δ and one interval of length $\theta\delta$ by points $x_i = x + i\delta$. Then $|f(x) - f(y)| \leq |f(x) - f(x_1)| + |f(x_1) - f(x_2)| + \cdots$.
4. The modulus of continuity of a function has the properties

$$\omega(\delta) \leq n\omega\left(\frac{\delta}{n}\right)$$

$$\omega(\alpha\delta) \leq ([\alpha]+1)\omega(\delta) \leq (\alpha+1)\omega(\delta)$$

where $\alpha > 0$, $\delta > 0$, n is an integer, and $[\alpha]$ denotes the largest integer in α. Interpret the first inequality graphically.
5. If $f \in C_{2\pi}$, then $E_n(f) \leq 2.5707\omega(1/n)$. *Hint*: In the proof of Jackson's Theorem III take a different choice of δ.
6. For continuously differentiable 2π-periodic f is it true that

$$E_n(f) \leq \frac{\pi}{2n+2} E_n(f') \quad \text{or that} \quad e_n(f) \leq \frac{\pi}{2n+2} e_n(f')$$

7. For a function f which satisfies $\|f^{(k)}\| \leq M$ for *all* k why can we not let $k \to \infty$ in Jackson's Theorem V and get $E_n(f) = 0$?
8. If f is a 2π-periodic function satisfying, for some $\alpha > 0$, the Lipschitz condition $|f(x) - f(y)| \leq L|x-y|^\alpha$, then the Fourier series of f converges uniformly to f. *Hint*: Use the Dini-Lipschitz test.
9. Deduce a weakened form of Jackson's Theorem II from III, and I from II.
10. For $\phi(x) = |x|$, prove that $E_n(\phi) \leq \pi/(2n+2)$.
11. If each derivative of f exists and does not change sign in $[a,b]$, then $E_0(f) > E_1(f) > \cdots$. *Hint*: If $E_n(f) = E_{n+1}(f)$, and if P_n is the polynomial of degree $\leq n$ which best approximates f, then $f - P_n$ alternates $n+3$ times. [Shohat, 1941]
12. We denote by Lip α the class of functions f which satisfy $|f(x) - f(y)| \leq k|x-y|^\alpha$ for some k. Prove that
    (*a*) f' is continuous $\Rightarrow f \in \text{Lip } 1$
    (*b*) $f \in \text{Lip } \alpha$, $\alpha > 1 \Rightarrow f = \text{constant}$
    (*c*) $f \in \text{Lip } \alpha \Rightarrow \omega_f(\delta)\delta^{-\alpha}$ is bounded
    (*d*) $f \in \text{Lip } \alpha$, $0 < \alpha \leq 1$, f 2π-periodic $\Rightarrow n^\alpha E_n(f)$ is bounded
13. If $f \in C[-1,1]$ and $f \sim \sum a_k T_k$, then the polynomials $S_n f = \sum_{k=0}^{n} a_k T_k$ are not bad substitutes for the polynomials of best approximation to f. Specifically, $E_n(f) \geq (4 + \log n)^{-1} \|f - S_n f\|$. Thus, for all n up to 400, we can secure at most one extra decimal place of accuracy in replacing $S_n f$ by the polynomial of best approximation.

Chapter 5

Rational approximation

1. Introduction

In the preceding chapters, our attention has been confined to problems of approximation in which the parameters to be determined have occurred *linearly*. In principle, the parameters in an approximation problem may enter in any manner whatsoever; indeed there need not be any parameters discernible. As a simple example of a nonlinear problem *with* parameters, we may cite the problem of determining optimum values of α and β in the approximation

$$f(x) \approx \cos\,[\alpha(\log x)^{\beta}]$$

A problem *without* parameters may arise if we seek an approximation which achieves a prescribed accuracy and is optimum in its expenditure of computing time (in a certain specified computer).

Nonlinear problems present us with such a diversity that it is almost inconceivable for any truly general theory to exist for them. At the present time it is certainly true that only very restricted types of approximation problems have any theory at all, and these special problems invariably retain some of the features of polynomial approximation. This is the situation with rational approximation. Some of the results will be strongly reminiscent of the linear theory in Chap. 3. The rational approximations known by the name of Padé also share this affinity with the linear problems—in this case the Taylor-series approximation.

As in earlier chapters, our interest focuses upon the Tchebycheff theory. The existence, unicity, characterization, and computation of best approximations will occupy several sections. The Padé theory and continued fractions are also discussed.

One of the strong motivations for studying rational approximations is the perennial and concrete problem of representing functions efficiently by easily computed expressions. In this capacity the rational functions

$$(1) \qquad R(x) = \frac{a_0 + a_1x + \cdots + a_nx^n}{b_0 + b_1x + \cdots + b_mx^m}$$

have been found to be extremely effective. In a loose manner of speaking, one may say that the "curve-fitting ability" of $R(x)$ is roughly equal to that of a polynomial of degree $n + m$. However, we shall see that in competing with the polynomial of degree $n + m$, $R(x)$ has an unsuspected advantage in that the computation of $R(x)$ for a given x does *not* require $n + m$ additions, $n + m - 1$ multiplications, and one division as might be surmised at first. By transforming $R(x)$ into a "continued fraction"

$$(2) \qquad R(x) = P_1(x) + \cfrac{c_2}{P_2(x) + \cfrac{c_3}{P_3(x) + \cfrac{\ddots}{+ \cfrac{c_k}{P_k(x)}}}}$$

(in which each P_j denotes a certain polynomial), we achieve the significant reduction in the number of "long" arithmetic operations (multiplications or divisions) to n or m.

Theorem. *The rational function* (1) *can be put into the continued-fraction form* (2), *and from this it can be evaluated for any x with at most* $\max\{n,m\}$ *long operations.*

Proof. The idea of the proof can be explained best by means of an example, such as

$$R(x) = \frac{2x^4 - 4x^3 - 2x^2 + 12x - 4}{x^3 - 2x^2 - x + 5}$$

This may be written in successive (self-explanatory) stages as

$$R(x) = 2x + \frac{2x - 4}{x^3 - 2x^2 - x + 5}$$

$$= 2x + \frac{2}{(x^3 - 2x^2 - x + 5)/(x - 2)}$$

$$= 2x + \frac{2}{x^2 - 1 + 3/(x - 2)}$$

In the general case, let the numerator and denominator be denoted by R_0 and R_1, respectively. Let ∂ stand for *degree of*, and assume first that $\partial R_0 \geq \partial R_1$. By successive division (of R_{j-1} by R_j) we obtain quotients Q_j and remainders R_{j+1} as follows:

$$R_0 = R_1Q_1 + R_2 \qquad (\partial R_2 < \partial R_1)$$

$$R_1 = R_2Q_2 + R_3 \qquad (\partial R_3 < \partial R_2)$$

etc.

Since the degrees ∂R_j form a decreasing sequence of nonnegative integers, we eventually reach a step in which $\partial R_k = 0$:

$$R_{k-2} = R_{k-1}Q_{k-1} + R_k \qquad (\partial R_k = 0)$$

$$R_{k-1} = R_kQ_k$$

From this schema, we have

$$R = \frac{R_0}{R_1} = Q_1 + \frac{1}{R_1/R_2}$$

$$= Q_1 + \cfrac{1}{Q_2 + \cfrac{1}{R_2/R_3}} \qquad \text{etc.}$$

$$= Q_1 + \cfrac{1}{Q_2 + \cfrac{1}{Q_3 + \cdots + \cfrac{1}{Q_{k-1} + 1/Q_k}}}$$

This can be written in the equivalent form (2), each P_j except P_1 being a *monic* polynomial (i.e., P_j has leading coefficient unity). To evaluate such a polynomial requires no more than $\partial P - 1$ multiplications since it can be written

$$x^\nu + a_{\nu-1}x^{\nu-1} + \cdots + a_0 = (\cdots((x + a_{\nu-1})x + a_{\nu-2})\cdots)x + a_0$$

The long operations necessary to calculate $R(x)$ from equation (2) are then the multiplications for P_j and the $k - 1$ divisions. The total number of these operations is

$$\begin{aligned}\partial P_1 + \partial P_2 + \cdots + \partial P_k & \\ &= \partial Q_1 + \cdots + \partial Q_k \\ &= (\partial R_0 - \partial R_1) + (\partial R_1 - \partial R_2) + \cdots + (\partial R_{k-1} - \partial R_k) \\ &= \partial R_0 \leq n\end{aligned}$$

Here we have used the inequalities $\partial R_{j+1} < \partial R_j$ to conclude that $\partial R_j = \partial R_{j+1} + \partial Q_{j+1}$. Note that if Q_1 is *monic*, then the number of operations is at most $n - 1$. Now, in the case that $\partial R_0 < \partial R_1$ we write

$$R = \frac{c_1}{c_1 R_1 / R_0}$$

where c_1 is selected so that $c_1 R_1$ and R_0 have the same leading coefficient. The preceding discussion now shows that $c_1 R_1 / R_0$ may be expressed as a continued fraction, any value of which may be computed with no more than $\partial R_1 - 1$ long operations. Hence in this case the evaluation of R requires at most $\partial R_1 \leq m$ long operations. ∎

2. Existence of Best Rational Approximations

Let us consider the following approximation problem. A function $f \in C[a,b]$ and a pair of integers $n \geq 0$, $m \geq 0$ are prescribed. We seek to approximate f by a function of the form $R \equiv P/Q$, where

$$\begin{aligned} P(x) &= a_0 + a_1 x + \cdots + a_n x^n \\ Q(x) &= b_0 + b_1 x + \cdots + b_m x^m \end{aligned}$$

We can always take for our function R a representation in the form P/Q which is *irreducible*: i.e., P and Q have no common factors other than constants. Then in order for $R \equiv P/Q$ to be bounded on $[a,b]$ it is necessary and sufficient that Q have no root on $[a,b]$. Thus, for approximating continuous functions in the uniform norm, there is no loss of generality in requiring that $Q(x) > 0$ on $[a,b]$. The resulting family of rational functions is denoted by $\mathbf{R}_m{}^n[a,b]$:

$$\mathbf{R}_m{}^n[a,b] = \left\{\frac{P}{Q} : \partial P \leq n,\ \partial Q \leq m,\ Q(x) > 0 \text{ on } [a,b]\right\}$$

Here ∂P denotes the *degree* of P, with the convention that $\partial 0 = -\infty$. We adopt the further convention that the irreducible representation of 0 is 0/1.

We face immediately the question of existence of best approximations in $\mathbf{R}_m{}^n$. A general technique that served to establish earlier existence theorems was to prove first that the point which we sought belonged to a compact set which could be prescribed on *a priori* grounds. For example, in the case of polynomials of degree $\leq n$, the best approximation to f surely must lie in the compact set

$$\{P: \partial P \leq n, \|f - P\| \leq \|f\|\}$$

since P must furnish an approximation to f at least as good as that furnished by zero.

The same technique is not effective in the present circumstance. In fact, the set

$$\{R \in \mathbf{R}_m^n: \|R - f\| \leq \|f\|\}$$

is not generally compact. (The norm here is the uniform norm.) A simple example in support of this assertion is given by the sequence of rational functions

$$R_k(x) = \frac{1}{kx + 1} \qquad (k = 1, 2, 3, \ldots)$$

On the interval $[0, 1]$ these have the property $\|R_k\| \leq 1$. If it were possible to extract from them a convergent subsequence, then the limit function would have to be continuous. But this is not possible since $R_k(0) = 1$ while $R_k(x) \to 0$ when $x > 0$. In spite of these remarks, compactness plays a crucial role in the existence theorem.

Existence Theorem. *To each function $f \in C[a,b]$ there corresponds at least one best rational approximation from the class $\mathbf{R}_m^n[a,b]$.*

Proof. Let $\delta = \text{dist}\,(f,\mathbf{R}_m^n)$, and let R_k be a sequence of elements in $\mathbf{R}_m^n$ such that $\|R_k - f\| \to \delta$. We may write $R_k = P_k/Q_k$ where $\partial P_k \leq n$, $\partial Q_k \leq m$, $\|Q_k\| = 1$, and $Q_k(x) > 0$ in $[a,b]$. By passing to a subsequence if necessary we may assume that $\|R_k - f\| \leq \delta + 1$ for all k. Consequently $\|R_k\| \leq \|R_k - f\| + \|f\| \leq \delta + 1 + \|f\| \equiv \theta$. Since $|P_k(x)| = |Q_k(x)|\,|R_k(x)| \leq \|Q_k\|\,\|R_k\| \leq \theta$, the pairs (P_k,Q_k) lie in the compact set defined by inequalities $\|P\| \leq \theta$ and $\|Q\| = 1$. By passing to a subsequence if necessary, we may assume that $P_k \to P$ and $Q_k \to Q$. Clearly $\|Q\| = 1$; thus there can be at most m points x_i where $Q(x_i) = 0$. At all other points $P(x)/Q(x)$ is well defined, and we have $P_k(x)/Q_k(x) \to P(x)/Q(x)$. Consequently for these points, $|P(x)/Q(x)| \leq \theta$, or $|P(x)| \leq \theta\,|Q(x)|$. By continuity, this last inequality is valid for all x in $[a,b]$. Consequently any zero of Q in $[a,b]$ is also a zero of P, and the linear factor corresponding to it may be canceled from P and Q. The removal of such a linear factor does not disturb the previous inequality, and so we may repeat this cancellation process until Q is free of zeros on $[a,b]$. Let R denote the resulting element of $\mathbf{R}_m^n$. Since $R_k \to R$ at all points where $Q(x) \neq 0$, $\|R - f\| = \delta$. ∎

If one attempts to extend the existence theorem to "generalized rational functions" of the form

$$\frac{a_0 g_0(x) + \cdots + a_n g_n(x)}{b_0 h_0(x) + \cdots + b_m h_m(x)}$$

then some difficulties will be encountered because the technique of factorization is no longer available. One may, however, proceed as follows.

Definition. *Let all the functions g_i and h_i be analytic on $[a,b]$. Thus at any point $x \in [a,b]$ each function possesses a Taylor's expansion which represents that function in a neighborhood of x. Let* **R** *denote the family of all continuous functions R on $[a,b]$ which satisfy an equation of the form*

$$R(x) \sum b_i h_i(x) = \sum a_i g_i(x) \qquad (\sum | b_i | \neq 0)$$

Theorem. *Each function in $C[a,b]$ possesses a best approximation in* **R**.

Proof. We may assume that the set $\{h_0, \ldots, h_m\}$ is linearly independent, for in the contrary case $\mathbf{R} = C[a,b]$, and the theorem is trivial. Select elements $R_k \in \mathbf{R}$ in such a way that $\| f - R_k \| \to \delta = \text{dist}\,(f, \mathbf{R})$. By the definition of R, there is for each k, a function P_k in the linear span of $\{g_0, \ldots, g_n\}$ and a function Q_k in the linear span of $\{h_0, \ldots, h_m\}$ such that $R_k Q_k = P_k$ and $Q_k \neq 0$. There is no loss of generality in supposing that $\| Q_k \| = 1$. Since $\| f - R_k \| \to \delta$, $\| R_k \|$ is bounded. Hence $\| P_k \|$ is bounded. By compactness, we may assume that $Q_k \to Q$ and $P_k \to P$. Clearly $\| Q \| = 1$. We must now stop to prove that Q can have at most a finite number of zeros on $[a,b]$. The reader to whom this fact is familiar should proceed to the next paragraph. Suppose first that Q vanishes identically in a subinterval $[\alpha, \beta]$. We may assume that $\beta - \alpha$ is a maximum. Since Q is not zero throughout $[a,b]$, our subinterval is properly contained in $[a,b]$. Let us say for example that $\alpha > a$. Let the Taylor's expansion of Q at α be $Q(x) = \sum c_k (x - \alpha)^k$, and let this equation be valid in a neighborhood, N, of α. Since $\beta - \alpha$ was maximal, there is a point of N at which $Q(x) \neq 0$; hence not all the coefficients c_k vanish. Let c_ν be the first nonzero one of these. Then

$$Q(x) = (x - \alpha)^\nu \{c_\nu + (x - \alpha)[c_{\nu+1} + c_{\nu+2}(x - \alpha) + \cdots]\}$$

From this equation we see that for all x near α but different from α, $Q(x) \neq 0$. Indeed, let B be an upper bound for the modulus of the expression in square brackets, as x varies in N. If $0 < | x - \alpha | B < | c_\nu |$ and $x \in N$, then $Q(x) \neq 0$. Thus we arrive at the contradiction that for some points in (α, β), $Q(x) \neq 0$. Now suppose that Q possesses an infinite number of zeros in $[a,b]$. By compactness we may find a convergent sequence of zeros, say $z_k \to z$. Since Q does not vanish throughout any interval, the Taylor series of Q at z is not identically zero. Proceeding as we did with the point α, we see that Q *cannot* vanish at all z_k.

Now at any point x where $Q(x) \neq 0$ we may define $R(x) = P(x)/Q(x)$, and R is continuous there. Furthermore $R(x) = \lim R_k(x)$ whence $|R(x) - f(x)| \leq \delta$. At any point z where $Q(z) = 0$ we write the Taylor series $Q(x) = \sum_{k \geq \nu} c_k(x - z)^k$ and $P(x) = \sum_{k \geq \mu} d_k(x - z)^k$, where $c_\nu d_\mu \neq 0$. Since $|R(x)|$ is bounded by $\delta + \|f\|$ for all x near z but different from z, we conclude that $\mu \geq \nu$. Thus the quotient $P(x)/Q(x)$ is well-defined in a neighborhood of z by the expression

$$R(x) = \frac{d_\mu(x - z)^{\mu-\nu} + d_{\mu+1}(x - z)^{\mu-\nu+1} + \cdots}{c_\nu + c_{\nu+1}(x - z) + c_{\nu+2}(x - z)^2 + \cdots}$$

Clearly R is continuous at z and is an element of $\mathbf{R}$. Since $|R(x) - f(x)| \leq \delta$ when $Q(x) \neq 0$, and R is continuous, this inequality is true also at points where $Q(x) = 0$. Hence $\|R - f\| \leq \delta$. ∎

We conclude this section with a theorem which guarantees the existence of best approximations by rational *trigonometric* functions

$$(1) \qquad R(\theta) = \frac{\sum_{j=0}^{n} (a_j \cos j\theta + b_j \sin j\theta)}{\sum_{j=0}^{m} (c_j \cos j\theta + d_j \sin j\theta)}$$

If the interval is taken to be $[-\pi,\pi]$, we can always find a best approximation in which the denominator is strictly positive. This is the crux of the matter. We require a lemma.

Lemma. *Let P and Q be two nonzero trigonometric polynomials with real coefficients such that $|P(\theta)| \leq |Q(\theta)|$ for all real θ. If Q has a real zero, then there exist nonzero trigonometric polynomials P^* and Q^* with real coefficients such that $\partial P^* < \partial P$, $\partial Q^* < \partial Q$, and $P^*Q = PQ^*$.*

Proof. It is clear that the lemma is true for *algebraic* polynomials. The proof for trigonometric polynomials will be effected by mapping to and from the domain of algebraic polynomials. We define by the following equation a map L_n from the space of real trigonometric polynomials of degree $\leq n$ into the space of real algebraic polynomials of degree $\leq 2n$:

$$(2) \qquad (L_n f)(x) = (1 + x^2)^n f(2 \tan^{-1} x)$$

That $L_n f$ is an algebraic polynomial of degree $\leq 2n$ whenever f is a trigonometric polynomial of degree $\leq n$ may be seen by starting with $\theta = 2 \tan^{-1} x$, so that $\sin \theta = 2x(1 + x^2)^{-1}$ and $\cos \theta = (1 - x^2)(1 + x^2)^{-1}$. Then

$(1 + x^2)^n \cos k\theta = (1 + x^2)^n T_k(\cos\theta) = (1 + x^2)^n T_k[(1 - x^2)(1 + x^2)^{-1}]$, which is an algebraic polynomial of degree $\leq 2n$. Here T_k denotes the kth Tchebycheff polynomial. Similarly, using the Tchebycheff polynomials U_k, we have

$$\begin{aligned}(1 + x^2)^n \sin k\theta &= (1 + x^2)^n \sin\theta U_{k-1}(\cos\theta)\\ &= (1 + x^2)^n 2x(1 + x^2)^{-1} U_{k-1}[(1 - x^2)(1 + x^2)^{-1}]\end{aligned}$$

which is an algebraic polynomial of degree $\leq 2n - 1$. It is true, but its verification is left to the problems, that L_n has an inverse given by the formula

$$(L_n^{-1}f)(\theta) = \left(\cos\frac{\theta}{2}\right)^{2n} f\left(\tan\frac{\theta}{2}\right)$$

Now let P and Q be as in the lemma. We may assume without loss of generality that $Q(\pi) \neq 0$, for otherwise we make a change of variable, $\theta \to \theta + \alpha$. Suppose that Q has a root $\theta_0 \in (-\pi,\pi)$. Since Q is periodic and continuous, it has another root in this same interval or has θ_0 as a double root. It is easy to see, then, that L_mQ has two real roots also. Since $|P(\theta)| \leq |Q(\theta)|$, we obtain $|(L_nP)(x)| \leq (1 + x^2)^{n-m} |(L_mQ)(x)|$. This shows that any real roots of L_mQ are present, with multiplicities at least as great, in L_nP. Thus L_mQ and L_nP share a quadratic factor corresponding to two real roots. The polynomials which result from removing this quadratic factor may be denoted by $L_{n-1}P^*$ and $L_{m-1}Q^*$, where P^* and Q^* are trigonometric polynomials of degrees $<n$ and $<m$, respectively. This argument requires the invertibility of the operators L_{n-1} and L_{m-1}. That $PQ^* = P^*Q$ follows from the fact that $L_nP/L_mQ = L_{n-1}P^*/L_{m-1}Q^*$ and from equation (2). ∎

Theorem. *To each $f \in C[-\pi,\pi]$ there corresponds a rational trigonometric function of the above form* (1) *which best approximates f. With no loss of generality this function may be assumed to have a positive denominator.*

Proof. By the theorem preceding the lemma, a rational trigonometric function, P/Q, of best approximation exists, but at points where the denominator vanishes, it must be defined as a limit. If this occurs, an inequality $|P(\theta)| \leq |kQ(\theta)|$ is clearly valid. Thus we may apply the lemma (perhaps repeatedly) to obtain other trigonometric polynomials P^* and Q^* such that $\partial P^* < \partial P$, $\partial Q^* < \partial Q$, $P/Q = P^*/Q^*$, and $Q^*(\theta) > 0$ on $[-\pi,\pi]$. ∎

Problems

1. Prove that best approximations of the form $ax/(b|x| + c)$ do not exist for all functions in $C[-1,1]$. *Hint*: Consider a piecewise-linear function f such that $f(-1) = f(1) = 0, f(\frac{1}{2}) = -f(-\frac{1}{2}) = 1$.

2. Prove that best approximations by rational functions do not always exist if we use a seminorm $\|f\| = \max |f(x_i)|$. *Hint*: Consider $x_0 = 0$, $x_1 = 1$, $f(x_0) = 1$, $f(x_1) = 0$, $r(x) = a/(bx + c)$.

*3. Let $g_0, \ldots, g_n$, $h_0, \ldots, h_m$ be elements of $C[X]$. Prove the existence of best approximations in $C[X]$ by functions of the form $(\sum a_i g_i)(\sum b_i h_i)$. [Boehm, 1964]

4. Prove that if K is any convex set in $C[a,b]$, then the coefficient vectors of the rational functions in $K \cap \mathbf{R}_m{}^n[a,b]$ form a convex set. Prove that $\mathbf{R}_m{}^n[a,b]$ is *not* convex.

5. Given $g_0, \ldots, g_n$, $g_0, \ldots, h_m$, define for any $\epsilon > 0$,

$$\mathbf{R}_\epsilon = \left\{ \frac{\sum a_i g_i}{\sum b_i h_i} : \sum |b_i| = 1 \quad \text{and} \quad \inf_x \sum b_i h_i(x) \geq \epsilon \right\}$$

Prove the existence of best approximations in $\mathbf{R}_\epsilon$.

6. Consider in the space $C[a,b]$ some other norm, for example, $\|f\| = \langle f,f \rangle^{1/2}$ where $\langle f,g \rangle = \int_a^b f(x)g(x)w(x)\,dx$. What can you say about the existence of best approximations out of $\mathbf{R}_m{}^n[a,b]$?

7. Verify the formula for $L_n{}^{-1}$ as given in the proof of the lemma. Prove also that $\partial L_n P = 2n - k$, where k is the multiplicity of π as a root of P. Prove that L_n is linear.

8. It is easy to prove (do so) that for each fixed m, $\bigcup_{n=0}^{\infty} \mathbf{R}_m{}^n[a,b]$ is *dense* in $C[a,b]$; that is, to each $f \in C[a,b]$ and to each $\epsilon > 0$ there corresponds an n and an $R \in \mathbf{R}_m{}^n[a,b]$ such that $\|R - f\| < \epsilon$. Show that $\bigcup_{m=0}^{\infty} \mathbf{R}_m{}^n[a,b]$ is dense in the cone

$$C_+[a,b] = \{f \in C[a,b] : f(x) > 0 \text{ for all } x \in [a,b]\}$$

More generally, show that if D is a dense set in $C[X]$, then $\{1/f : f \in D, f > 0\}$ is dense in $C_+[X]$. For further results see [Boehm, 1964].

3. The Characterization of Best Approximations

Our immediate objective is to establish for generalized rational approximations the analogue of the characterization theorem given on page 73. This in turn will lead to a characterization of best approximations by means of the alternations in the error curve. In this section we adopt a setting which is much more general than the one which was required by the existence theorem of the preceding section. We suppose now that two finite-dimensional subspaces $\mathbf{P}$ and $\mathbf{Q}$ have been prescribed in $C[X]$. The set X may be any compact metric space, although later X will be required to be an interval. It is assumed that $\mathbf{Q}$ contains at least one function which is positive throughout X. Our approximating family is then the class $\mathbf{R}$ of all functions $R = P/Q$, where $P \in \mathbf{P}$, $Q \in \mathbf{Q}$, and $Q(x) > 0$ in X. Such functions R will be termed *generalized rational functions.*

If f is a given element of $C[X]$, there may or may not exist in $\mathbf{R}$ an element of best approximation to f. One case in which existence has already been proved is that in which **P** consists of *all* polynomials of degree $\le n$, **Q** consists of *all* polynomials of degree $\le m$, and X is an interval. Another case is that in which **Q** has dimension 1 and **P** is arbitrary, this being the purely *linear* problem of Chap. 3. Even in the general case, where an existence theorem is lacking, we will be able to *characterize* best approximations from **R**. Given a fixed element R in **R**, we shall write

$$\mathbf{P} + R\mathbf{Q} = \{P + RQ\colon P \in \mathbf{P} \text{ and } Q \in \mathbf{Q}\}$$

This is a linear subspace of $C[a,b]$. If $\{g_1, \ldots, g_n\}$ is a basis for **P** and if $\{h_1, \ldots, h_m\}$ is a basis for **Q**, then $\mathbf{P} + R\mathbf{Q}$ is *spanned* by

$$\{g_1, \ldots, g_n, Rh_1, \ldots, Rh_m\}$$

Even when $R \neq 0$, the latter is *not* a basis. Indeed, if $R = \sum a_i g_i / \sum b_i h_i$, then we have the linear dependence

$$\sum a_i g_i - \sum b_i R h_i = 0$$

Thus $\mathbf{P} + R\mathbf{Q}$ can have dimension at most $n + m - 1$.

Characterization Theorem. *An element $R \in \mathbf{R}$ is a best approximation to $f \notin \mathbf{R}$ if and only if no element $\phi \in \mathbf{P} + R\mathbf{Q}$ has the same signs as $f - R$ on the set of critical points*

$$Y = \{y\colon |f(y) - R(y)| = \|f - R\|\}$$

Proof. If R is not a best approximation to f, select a better one, $R^* = P^*/Q^* \in \mathbf{R}$. Put $\phi = Q^*(R^* - R)$. This is an element of $\mathbf{P} + R\mathbf{Q}$. Furthermore, if $y \in Y$ and if $\sigma(y) = \operatorname{sgn}(f - R)(y)$, then

$$\sigma(y)(f - R^*)(y) \le \|f - R^*\| < \|f - R\| = \sigma(y)(f - R)(y)$$

whence $\sigma(y)(R^* - R)(y) > 0$ and $\sigma(y)\phi(y) > 0$.

For the converse, let ϕ agree in sign with $f - R$ on Y. Write $\phi = P_0 + RQ_0$, $R = P/Q$, and

$$R_\lambda = \frac{P + \lambda P_0}{Q - \lambda Q_0}$$

The remainder of the proof is devoted to showing how to select λ so that $\|f - R_\lambda\| < \|f - R\|$. Define

$$\delta = \inf_{x \in Y} \sigma(x)\phi(x)$$

By continuity and compactness, $\delta > 0$. Let $e = f - R$, and define sets

$$X_1 = \{x \in X\colon \sigma(x)\phi(x) > \tfrac{1}{2}\delta \text{ and } |e(x)| > \tfrac{1}{2}\|e\|\}$$

$$X_2 = X \sim X_1$$

It is clear that X_1 contains Y and that X_2 is a compact set containing no points of Y. Hence there is a number μ satisfying the inequalities

$$|e(x)| \leq \mu < \|e\| \qquad (x \in X_2)$$

In order to determine the appropriate restrictions on λ, we must perform some computations. First, for $x \in X_2$ we have

$$\begin{aligned} |f(x) - R_\lambda(x)| &\leq |f(x) - R(x)| + |R(x) - R_\lambda(x)| \\ &\leq \mu + \|R - R_\lambda\| \end{aligned}$$

Since $\|R - R_\lambda\| \to 0$ as $\lambda \to 0$, this last term is less than $\|e\|$ for all sufficiently small λ. Now we take λ so small that $f(x) - R_\lambda(x)$ has the same sign as $f(x) - R(x)$ on X_1. Then for $x \in X_1$

$$\begin{aligned} |f(x) - R_\lambda(x)| &= \sigma(x)(f - R)(x) + \sigma(x)(R - R_\lambda)(x) \\ &\leq \|e\| - \frac{\lambda\sigma(x)\phi(x)}{(Q - \lambda Q_0)(x)} \\ &\leq \|e\| - \frac{\lambda\delta}{2\|Q - \lambda Q_0\|} < \|e\| \end{aligned}$$

In this argument it is necessary to restrict λ to small positive values such that $Q - \lambda Q_0$ is positive throughout X. ∎

The characterization theorem just proved can be stated in other equivalent forms. For example:

Theorem. *An element $R \in \mathbf{R}$ is a best approximation to f if and only if there exist points $x_i \in X$ and scalars $\lambda_i \neq 0$ such that*

(i) $f(x_i) - R(x_i) = (\operatorname{sgn} \lambda_i)\|f - R\|$

(ii) $\sum \lambda_i \phi(x_i) = 0$ *for every* $\phi \in \mathbf{P} + R\mathbf{Q}$

Theorem. *An element $R \in \mathbf{R}$ is a best approximation to f if and only if the origin of n space lies in the convex hull of the set*

$$\{\sigma(x)\hat{x}: |f(x) - R(x)| = \|f - R\|\}$$

where $\sigma(x) = \operatorname{sgn}[f(x) - R(x)]$, $\hat{x} = [\phi_1(x),\ldots,\phi_n(x)]$, *and* $\{\phi_1,\ldots,\phi_n\}$ *is any basis for* $\mathbf{P} + R\mathbf{Q}$.

The reader will recall that in the linear theory of Chap. 3 the characterization of best approximations by the oscillations of the error function was made possible by adopting the hypothesis of the *Haar conditions* on the approximating family. In the present circumstances, similar hypotheses will be useful. The results that we will obtain contain the linear theory as a special case, and the case in which the Haar conditions are fulfilled as a special *subcase*.

We shall speak henceforth of a *Haar subspace* of $C[a,b]$ as a finite-dimensional subspace which has a basis satisfying the Haar condition. Thus M is a Haar subspace provided that there is a basis $\{g_1,\ldots,g_n\}$ for M such that each determinant

$$\begin{vmatrix} g_1(x_1) & \cdots & g_n(x_1) \\ \cdots & \cdots & \cdots \\ g_1(x_n) & \cdots & g_n(x_n) \end{vmatrix}$$

(formed with distinct points $a \leq x_i \leq b$) is nonzero. That this property of M is independent of the basis follows immediately from Prob. 1 of Chap. 3, Sec. 4: M is a Haar subspace of dimension n if and only if 0 is the only function in M which has n or more roots in $[a,b]$.

Whether or not the subspace M is a *Haar* subspace, the number of changes in sign which its members may possess has a least upper bound which depends only on M. Let this be denoted by $\nu(M) - 1$. We admit the possibility that for some subspaces M, $\nu(M)$ may be $+\infty$. Next let us denote by $\delta(M)$ the *dimension* of M. It follows that every Haar subspace M satisfies the equation $\delta(M) = \nu(M)$. Finally, in any subspace M we may look for Haar subspaces. Let $\eta(M)$ be the maximum dimension among these. Thus M itself is a Haar subspace if and only if $\delta(M) = \eta(M)$. In summary,

$$\begin{aligned} \delta(M) &= \text{dimension of } M \\ \nu(M) &= 1 + \text{maximum number of variations} \\ &\qquad \text{in sign possessed by members of } M \\ \eta(M) &= \max\,\{\delta(H): H \text{ is a Haar subspace of } M\} \end{aligned}$$

Given an element R of $\mathbf{R}$, we again form the subspace $\mathbf{P} + R\mathbf{Q}$. The indices $\nu(\mathbf{P} + R\mathbf{Q})$ and $\eta(\mathbf{P} + R\mathbf{Q})$ depend now only on R. Recall that a function e is said to *have k alternations* if there exist points $x_1 < \cdots < x_k$ such that $e(x_i) = (-1)^i\lambda$, with $|\lambda| = \|e\|$.

Alternation Theorem. *If the error function $e = f - R$ has at least $1 + \nu(\mathbf{P} + R\mathbf{Q})$ alternations, then R is a best approximation to f from $\mathbf{R}$. If R is a best approximation to f, then e has at least $1 + \eta(\mathbf{P} + R\mathbf{Q})$ alternations.*

Proof. If R is not a best approximation to f, then by the above characterization theorem, we may find $\phi \in \mathbf{P} + R\mathbf{Q}$ such that

$$|e(x)| = \|e\| \Rightarrow \phi(x)e(x) > 0$$

This shows that if e alternates k times, then ϕ has $k - 1$ variations in sign. Hence e can alternate no more than $\nu(\mathbf{P} + R\mathbf{Q})$ times.

Suppose now that R is a best approximation to f. In $\mathbf{P} + R\mathbf{Q}$ we may select a Haar subspace M of dimension $n = \eta(\mathbf{P} + R\mathbf{Q})$. Let $Y = \{x: |e(x)| = \|e\|\}$. By the characterization theorem, there is no $\phi \in M$ such that $\phi(x)e(x) > 0$ on Y. If $\{\phi_1, \ldots, \phi_n\}$ is a basis for M, then the system of inequalities

$$e(x) \sum c_i \phi_i(x) > 0 \qquad (x \in Y)$$

is inconsistent, and thus the origin of n space lies in the convex hull of the point set

$$\{e(x)\hat{x}: x \in Y\}$$

where $\hat{x}$ denotes $[\phi_1(x), \ldots, \phi_n(x)]$. By Carathéodory's theorem, and by the Haar condition the origin lies in the convex hull of some set of precisely $n + 1$ such points, $e(x_i)\hat{x}_i$. By the lemma of Chap. 3, Sec. 4, the numbers $e(x_i)$ must alternate in sign if $x_1 < x_2 < \cdots$. Hence e alternates $n + 1$ times. ∎

The preceding theorem gives a *complete* characterization of best generalized rational approximations only when the two indices η and ν are the same for the subspace $\mathbf{P} + R\mathbf{Q}$. Fortunately this turns out to be true for ordinary rational approximations, in accordance with the following lemma.

Lemma. *Let $\mathbf{P}$ and $\mathbf{Q}$ be the spaces of polynomials of degree $\leq n$ and $\leq m$, respectively. Let $R = P/Q$, with $P \in \mathbf{P}$, $Q \in \mathbf{Q}$, $Q > 0$ on $[a,b]$, and P/Q irreducible. Then $\mathbf{P} + R\mathbf{Q}$ is a Haar subspace in $C[a,b]$ of dimension $1 + \max\{n + \partial Q, m + \partial P\}$.*

Proof. We begin by showing that the dimension of $\mathbf{P} + R\mathbf{Q}$ is $k \equiv 1 + \max\{n + \partial Q, m + \partial P\}$. If $R = 0$, then by our conventions, $P = 0$, $Q = 1$, and $\partial P = -\infty$ so that $k = 1 + n$, and this is the dimension of $\mathbf{P}$. In the other case ($R \neq 0$), we use the equation

$$\delta(\mathbf{P} + R\mathbf{Q}) = \delta(\mathbf{P}) + \delta(R\mathbf{Q}) - \delta(\mathbf{P} \cap R\mathbf{Q})$$

The dimension of $\mathbf{P}$ is $n + 1$, and the dimension of $R\mathbf{Q}$ is $m + 1$. Now $R\mathbf{Q} = \{(P/Q)Q_1: \partial Q_1 \leq m\}$, and an element $(P/Q)Q_1$ will belong also to $\mathbf{P}$ if and only if Q divides Q_1, leaving a quotient of degree $\leq n - \partial P$. In this event, Q_1 must be of the form QQ_2 with $\partial Q_2 \leq n - \partial P$. Since $\partial Q_1 \leq m$, we must also have $\partial Q_2 \leq m - \partial Q$. Thus $\delta(\mathbf{P} \cap R\mathbf{Q}) = 1 + \min\{m - \partial Q, n - \partial P\} = m + n + 2 - k$, whence $\delta(\mathbf{P} + R\mathbf{Q}) = k$.

Now to prove that $\mathbf{P} + R\mathbf{Q}$ is a Haar subspace we need only establish that its nontrivial elements can have at most $k - 1$ roots in $[a,b]$. If one of its elements, say $P_1 + RQ_1$, has k roots, then $P_1Q + PQ_1$ also has k roots. But this is not possible since this polynomial is of degree at most

$$\max\{n + \partial Q, m + \partial P\} \equiv k - 1$$

∎

Corollary. *In order that the irreducible rational function P/Q be a best approximation to f from the class $\mathbf{R}_m{}^n[a,b]$, it is necessary and sufficient that the error have at least $2 + \max\{n + \partial Q, m + \partial P\}$ alternations.*

Problems

1. For any finite-dimensional subspace M of $C[a,b]$ the inequality $\nu(M) \geq \delta(M) \geq \eta(M)$ is valid.

2. On the interval $[1,2]$ let $\mathbf{P}$ have the basis $\{1,x^2,\ldots,x^{2n}\}$, and let $\mathbf{Q}$ have the basis $\{1,x^2,\ldots,x^{2m}\}$. Show that if $R = P/Q$ is an irreducible element of $\mathbf{R}$, then $\eta(\mathbf{P} + R\mathbf{Q}) = \max\{n + \frac{1}{2}\partial Q, m + \frac{1}{2}\partial P\}$.

3. A best rational approximation on an interval is not necessarily a best approximation on the set of "critical" points, as it must be in the linear case. For example, let $f(x) = 2x + 1$ on $[-1,\frac{1}{2}]$, and let $f(x) = 1/x$ on $[\frac{1}{2},\frac{3}{2}]$. The best approximation of f of the form $a/(bx + c)$ is $(2 - x)^{-1}$, as can be verified by checking that the maximum deviation occurs with alternating sign at the points -1, $\frac{1}{2}$, and $\frac{3}{2}$. But at these three points we can approximate f exactly by $1/x$. [P. C. Curtis]

4. In the class $\mathbf{R}_1{}^0[0,1]$ the best approximation of $f(x) = \frac{1}{2} - x$ is 0, and the minimax error is $\frac{1}{2}$. The rational functions $R_\lambda(x) = \lambda(x + \lambda)^{-1}$ have the property that $\|f - R_\lambda\| \downarrow \frac{1}{2}$ as $\lambda \downarrow 0$, but do not converge uniformly. They converge (pointwise) to a discontinuous function.

5. Prove the two alternative forms of the characterization theorem.

6. For an arbitrary finite-dimensional subspace M of $C[a,b]$ is it true that $\nu(M) = \delta(M) \Rightarrow \eta(M) = \delta(M)$? What about the converse implication?

7. On $[-1,1]$ let $\mathbf{P}$ be spanned by the single function $g(x) = x$. What are $\delta(\mathbf{P})$, $\nu(\mathbf{P})$, and $\eta(\mathbf{P})$? Give an $f_1 \in C[a,b]$, together with its best approximation $P_1 \in \mathbf{P}$, such that $f_1 - P_1$ has fewer than 2 alternations. Give an $f_2 \in C[a,b]$ and a $P_2 \in \mathbf{P}$ such that $f_2 - P_2$ has two alternations, yet P_2 is not a best approximation to f.

8. Prove the following generalization of de La Vallée Poussin's theorem (Chap. 3, Sec. 4): If R^* is an element of $\mathbf{R}$ such that $f - R^*$ is alternately positive and negative at the points $x_1 < \cdots < x_k$, with $k > \nu(\mathbf{P} + R^*\mathbf{Q})$, then
$$\inf_{R\in\mathbf{R}} \|f - R\| \geq \min_i |(f - R^*)(x_i)|$$

9. A generalization of de La Vallée Poussin's theorem to the general case, when X is not necessarily an interval, is as follows. If R^* is an element of $\mathbf{R}$ and if Z is a subset of X such that $0 \geq \inf_{z\in Z} \phi(z)(f - R^*)(z)$ for all $\phi \in \mathbf{P} + R^*\mathbf{Q}$, then
$$\inf_{R\in\mathbf{R}} \|f - R\| \geq \inf_{z\in Z} |(f - R^*)(z)|$$

10. If $\mathbf{Q}$ is a linear subspace of $C[X]$, then let $\mathbf{Q}_+$ ("the positive part of $\mathbf{Q}$") denote the set of elements in $\mathbf{Q}$ which take positive values at each point $x \in X$. Prove that $\mathbf{Q}_+$ is a "cone": it is closed under addition and under multiplication by *positive* scalars. Prove that if $\mathbf{Q}_+$ is nonempty, then it contains a basis for $\mathbf{Q}$. Under the same hypothesis, each element of $\mathbf{Q}$ may be written as the difference of two elements from $\mathbf{Q}_+$.

4. Unicity; Continuity of Best-approximation Operators

We retain the setting of the previous section. Thus our approximating family is the set $\mathbf{R}$ of all functions $R = P/Q$, where P varies in one finite-dimensional subspace $\mathbf{P}$ of $C[a,b]$, and Q varies in another finite-dimensional subspace $\mathbf{Q}$, but subject to the restriction $Q(x) > 0$ in $[a,b]$. Given $R \in \mathbf{R}$, we form the subspace $\mathbf{P} + R\mathbf{Q}$ consisting of all $P_1 + RQ_1$ with $P_1 \in \mathbf{P}$ and $Q_1 \in \mathbf{Q}$. The indices η, δ, and ν from page 161 will again play a role. The first result is a lemma which strengthens the characterization theorem in the case that $\mathbf{P} + R\mathbf{Q}$ is a Haar subspace.

Lemma 1. *Let R be a best approximation in $\mathbf{R}$ to a function $f \notin \mathbf{R}$. If $\mathbf{P} + R\mathbf{Q}$ is a Haar subspace, then 0 is the only element ϕ of $\mathbf{P} + R\mathbf{Q}$ having the property $\phi(y)(f - R)(y) \geq 0$ for all y in the set of critical points, $Y = \{y: |f(y) - R(y)| = \|f - R\|\}$.*

Proof. Let $\{\phi_1, \ldots, \phi_n\}$ be a basis for $\mathbf{P} + R\mathbf{Q}$. As in the alternation theorem (Sec. 3), we infer that the origin of n space lies in the convex hull of the set

$$\{e(x)\hat{x}: x \in Y\}$$

where $\hat{x} = [\phi_1(x), \ldots, \phi_n(x)]$ and $e = f - R$. Let us write $0 = \sum_{i=0}^{k} \theta_i e(x_i)\hat{x}_i$, with $x_i \in Y$, and $\theta_i > 0$. By the Haar conditions, $k \geq n$. If ϕ is any nonzero element of $\mathbf{P} + R\mathbf{Q}$, then $0 = \sum_{i=0}^{k} \theta_i e(x_i)\phi(x_i)$. By the Haar conditions, at most $n - 1$ of the numbers $e(x_i)\phi(x_i)$ can vanish. Hence at least one of these is positive and at least one is negative. ∎

Unicity Theorem. *Let R be a best approximation from $\mathbf{R}$ to the function f. If $\mathbf{P} + R\mathbf{Q}$ is a Haar subspace, then R is unique.*

Proof. Suppose on the contrary that $R_0 \equiv P_0/Q_0$ is another best approximation. The function $\phi = Q_0(R_0 - R)$ belongs to $\mathbf{P} + R\mathbf{Q}$, and from the identity

$$\phi = Q_0(R_0 - R) = Q_0[(f - R) - (f - R_0)]$$

it follows that for all critical points y (of the function $f - R$) we must have $\phi(y)(f - R)(y) \geq 0$. Since $f \notin \mathbf{R}$, Lemma 1 may be applied, with the conclusion that $\phi = 0$ and $R = R_0$. ∎

Corollary. *(Unicity for ordinary rational approximation) Best approximations in $\mathbf{R}_m^n[a,b]$ are always unique.*

Proof. In the preceding section, we have established that in the case of ordinary rational approximation, $\mathbf{P} + R\mathbf{Q}$ is always a Haar subspace. Unicity follows then from the theorem. ∎

It should be noted that even if a best approximation in $\mathbf{R}$ is unique, its *representation* P/Q is never unique. We may, for example, multiply numerator and denominator by any positive scalar. In some cases the possibility exists of multiplying numerator and denominator by a function other than a constant. For example, if $P/Q \in \mathbf{R}_m{}^n[a,b]$, if $\partial P < n$, and if $\partial Q < m$, then P and Q may be multiplied by any factor $x - c$ with $c < a$.

Before giving the analogue of the strong unicity theorem (Chap. 3, Sec. 5) in the case of rational approximation, it is convenient to extract a lemma which incorporates certain dimension arguments.

Lemma 2. *Let $R^* \equiv P^*/Q^*$ be an element of $\mathbf{R}$ such that $\delta(\mathbf{P} + R^*\mathbf{Q}) = \delta(\mathbf{P}) + \delta(\mathbf{Q}) - 1$. If $P \in \mathbf{P}, Q \in \mathbf{Q}, \| P \| + \| Q \| = \| P^* \| + \| Q^* \|$, $P = R^*Q$, and $Q(x) \geq 0$ on $[a,b]$, then $P = P^*$ and $Q = Q^*$.*

Proof. If $R^* = 0$, then $P^* = 0$ and $P = 0$. Furthermore, $\delta(\mathbf{Q}) = 1$. Since $\| Q \| = \| Q^* \|$ and $Q(x)Q^*(x) \geq 0$, it follows that $Q = Q^*$.

If $R^* \neq 0$, then the equations $P = R^*Q$ and $P^* = R^*Q^*$ show that P and P^* are nonzero elements of $\mathbf{P} \cap R^*\mathbf{Q}$. But the inequality

$$\delta(\mathbf{P} + R^*\mathbf{Q}) \leq \delta(\mathbf{P}) + \delta(\mathbf{Q}) - \delta(\mathbf{P} \cap R^*\mathbf{Q})$$

shows that $\delta(\mathbf{P} \cap R^*\mathbf{Q}) \leq 1$. Hence P is a scalar multiple of P^*. By the remaining conditions we see that $P = P^*$ and $Q = Q^*$. ∎

Strong Unicity Theorem. *Let R^* be a best approximation in $\mathbf{R}$ to f. If $\eta(\mathbf{P} + R^*\mathbf{Q}) = \delta(\mathbf{P}) + \delta(\mathbf{Q}) - 1$, then there exists a constant $\gamma > 0$ such that for all $R \in \mathbf{R}$,*

$$\| f - R \| \geq \| f - R^* \| + \gamma \| R - R^* \|$$

Proof. The theorem is trivial in the case that $f \in \mathbf{R}$. We therefore assume the contrary. For $R \in \mathbf{R}$ and $R \neq R^*$ let us define

$$\gamma(R) = \frac{\| f - R \| - \| f - R^* \|}{\| R - R^* \|}$$

Our task is to prove that $\gamma(R)$ is bounded away from zero. Suppose on the contrary that it is possible to find a sequence $R_k \in \mathbf{R}$ such that $R_k \neq R^*$ and $\gamma(R_k) \to 0$. Put $R_k = P_k/Q_k$ with $P_k \in \mathbf{P}$ and $Q_k \in \mathbf{Q}$. We may assume that $\| P_k \| + \| Q_k \| = 1$. Likewise, if $R^* = P^*/Q^*$, we may assume that $\| P^* \| +$

$\| Q^* \| = 1$. By compactness we may assume that $P_k \to P$, and $Q_k \to Q$. Since $\gamma(R_k) \to 0$, $\| R_k \|$ and $\| R_k - R^* \|$ remain bounded.

We show first that $P = R^*Q$. Let $\sigma(x) = \text{sgn}\,(f - R^*)(x)$, and let y denote an arbitrary critical point of $f - R^*$; i.e.,

$$y \in Y = \{x: |(f - R^*)(x)| = \|f - R^*\|\}$$

Then it follows that

$$\begin{aligned}
(1) \qquad \gamma(R_k) \| R^* - R_k \| &= \|f - R_k\| - \|f - R^*\| \\
&\geq \sigma(y)(f - R_k)(y) - \sigma(y)(f - R^*)(y) \\
&= \sigma(y)(R^* - R_k)(y) \\
&= \frac{\sigma(y)(R^*Q_k - P_k)(y)}{Q_k(y)}
\end{aligned}$$

By passing to the limit we obtain

$$\sigma(y)(P - R^*Q)(y) \geq 0 \qquad (y \in Y)$$

By Lemma 1, this implies that $P = R^*Q$. (Observe that $\mathbf{P} + R^*\mathbf{Q}$ is a Haar subspace since the two indices η and δ are the same for it.) By Lemma 2, it follows that $P = P^*$ and $Q = Q^*$.

Since $Q^*(x) > 0$ in $[a,b]$, we may pass to a subsequence such that for some $\epsilon > 0$ and for all k and x, $Q_k(x) \geq \epsilon$. Now define

$$c = \inf_{\substack{\phi \in \mathbf{P} + R^*\mathbf{Q} \\ \|\phi\| = 1}} \max_{y \in Y} \sigma(y)\phi(y)$$

Lemma 1 implies that $c > 0$. From the definition of c and from inequality (1) it follows that there exists a $y \in Y$ with the property

$$\begin{aligned}
\gamma(R_k) \| R_k - R^* \| &\geq \frac{\sigma(y)(R^*Q_k - P_k)(y)}{Q_k(y)} \\
&\geq \sigma(y)(R^*Q_k - P_k)(y) \\
&\geq c \| R^*Q_k - P_k \| \\
&\geq c\epsilon \| R^* - R_k \|
\end{aligned}$$

We have reached a contradiction, since $R_k \neq R^*$ and $\gamma(R_k) \to 0$. ∎

Corollary. *Let $R^* \equiv P^*/Q^*$ be the best approximation to f from the class $\mathbf{R}_m{}^n[a,b]$. If $\min\{n - \partial P^*, m - \partial Q^*\} = 0$, then there exists a $\gamma > 0$ such that for all $R \in \mathbf{R}_m{}^n[a,b]$,*

$$\|f - R\| \geq \|f - R^*\| + \gamma \|R - R^*\|$$

As in the linear theory, we are led to introduce at this juncture a *best-approximation operator*—that is, an operator $\mathfrak{J}$ which picks out from a class **R** of generalized rational functions the element of best approximation to a given function. Since existence and unicity of best approximations are ensured only by rather stringent hypotheses, we adopt a slightly different approach. Given **R**, we define $\mathfrak{J}$ as follows: For any f, $\mathfrak{J}f$ is the *set* of all best approximations to f in the class **R**.

$$\mathfrak{J}f = \{R \in \mathbf{R}: \|f - R\| = \min\}$$

This set may be empty. In the case of ordinary rational approximation, $\mathbf{R} = \mathbf{R}_m^n[a,b]$, $\mathfrak{J}f$ is never empty, and indeed always contains exactly one element. However, even in this ideal case the operator $\mathfrak{J}$ is discontinuous at some points of $C[a,b]$. This discovery is due to Maehly and Witzgall.

Let us pause here to give an example of this phenomenon. One may start with the rational functions

$$R_\lambda(x) = \frac{\lambda}{\lambda + x} \qquad (\lambda > 0) \qquad R_0(x) = 0$$

considered as elements of $\mathbf{R}_1^0[0,1]$. These have the property that

$$\| R_\lambda - R_0 \| = 1$$

for $\lambda > 0$, since $R_\lambda(0) = 1$. Now let us determine a continuous function f_λ for which R_λ is the best approximation in the considered class. By the alternation theorem it will be enough to give $f_\lambda - R_\lambda$ *three* points of alternation. For

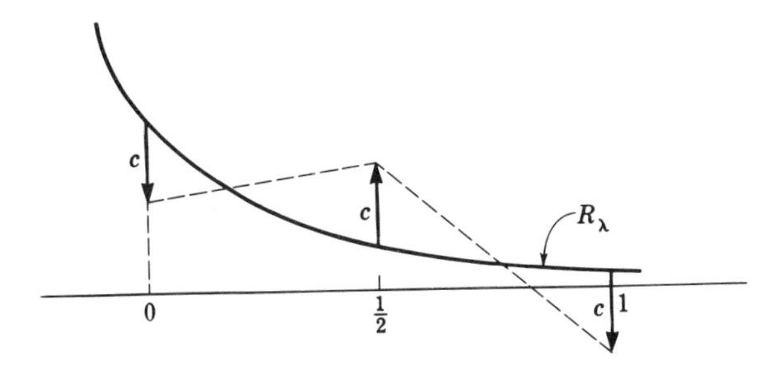

example, let f_λ be defined by the dotted line in the sketch, c being any constant in the interval $(\frac{1}{2},1)$. As $\lambda \to 0$, f_λ converges uniformly to a continuous function f_0 whose best approximation is $R_0 \equiv 0$. But as we have noted, R_λ does *not* converge to R_0. Hence the best-approximation operator $\mathfrak{J}$ is discontinuous at f_0. Note that in order for R_0 to be the best approximation of f_0, only *two* points of alternation are required.

For generalized rational approximations, including the ordinary rational case, the continuity of $\mathfrak{T}$ is governed by the following theorem.

Continuity Theorem. *If $R_0 \in \mathfrak{T}f_0$ and if $\eta(\mathbf{P} + R_0\mathbf{Q}) = \delta(\mathbf{P}) + \delta(\mathbf{Q}) - 1$, then $\mathfrak{T}f$ is nonempty for all f in a neighborhood of f_0, and $\mathfrak{T}$ is "continuous" at f_0: there is a $\beta > 0$ such that $\| R_0 - R \| < \beta \| f_0 - f \|$ whenever $R \in \mathfrak{T}f$.*

Proof. The search for a best approximation to f may clearly be confined to those $R \in \mathbf{R}$ for which $\| R - f \| \leq \| R_0 - f \|$. By the strong unicity theorem, such an R satisfies the inequalities

$$\begin{aligned} \gamma \| R - R_0 \| &\leq \| f_0 - R \| - \| f_0 - R_0 \| \\ &\leq \| f_0 - f \| + \| f - R \| - \| f_0 - R_0 \| \\ &\leq \| f_0 - f \| + \| f - R_0 \| - \| f_0 - R_0 \| \\ &\leq \| f_0 - f \| + \| f - f_0 \| \end{aligned}$$

Hence a best approximation of f, if it exists, must satisfy $\| R - R_0 \| \leq 2\gamma^{-1} \| f - f_0 \|$. Thus β in the theorem may be taken to be $2\gamma^{-1}$. That $\mathfrak{T}f$ is nonempty remains to be proved.

Write $R_0 = P_0/Q_0$, and assume that $\| P_0 \| + \| Q_0 \| = 1$. The number $2\epsilon_1 \equiv \inf Q_0(x)$ is positive. Now select $\epsilon_2 > 0$ such that

$$\left. \begin{aligned} &\| P \| + \| Q \| = 1 \\ &R = P/Q \in \mathbf{R} \\ &\| R - R_0 \| < \epsilon_2 \end{aligned} \right\} \Rightarrow \| Q - Q_0 \| < \epsilon_1$$

In order to see that this is possible, suppose on the contrary that there exists a sequence $R_k = P_k/Q_k \in \mathbf{R}$ such that $\| P_k \| + \| Q_k \| = 1$, $R_k \to R_0$, and $\| Q_k - Q_0 \| \geq \epsilon_1$. By compactness we may assume that $P_k \to P$ and $Q_k \to Q$. Since $R_k \to R_0$, $P = R_0Q$. By Lemma 2, $P = P_0$ and $Q = Q_0$, a contradiction.

Now to complete the proof, let us suppose that $\| f - f_0 \| < \frac{1}{2}\gamma\epsilon_2$. Then the best approximation R of f (if it exists) must satisfy $\| R - R_0 \| < \epsilon_2$. If we normalize $R = P/Q$ by setting $\| P \| + \| Q \| = 1$, then it will follow that $\| Q - Q_0 \| < \epsilon_1$. Since $Q_0(x) \geq 2\epsilon_1$, $Q(x) \geq \epsilon_1$. Thus our search for R is confined to

$$\{P/Q \colon P \in \mathbf{P}, Q \in \mathbf{Q}, \| P \| + \| Q \| = 1, Q(x) \geq \epsilon_1 \text{ on } [a,b]\}$$

It is elementary to prove that this is a compact set, and it must therefore contain a best approximation to f. ∎

Problems

1. Let $\mathfrak{J}$ be the best-approximation operator for the approximating family $\mathbf{R}_m{}^n[a,b]$. If $\mathfrak{J}f = P/Q$ and $\partial P_1 \le n + \partial Q - m$, then $\mathfrak{J}(f + P_1/Q) = \mathfrak{J}f + \mathfrak{J}(P_1/Q)$.
2. Fixing n and m, define a relation $\sim$ in $C[a,b]$ as follows: $f \sim g$ iff f and g have the same best approximation in $\mathbf{R}_m{}^n[a,b]$. Prove that $\sim$ is an equivalence relation: (*a*) $f \sim f$; (*b*) $f \sim g \Rightarrow g \sim f$; (*c*) $(f \sim g$ and $g \sim h) \Rightarrow f \sim h$. Prove also: (*d*) $f \sim g \Rightarrow \lambda f \sim \lambda g$; (*e*) $f \sim g$ is false when f and g are distinct elements of $\mathbf{R}_m{}^n[a,b]$; (*f*) to each f there corresponds a unique $R \in \mathbf{R}_m{}^n[a,b]$ such that $f \sim R$; and (*g*) to each $R \in \mathbf{R}_m{}^n[a,b]$ there correspond many f such that $f \sim R$.
3. $\mathbf{P} + R\mathbf{Q}$ is generally not a Haar subspace even though $\mathbf{P}$ and $\mathbf{Q}$ are. For example, on $[0,3]$ let $\mathbf{P}$ be spanned be $\{1,x^2\}$, and let $\mathbf{Q}$ be spanned by $\{1,x\}$. Take $R(x) = (1 + x^2)/(1 + x)$, and consider $\phi(x) = 6 + x^2 - 6R(x)$.
4. Let ϕ be an element of an n-dimensional Haar subspace in $C[a,b]$. Let $a \le x_0 < x_1 < \cdots < x_n \le b$, and suppose that $(-1)^i\phi(x_i) \ge 0$. Then $\phi = 0$.
5. Prove formally that if $(\|f - R_k\| - \|f - R^*\|)\,\|R_k - R^*\|^{-1} \to 0$, then $\|R_k\|$ remains bounded.
6. The set $\{P/Q\colon P \in \mathbf{P}, Q \in \mathbf{Q}, \|P\| + \|Q\| = 1, Q(x) \ge \epsilon$ on $[a,b]\}$ is compact in $C[a,b]$. (The subspaces $\mathbf{P}$ and $\mathbf{Q}$ are finite-dimensional.)
*7. The example in the text (showing the discontinuity of $\mathfrak{J}$) used a function f_0 whose best approximation produced *fewer* than $n + m + 2$ points of alternation. Here is an example where this does not happen. Let f_λ be defined as in the sketch, with $f_\lambda(0) = -1 - \lambda$,

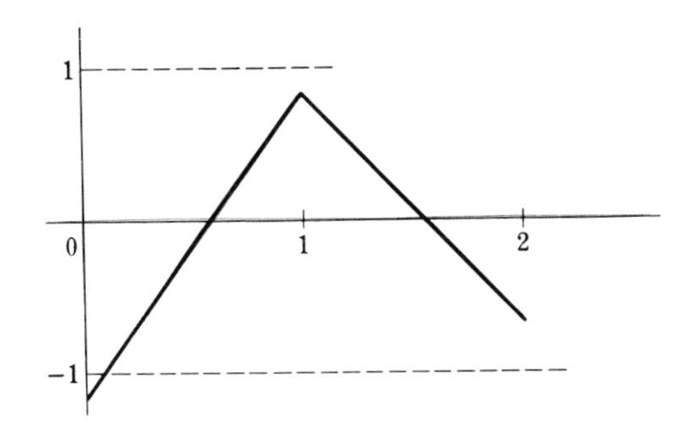

$f_\lambda(1) = 1 - \lambda$, $f_\lambda(2) = -1 + 2\lambda$. Determine the best approximation of f_λ in $\mathbf{R}_1{}^0[0,2]$, and show that $\mathfrak{J}f_\lambda \not\to \mathfrak{J}f_0$. *Hint*: One of the points of alternation lies in the interval $(0,1)$. [Witzgall]
8. Let M be a subset of $C[X]$ such that every $f \in C[X]$ has a unique best approximation, $\mathfrak{J}f$, in M. Show that $\mathfrak{J}$ is continuous at every point of M. What can you say if best approximations exist but are not always unique? (Cf. page 23.)

5. Algorithms

We consider again the problem of approximating a given function $f \in C[a,b]$ by a generalized rational function of the form

$$R = \frac{P}{Q} = \frac{a_0g_0 + \cdots + a_ng_n}{b_0h_0 + \cdots + b_mh_m}$$

in which $Q(x) > 0$ on $[a,b]$. The class of all such R, obtained by varying the parameters a_i and b_i, is denoted as usual by **R**.

One of the simplest of algorithms can be based upon the use of systems of linear inequalities. Suppose that for a certain positive value of ϵ we ask for an $R = P/Q$ satisfying $|f(x) - R(x)| \leq \epsilon$. We may arrange that $Q(x) \geq 1$ on $[a,b]$ by multiplying numerator and denominator by an appropriate positive number. The inequality $-\epsilon \leq f(x) - R(x) \leq \epsilon$ may be written in the equivalent form

$$-\epsilon Q(x) \leq f(x)Q(x) - P(x) \leq \epsilon Q(x)$$

and thus the conditions to be imposed on P and Q are just these:

$$\text{(1)} \qquad \left\{ \begin{array}{c} -Q(x) \leq -1 \\ f(x)Q(x) - P(x) \leq \epsilon Q(x) \\ -f(x)Q(x) + P(x) \leq \epsilon Q(x) \end{array} \right\} \qquad (a \leq x \leq b)$$

This system of linear inequalities on the coefficient vector

$$[a_0,\ldots,a_n,b_0,\ldots,b_m]$$

is either *consistent* or *inconsistent*. In the former case any solution of the system provides the approximation sought. In the latter case, ϵ is so small that no element of **R** exists within distance ϵ from f. If a *best* approximation is desired, then the value of ϵ must be adjusted until a minimum ϵ is found for which the system (1) is consistent. The test for consistency of (1) may be made by seeking the minimum of the following convex function:

$$\delta = \max_x \max \{1 - Q, fQ - P - \epsilon Q, P - fQ - \epsilon Q\}$$

Here δ is a function of the coefficient vector $c = [a_0,\ldots,a_n,b_0,\ldots,b_m]$, while P, f, and Q are functions of x in the range $[a,b]$. The methods of Chap. 2 can be used to find that value of c for which $\delta(c)$ is a minimum. If $\delta(c) \leq 0$, then c is a solution of the system (1). If $\delta(c) > 0$, then (1) is inconsistent and ϵ was chosen too small. This algorithm is known as the *linear inequality method.*

Another procedure which is easily explained may be termed the *weighted minimax algorithm*. The function whose minimum we seek is written in the form

$$\Delta = \max_{a \leq x \leq b} \left| f(x) - \frac{P(x)}{Q(x)} \right| = \max_{a \leq x \leq b} \frac{1}{Q(x)} |f(x)Q(x) - P(x)|$$

which immediately suggests applying the method of *iteration*. Thus we introduce indices in the expression appearing above and minimize instead the function

$$\delta_k = \max_{a \le x \le b} \frac{1}{Q_{k-1}(x)} \, | f(x) Q_k(x) - P_k(x) |$$

In the kth step of the process, $1/Q_{k-1}(x)$ is held fixed at the value determined from the preceding step while Q_k and P_k are varied to make δ_k a minimum. In order to avoid a trivial solution, one of the coefficients in P or Q is set equal to a constant. The minimization of δ is carried out by the methods of Chap. 2. Although this method is easily programmed and works quite well in practice, a convergence theorem for it is lacking.

The next algorithm to be considered is similar in spirit, is slightly more difficult to carry out, but possesses a convergence theorem. At the kth step of this algorithm an approximation $R_k = P_k/Q_k$ will be available from the preceding step. We compute then the number

$$\Delta_k = || f - R_k ||$$

Now define an auxiliary function

$$\delta_k(R) = \max_x \{ | f(x) Q(x) - P(x) | - \Delta_k Q(x) \}$$

We select $R_{k+1} = P_{k+1}/Q_{k+1}$ so as to minimize δ_k under the constraint that $|| Q_{k+1} || = 1$. It is clear that if $|| P ||$ is large, then $\delta_k(R)$ is large, so that in the minimization of δ_k it is not necessary to constrain $|| P_{k+1} ||$. In the beginning, R_0 may be arbitrary except that its denominator should be positive in $[a,b]$. If $\delta_k(R_{k+1}) \geq 0$, then we stop, and R_k is a best approximation of f. This procedure is known as the *differential correction algorithm*. We now prove that it is *effective*.

Theorem. *In the differential correction algorithm, $\Delta_k \downarrow \Delta^* = \inf \Delta$. If a best approximation exists, then the convergence is at least linear: $\Delta_{k+1} - \Delta^* \leq \theta(\Delta_k - \Delta^*)$ with $\theta < 1$.*

Proof. If the denominator, Q, of R is positive throughout $[a,b]$, then $\delta_k(R)$ may be written in the following form:

$$\delta_k(R) = \max_x \{ [| f(x) - R(x) | - \Delta_k] Q(x) \} \tag{2}$$

Now suppose that there exists an index k such that $\inf_{a \le x \le b} Q_{k+1}(x) \leq 0$. We may take k to be the first such index. Since $Q_0(x) > 0$, $k \geq 0$. Also $R_k \in \mathbf{R}$. We shall prove that R_k is a best approximation to f. In the contrary case there exists $R = P/Q \in \mathbf{R}$ such that $|| Q || = 1$ and $\Delta(R) < \Delta(R_k)$. Thus $| f(x) - R(x) | < \Delta_k$ for all x, and consequently from (2), $\delta_k(R_{k+1}) \leq$

$\delta_k(R) < 0$. But if $Q_{k+1}(x_0) \le 0$, we obtain the contradictory inequality $\delta_k(R_{k+1}) \ge |f(x_0)Q_{k+1}(x_0) - P_{k+1}(x_0)| - \Delta_k Q_{k+1}(x_0) \ge 0$. Thus unless the algorithm produces a solution in finitely many steps, we may assume that $Q_k(x) > 0$ for all k and x. Now we can prove that $\delta_k(R_{k+1}) \le 0$, equality occurring only if R_k is a best approximation. Indeed, $\delta_k(R_{k+1}) \le \delta_k(R_k) = 0$ from (2), whereas if R_k is not a best approximation, then as above we may show that $\delta_k(R_{k+1}) < 0$. Next we establish that $\Delta_0 > \Delta_1 > \cdots$. Actually, $0 > \delta_k(R_{k+1}) \ge \max_x \{|f(x) - R_{k+1}(x)| - \Delta_k\} = \Delta_{k+1} - \Delta_k$, where we have used the fact that $\| Q_{k+1} \| = 1$ in equation (2). Thus the sequence $\{\Delta_k\}$ converges downward to a limit L. Let $\Delta^* = \inf_{R \in \mathbf{R}} \Delta(R)$. It remains to be shown that $\Delta^* = L$. If not, then there exists an $R = P/Q \in \mathbf{R}$ such that $\Delta(R) < L$. We may assume that $\| Q \| = 1$. Then $|f(x) - R(x)| \le \Delta(R) < L \le \Delta_k$, whence $\delta_k(R_{k+1}) \le \delta_k(R) \le \alpha \max_x \{|f(x) - R(x)| - \Delta_k\} = \alpha[\Delta(R) - \Delta_k]$ with $\alpha = \min_x Q(x)$. Thus $\Delta_{k+1} \le \delta_k(R_{k+1}) + \Delta_k \le \alpha[\Delta(R) - \Delta_k] + \Delta_k$. Now letting $k \to \infty$, we obtain $L \le \alpha[\Delta(R) - L] + L$, which is a contradiction. The assertion about the linear convergence is proved as follows. Let R be a best approximation to f. Then by the preceding argument we have $\Delta_{k+1} - \Delta_k \le \alpha(\Delta^* - \Delta_k)$. Hence $\Delta_{k+1} - \Delta^* = (\Delta_k - \Delta^*) + (\Delta_{k+1} - \Delta_k) \le \Delta_k - \Delta^* + \alpha(\Delta^* - \Delta_k) = (1 - \alpha)(\Delta_k - \Delta^*)$. Since $\| Q \| = 1$, $0 \le 1 - \alpha < 1$. ∎

Corollary. *Let R^* be a best approximation in $\mathbf{R}$ to f. If $\eta(\mathbf{P} + R^*\mathbf{Q}) = n + m + 1$, then the rational functions produced by the differential correction algorithm converge at least linearly to R^*:*

$$\| R_k - R^* \| \le A\theta^k \qquad (\theta < 1)$$

Proof. By the strong unicity theorem (Sec. 4),

$$\| R_k - R^* \| \le \gamma^{-1}[\| f - R_k \| - \| f - R^* \|] = \gamma^{-1}(\Delta_k - \Delta^*)$$

By the preceding theorem this is majorized by $\gamma^{-1}(1 - \alpha)^k(\Delta_0 - \Delta^*)$. ∎

In the practical application of the differential correction algorithm, the constraint $\| Q \| = 1$ is less convenient than $|b_i| \le 1$, where $Q = \sum_{i=0}^{m} b_i h_i$. The proof of convergence remains valid if we write $0 > \delta_k(R_{k+1}) \ge \beta(\Delta_{k+1} - \Delta_k)$, with $\beta = \max_{|b_i| \le 1} \| Q \|$. The constant $1 - \alpha$ is replaced by $1 - \alpha\beta^{-1}$ at the end, and this also is a number in the interval $[0,1)$. The minimization of δ_k is a problem of "convex programming," which can be solved by the method of Chap. 2, Sec. 8, or its extension to the case of infinitely many functions r_i.

An algorithm is also known in which the successive rational approximations converge *quadratically* (under certain conditions) to the best approximation. This is the analogue of Remes's second method described in Chap. 3, Sec. 8, for the linear problem. The interested reader may refer to the analysis of H. Werner published in a series of articles [1962, a, 1963].

6. Padé Approximation and Its Generalizations

The Padé approximation is the rational-function analogue of the Taylor polynomial approximation. Before proceeding to a formal definition, let us recall a few facts from elementary calculus, and interpret these as results in approximation theory.

Suppose that a function to be approximated is defined by a power series in an interval about the origin:

$$f(x) = \sum_{k=0}^{\infty} a_k x^k \qquad (-b < x < b)$$

A polynomial which results from truncating this series, $P(x) = \sum_{k=0}^{n} a_k x^k$, is called a *Taylor polynomial of f* and enjoys the remarkable interpolating property

$$P(0) = f(0),\ P'(0) = f'(0),\ \ldots,\ P^{(n)}(0) = f^{(n)}(0) \tag{1}$$

Thus among all polynomials of degree $\leq n$, P is a best approximation to f in the sense of the seminorm

$$\| g \| = | g(0) | + | g'(0) | + \cdots + | g^{(n)}(0) |$$

A description of this approximation may also be given in a manner which avoids the explicit assumption that f possesses a certain number of continuous derivatives. Thus we might ask simply that P should satisfy an inequality of the form

$$| f(x) - P(x) | \leq M \, | x^\nu | \qquad (-b < x < b) \tag{2}$$

with as high a value of ν as possible. If f possesses $n + 1$ continuous derivatives in $(-b,b)$, then we may take $\nu = n + 1$, because Taylor's theorem provides a polynomial with the desired property:

$$f(x) = \sum_{k=0}^{n} \frac{1}{k!} f^{(k)}(0) x^k + \frac{1}{(n+1)!} f^{(n+1)}(\xi) x^{n+1}$$

Here ξ depends on x and lies in $(-b,b)$, so that with $\nu = n + 1$, inequality (2) is satisfied by taking

$$(3) \qquad P(x) = \sum_{k=0}^{n} \frac{1}{k!} f^{(k)}(0)x^k \qquad M = \frac{\| f^{(n+1)} \|}{(n+1)!}$$

The norm here is the Tchebycheff norm on $(-b,b)$. Actually, a polynomial P of degree $\leq n$ satisfying inequality (2) with $\nu = n + 1$ is *unique*; this fact may be proved without any assumptions on f (cf. Prob. 1). In this case, inequality (2) is equivalent to the $n + 1$ equations in (1) or to the $n + 1$ inequalities

$$|f(x) - P(x)| \leq M |x^k| \qquad (k = 1, \ldots, n+1)$$

so that the number of coefficients at our disposal is equal to the number of conditions imposed.

In analogy with the above, it would be natural (at least for a function possessing a power series) to seek rational approximations $P_n(x)/Q_m(x) = \sum_{k=0}^{n} p_k x^k / \sum_{k=0}^{m} q_k x^k$ satisfying the inequality

$$(4) \qquad \left| f(x) - \frac{P_n(x)}{Q_m(x)} \right| \leq M |x^\nu| \qquad (-b \leq x \leq b)$$

with $\nu = n + m + 1$. However, this requirement is sometimes too stringent, *even for functions which possess power series*, and we ask instead only that the coefficients in our rational function should be so chosen as to render inequality (4) valid with the highest possible value of ν. (The possibility $\nu = +\infty$ is admitted.) Typically, this "highest possible value" will be $n + m + 1$. The proof of the next theorem will disclose how it can happen that $\nu \leq n + m$. In any case, the rational function P_n/Q_m that results is called a *Padé approximation of order* (n,m) *to the function* f (at the point 0). It is clear that every bounded function possesses a Padé approximation for each pair (n,m), since if necessary we may take $\nu = 0$.

Theorem 1. *If f possesses $n + m + 1$ continuous derivatives in some neighborhood of 0, then f possesses a Padé approximation of order (n,m), with $\nu > n$. If $\nu \leq n + m + 1$, then the coefficients p_i and q_i satisfy the homogeneous equations*

$$(5) \qquad \sum_{j=0}^{k} a_j q_{k-j} = p_k \qquad (k = 0, \ldots, \nu - 1)$$

wherein $a_k = f^{(k)}(0)/k!$, and $p_{n+i} = q_{m+i} = 0$ if $i \geq 1$.

Proof. We can secure inequality (4) with $\nu = n + 1$ simply by taking $Q_m(x) =$

1 and letting P_n be the Taylor polynomial in (3). Now in the approximation sought we may assume that $Q_m(0) \neq 0$, for in the contrary case the rational function either is *reducible* or *infinite* at 0. Inequality (4) is therefore equivalent to the following inequality throughout some neighborhood of 0:

$$|f(x)Q_m(x) - P_n(x)| \leq M|x^\nu Q_m(x)| \leq M_1|x^\nu| \tag{6}$$

By Taylor's theorem we may write next

$$f(x) = f^*(x) + r(x)$$

with $f^*(x) = a_0 + a_1x + \cdots + a_{n+m}x^{n+m}$ and

$$r(x) = \frac{f^{(n+m+1)}(\xi)x^{n+m+1}}{(n+m+1)!}$$

Here again ξ depends on x and satisfies $|\xi| < |x|$. Inequality (6) now reads

$$|f^*(x)Q_m(x) + r(x)Q_m(x) - P_n(x)| \leq M_1|x^\nu|$$

If $\nu \leq n + m + 1$, then it is sufficient to secure the inequality

$$|f^*(x)Q_m(x) - P_n(x)| \leq M_2|x^\nu|$$

since the ignored term $r(x)Q_m(x)$ already satisfies such an inequality. For convenience, let the polynomials f^*, P_n, and Q_m be written as infinite series by furnishing zero coefficients. Then we have

$$\left|\left(\sum_0^\infty a_kx^k\right)\left(\sum_0^\infty q_kx^k\right) - \sum_0^\infty p_kx^k\right| \leq M_2|x^\nu| \tag{7}$$

If we multiply the two series together by the Cauchy rule, this last may be written as

$$\left|\sum_{k=0}^\infty \left(\sum_{j=0}^k a_jq_{k-j} - p_k\right)x^k\right| \leq M_2|x^\nu|$$

This condition implies the equations (5) of the theorem. ∎

Observe that the homogeneous system of equations (5) *always* has a nontrivial solution if $\nu \leq n + m + 1$, because we are asking only that the vector $[p_0,\ldots,p_n,q_0,\ldots,q_m]$ in $(n + m + 2)$ space be orthogonal to ν other vectors (the rows of the matrix). However, it is not enough to find an arbitrary solution of (5); we must have a solution in which $q_0 \neq 0$.

In order to illustrate the general method of computing Padé approximations and also to furnish an example in which $\nu \leq n + m$, let us consider the Bessel function,

$$J_0(2x) = 1 - x^2 + \frac{x^4}{4} - \frac{x^6}{36} + \cdots = \sum_{k=0}^\infty (-1)^k \left(\frac{x^k}{k!}\right)^2$$

and let $(n,m) = (1,1)$. Inequality (7) reads now

$$\left| \left(1 - x^2 + \frac{x^4}{4} - \frac{x^6}{36} + \cdots\right)(q_0 + q_1x) - (p_0 + p_1x) \right| \leq M \mid x^\nu \mid$$

$$\mid (q_0 - p_0) + (q_1 - p_1)x - q_0x^2 + \cdots \mid \leq M \mid x^\nu \mid$$

It is clear that if $q_0 \neq 0$, then ν can be at most 2 because of the presence of the term $-q_0x^2$ on the left. Taking $q_0 = p_0 = 1$ and $q_1 = p_1$, we simply arrive at the approximation $J_0(2x) \approx 1$ as the Padé approximation of order (1,1). Thus ν is one less than the expected value $n + m + 1$, and the approximation is not very good. This result might have been anticipated from the observation that a rational function $(p_0 + p_1x)/(q_0 + q_1x)$ cannot conform nontrivially to the *evenness* of the function $J_0(2x)$.

Let us continue with the same function, but take $(n,m) = (2,4)$. Instead of writing $\mid f(x)Q_m(x) - P_n(x) \mid \leq M \mid x^\nu \mid$, let us write

$$f(x)Q_m(x) - P_n(x) = c_0 + c_1x + \cdots + c_{\nu-1}x^{\nu-1} + c_\nu x^\nu + \cdots$$

and seek to make $c_0 = c_1 = \cdots = c_{\nu-1} = 0$ with a maximum value of ν. This procedure is justified whenever the function f is given by a power series, and it permits us to write down *explicitly* the error in the approximation, once the c's are known:

$$f(x) - \frac{P_n(x)}{Q_m(x)} = \frac{c_\nu x^\nu + \cdots}{Q_m(x)} \tag{8}$$

For the function $J_0(2x)$ we have

$$\left(1 - x^2 + \frac{x^4}{4} - \frac{x^6}{36} + \frac{x^8}{576} - \cdots\right)(q_0 + q_1x + q_2x^2 + q_3x^3 + q_4x^4)$$
$$- (p_0 + p_1x + p_2x^2) = c_0 + c_1x + \cdots + c_{\nu-1}x^{\nu-1} + c_\nu x^\nu + \cdots$$

By collecting terms on the left we obtain

$$(q_0 - p_0) + (q_1 - p_1)x + (q_2 - q_0 - p_2)x^2 + (q_3 - q_1)x^3$$
$$+ \left(q_4 - q_2 + \frac{q_0}{4}\right)x^4 + \left(\frac{q_1}{4} - q_3\right)x^5 + \left(-\frac{q_0}{36} + \frac{q_2}{4} - q_4\right)x^6$$
$$+ \left(-\frac{q_1}{36} + \frac{q_3}{4}\right)x^7 + \left(\frac{q_0}{576} - \frac{q_2}{36} + \frac{q_4}{4}\right)x^8 + \cdots$$

Taking $q_0 = p_0 = 1$, $q_1 = p_1 = q_3 = 0$, $q_2 = \frac{8}{27}$, $p_2 = -\frac{19}{27}$, $q_4 = \frac{5}{108}$, we can make all the terms in $x^0, x^1, \ldots, x^7$ vanish. The coefficient c_8 is then

$$c_8 = \frac{q_0}{576} - \frac{q_2}{36} + \frac{q_4}{4} = \frac{79}{15{,}552} \approx \frac{1}{200}$$

and we can compute as many additional c's as are desired. The Padé approximation of order (2,4) can be written then, with an indication of the error, as

$$J_0(2x) = \frac{1 - \frac{19}{27}x^2}{1 + \frac{8}{27}x^2 + \frac{5}{108}x^4} + \frac{(79/15{,}552)x^8 + \cdots}{1 + \frac{8}{27}x^2 + \frac{5}{108}x^4}$$

As an interesting application of this approximation, we may approximate the first root of $J_0(2x)$—which is known to be ± 1.202—by solving the simple equation

$$1 - \tfrac{19}{27}x^2 = 0$$

This gives $x = \pm 1.192$, whereas the first three terms of the Taylor series give 1.414. (One might contemplate a systematic determination of this root by the use of successive Padé approximations in which the numerators *remained quadratic.*)

The procedure just illustrated is of general applicability and leads us to the following result.

Theorem 2. *If P_n/Q_m is a Padé approximation of order (n,m) to the function*

$$f(x) = \sum_{k=0}^{\infty} a_k x^k, \text{ then the error is } \frac{\sum_{k=n+1}^{\infty} c_k x^k}{Q_m(x)}, \text{ where } c_k = \sum_{i=0}^{m} a_{k-i} q_i.$$

Proof. As noted above, the Padé approximation satisfies

$$\left(\sum_{k=0}^{\infty} a_k x^k\right)\left(\sum_{k=0}^{m} q_k x^k\right) - \sum_{k=0}^{n} p_k x^k = \sum_{k=0}^{\infty} c_k x^k$$

where $c_0 = \cdots = c_{\nu-1} = 0$. By Theorem 1, $\nu > n$, so that simply by comparing the coefficients of x^k in this equation we obtain the asserted formula for c_k. ∎

The Padé approximations may be generalized in such a way as to enable us to obtain rational approximations which are more nearly optimal in the *uniform* norm or a least-squares norm. In the previous discussion we have only to replace the sequence of monomials $1, x, x^2, \ldots$ by a sequence of other polynomials $\phi_0, \phi_1, \ldots$, where ϕ_k has (exact) degree k. For example, interesting special cases arise if we choose ϕ_k to be the kth Tchebycheff polynomial or the kth Legendre polynomial. Once the sequence $\phi_0, \phi_1, \ldots$ has been prescribed, it is necessary to determine a set of coefficients A_{ijk} with the property

$$\phi_i \phi_j = \sum_{k=0}^{i+j} A_{ijk} \phi_k$$

The existence of such coefficients follows from two facts: (*i*) that the product $\phi_i\phi_j$ is a polynomial of degree $i + j$, and (*ii*) that $\{\phi_0, \ldots, \phi_{i+j}\}$ is a basis for the linear space of all polynomials having degree $\leq i + j$. In this connection, recall Theorem 1 of Chap. 4, Sec. 2. In the classical Padé theory, the equation above reads simply $x^i x^j = x^{i+j}$. In the case of the Tchebycheff polynomials, the coefficients A_{ijk} are also simple. (See Prob. 8.)

Now assume that we have a function of the form $f = \sum_{k=0}^{\infty} a_k\phi_k$ which is to be approximated by a rational function, P_n/Q_m. Setting $P_n = \sum_{j=0}^{n} p_j\phi_j$ and $Q_m = \sum_{j=0}^{m} q_j\phi_j$ as in the classical Padé theory, we may seek to satisfy an equation of the following form, with ν as large as possible:

$$\Big(\sum_{k=0}^{\infty} a_k\phi_k\Big)\Big(\sum_{k=0}^{m} q_k\phi_k\Big) - \sum_{k=0}^{n} p_k\phi_k = \sum_{k>\nu}^{\infty} c_k\phi_k$$

An approximation obtained in this manner may be termed a *generalized Padé approximation.* If the polynomial Q_m does not vanish on the interval of interest, then the error of the approximation is easily estimated as $\sum c_k\phi_k/Q_m$. If ϕ_k is, for example, the kth Tchebycheff polynomial, the uniform norm of the error is likely to be much lower than that of the classical Padé theory. Again, if $\{\phi_k\}$ is the orthogonal sequence with respect to an inner product

$$\langle f,g\rangle = \int_a^b f(x)g(x)w(x)\,dx$$

then the generalized Padé approximation is likely to be a good approximation to f in the norm $\langle g,g\rangle^{1/2}$. The theory may even be generalized to embrace functions ϕ_k which are not polynomials. The proof of the following theorem is left to the reader.

Theorem 3. *If f is defined by a finite series $f = \sum a_i\phi_i$, where the functions ϕ_k are arbitrary except for satisfying relations $\phi_i\phi_j = \sum_k A_{ijk}\phi_k$, then for each (n,m) a formal generalized Padé approximation $P_n/Q_m = \sum_{i=0}^{n} p_i\phi_i / \sum_{i=0}^{m} q_i\phi_i$ exists whose coefficients are determined by the linear equations $\sum_j \sum_i A_{ijk}a_iq_j = p_k$, with $p_k = 0$ for $k > n$ and $q_k = 0$ for $k > m$.*

If the functions $\phi_0, \phi_1, \ldots$ form an orthonormal system, the generalized Padé approximation can be described as follows.

Theorem 4. *In $C[a,b]$, consider an orthonormal system $\{\phi_n\}$ and an element f. For every (n,m), f possesses a formal generalized Padé approximation $P/Q = \sum_{i=0}^{n} p_i\phi_i / \sum_{i=0}^{m} q_i\phi_i$, in which Q is determined by the homogeneous equations $Q \perp f\phi_k$ $(k = n+1, \ldots, n+m)$. Then $p_i = \langle fQ, \phi_i \rangle$ and the error is r/Q, where*

$$r = \sum_{k>n+m} \langle fQ, \phi_k \rangle \phi_k.$$

Proof. We start from the equation $fQ - P = \sum_{i>n+m} c_i\phi_i$. Taking inner products with ϕ_k leads to three sets of equations:

$$\langle fQ, \phi_k \rangle = 0 \qquad (k = n+1, \ldots, n+m)$$

$$\langle fQ - P, \phi_k \rangle = 0 \qquad (k = 0, \ldots, n)$$

$$\langle fQ - P, \phi_k \rangle = c_k \qquad (k > n+m)$$

The first set of m equations can be written $\langle Q, f\phi_k \rangle = 0$. It admits a nontrivial solution for the $m+1$ coefficients $q_0, \ldots, q_m$. The second set of equations then gives the generalized Fourier coefficients of P. Owing to the form of P, the last equation becomes $\langle fQ, \phi_k \rangle = c_k$. ∎

Problems

1. Let f be any function defined on $[-1,1]$ such that some polynomial P_n of degree $\leq n$ exists satisfying

$$|f(x) - P_n(x)| \leq M\,|x^{n+1}| \qquad (-a \leq x \leq a)$$

Show that P_n is unique. *Hint*: If Q_n is another such polynomial, then what can be said about $|P_n(x) - Q_n(x)|$?

2. Obtain the following Padé approximation with its error

$$e^x = \frac{1 + \frac{1}{3}x}{1 - \frac{2}{3}x + \frac{1}{6}x^2} + \frac{\frac{1}{72}x^4 + \frac{1}{120}x^5 + \cdots}{1 - \frac{2}{3}x + \frac{1}{6}x^2}$$

Show that the coefficients in the error term are given recursively by $c_{k+1} = (k-1)c_k/(k+1)(k-3)$ and explicitly by $c_k = [6k(k-1)(k-4)!]^{-1}$. Prove that with $|x| < 0.1$ the given approximation is accurate to within 1.7×10^{-6}. Rearrange the rational function so that it may be evaluated with two divisions and no multiplications.

3. If $\nu = n + m + 1$ in the classical Padé theory, the vector $q = (q_0, \ldots, q_m)$ should be chosen to be orthogonal to the rows of the $m \times (m + 1)$ matrix

$$\begin{bmatrix} a_{n+1} & a_n & \cdots & a_{n-m+1} \\ a_{n+2} & a_{n+1} & \cdots & a_{n-m+2} \\ \cdot & \cdot & \cdots & \cdot \\ a_{n+m} & a_{n+m-1} & \cdots & a_n \end{bmatrix}$$

in which $a_k = 0$ for $k < 0$. After obtaining q, we get p from $p_k = \sum_{i=0}^{k} a_{k-i}q_i$, in which $q_i = 0$ for $i > m$.

4. Derive the Padé approximation

$$\sin x \approx \frac{60x - 7x^3}{60 + 3x^2}$$

and assess the error. *Ans.*: At $x = 0$, the error behaves like $(11/50{,}400)\ x^7$.

5. Obtain a Padé approximation for $\sqrt[3]{x + 8}$ of order $(2,2)$, and assess the error.

6. Prove Theorem 3.

7. Prove that if P/Q is the (classical) Padé approximation of e^x of order (n,n) at $x = 0$, then $Q(x) = P(-x)$. For example, if $n = 4$,

$$e^x \approx \frac{1{,}680 + 840x + 180x^2 + 20x^3 + x^4}{1{,}680 - 840x + 180x^2 - 20x^3 + x^4}$$

8. Prove that the Tchebycheff polynomials have the property $T_iT_j = \frac{1}{2}T_{i+j} + \frac{1}{2}T_{|i-j|}$. How does this generalize the standard recurrence relation for T_n?

9. The equations from which a generalized Padé approximation is obtained when ϕ_k is the kth Tchebycheff polynomial read as follows:

$$\tfrac{1}{2}\sum_{i=0}^{m} (a_{k+i}q_i + a_iq_{k+i} + a_iq_{k-i}) - p_k = c_k \qquad (k = 1, 2, \ldots)$$

$$\tfrac{1}{2}(a_0q_0 + \sum_{i=0}^{m} a_iq_i) - p_0 = c_0$$

Here $f = \sum a_iT_i = P/Q + R/Q$ with $P = \sum p_iT_i$, $Q = \sum q_iT_i$, $R = \sum c_iT_i$ and $c_{n+m+1-i} = p_{n+i} = q_{m+i} = q_{-i} = p_{-i} = a_{-i} = c_{-i} = 0$ for all $i = 1, 2, \ldots$.

7. Continued Fractions

The theory of continued fractions is a vast one, and is developed most efficiently with the aid of complex variables. In this section the reader is introduced to some of the basic parts of the theory, all within the field of the real numbers.

Let us begin by writing down an example which will indicate what is meant by a continued fraction and why the concept is important in approximation theory. A special case of a formula due to Gauss is the following:

$$\tan z = \cfrac{z}{1 - \cfrac{z^2}{3 - \cfrac{z^2}{5 - \cfrac{z^2}{7 - \cdots}}}}$$

A more convenient notation for the same thing is

$$\tan z = \frac{z}{1 -}\,\frac{z^2}{3 -}\,\frac{z^2}{5 -}\,\frac{z^2}{7 - \cdots}$$

It is known that the formula is valid for all values of the complex variable z. But what meaning do we assign to the nonterminating expression on the right? Or what meaning do we give to the equation as a whole? It is simply this: that if we form a *sequence* of functions by truncating the above expression,

$$f_n(z) = \frac{z}{1 -}\,\frac{z^2}{3 -}\,\frac{z^2}{5 - \cdots}\,\frac{z^2}{2n - 1}$$

then the equation asserts that $\lim_{n\to\infty} f_n(z) = \tan z$ for all z. An appreciation of the accuracy of the approximations f_n to $\tan z$ may be gained by comparing $\tan \pi/4 \equiv 1$ to the values of $f_n(\pi/4)$. It turns out that

$$f_1\left(\frac{\pi}{4}\right) = 0.785398163397 \qquad f_4\left(\frac{\pi}{4}\right) = 0.999997868416$$

$$f_2\left(\frac{\pi}{4}\right) = 0.988689239934 \qquad f_5\left(\frac{\pi}{4}\right) = 0.999999986526$$

$$f_3\left(\frac{\pi}{4}\right) = 0.999787680915 \qquad f_6\left(\frac{\pi}{4}\right) = 0.999999999941$$

Proceeding now to some formal definitions, we say that any expression

$$\frac{a_1}{b_1 +}\,\frac{a_2}{b_2 +}\,\frac{a_3}{b_3 + \cdots}$$

is a *continued fraction*. Its *nth convergent* is the expression

$$c_n = \frac{a_1}{b_1 +} \frac{a_2}{b_2 +} \cdots \frac{a_n}{b_n}$$

The number c_n cannot be defined if any of the denominators encountered in the successive divisions vanish. If all but a finite number of the c_n are well-defined and if $c \equiv \lim_{n\to\infty} c_n$ exists, then we write

$$c = \frac{a_1}{b_1 +} \frac{a_2}{b_2 +} \frac{a_3}{b_3 + \cdots}$$

The infinite process which the continued fraction denotes is similar in some respects to an infinite sum $\sum a_k$ or an infinite product $\prod a_k$. For these two processes, successive convergents may be easily calculated with the aid of two simple formulas:

$$\sum_{k=1}^{n} a_k = a_n + \sum_{k=1}^{n-1} a_k$$

$$\prod_{k=1}^{n} a_k = a_n \prod_{k=1}^{n-1} a_k$$

For continued fractions, the following theorem supplies the analogous computational technique.

Theorem 1. *For the nth convergent of a continued fraction,*

$$\frac{A_n}{B_n} = \frac{a_1}{b_1 +} \frac{a_2}{b_2 +} \cdots \frac{a_n}{b_n}$$

the following formulas are valid: $A_0 = 0$, $A_1 = a_1$, $B_0 = 1$, $B_1 = b_1$, $A_k = b_k A_{k-1} + a_k A_{k-2}$, $B_k = b_k B_{k-1} + a_k B_{k-2}$, *for* $k = 2, 3, \ldots$.

Proof. Let $f_n(z) = \dfrac{a_1}{b_1 +} \dfrac{a_2}{b_2 +} \cdots \dfrac{a_n}{b_n + z}$. It will be sufficient to establish

$$f_n(z) = \frac{A_{n-1}z + A_n}{B_{n-1}z + B_n} \tag{1}$$

since we can take $z = 0$ to get the formula in the theorem. For $n = 1$, (1) reads

$$\frac{a_1}{b_1 + z} = \frac{A_0 z + A_1}{B_0 z + B_1}$$

and this is clearly true. Now assume (1) for the index $n - 1$. Then

$$f_{n-1}(z) = \frac{A_{n-2}z + A_{n-1}}{B_{n-2}z + B_{n-1}}$$

In this equation replace z by $a_n/(b_n + z)$. The result is

$$f_{n-1}\left(\frac{a_n}{b_n + z}\right) = \frac{A_{n-2}a_n/(b_n + z) + A_{n-1}}{B_{n-2}a_n/(b_n + z) + B_{n-1}}$$

$$\frac{a_1}{b_1 +} \frac{a_2}{b_2 +} \cdots \frac{a_{n-1}}{b_{n-1} + a_n/(b_n + z)} = \frac{A_{n-2}a_n + A_{n-1}b_n + A_{n-1}z}{B_{n-2}a_n + B_{n-1}b_n + B_{n-1}z}$$

$$f_n(z) = \frac{A_n + A_{n-1}z}{B_n + B_{n-1}z}$$

■

Corollary 1. *Two successive convergents of the above continued fraction are related by the formula*

$$\frac{A_n}{B_n} - \frac{A_{n-1}}{B_{n-1}} = \frac{(-1)^{n-1}a_1a_2\cdots a_n}{B_nB_{n-1}}$$

Proof. Inspection of the formula shows that we have only to prove

$$B_{n-1}A_n - A_{n-1}B_n = (-1)^{n-1}a_1a_2\cdots a_n$$

and this may be done by induction on n. For $n = 1$ we have $B_0A_1 - A_0B_1 = a_1$ as required. If the formula is valid for n, then it is valid for $n + 1$ because

$$\begin{aligned} B_nA_{n+1} - A_nB_{n+1} &= B_n(b_{n+1}A_n + a_{n+1}A_{n-1}) - A_n(b_{n+1}B_n + a_{n+1}B_{n-1}) \\ &= -a_{n+1}(B_{n-1}A_n - A_{n-1}B_n) \\ &= (-1)^n a_1a_2\cdots a_{n+1} \end{aligned}$$

■

Corollary 2. *The nth convergent of the above continued fraction is expressible as*

$$\frac{A_n}{B_n} = \frac{a_1}{B_0B_1} - \frac{a_1a_2}{B_1B_2} + \frac{a_1a_2a_3}{B_2B_3} - \cdots + (-1)^{n+1}\frac{a_1a_2\cdots a_n}{B_{n-1}B_n}$$

Proof. This follows from the preceding corollary if we first write

$$\frac{A_n}{B_n} = \left(\frac{A_n}{B_n} - \frac{A_{n-1}}{B_{n-1}}\right) + \left(\frac{A_{n-1}}{B_{n-1}} - \frac{A_{n-2}}{B_{n-2}}\right) + \cdots + \left(\frac{A_1}{B_1} - \frac{A_0}{B_0}\right)$$

■

We come now to one of the most elegant convergence theorems in the subject of continued fractions, a theorem due to Seidel.

Theorem 2. *Let all the b_n be positive in the continued fraction*

$$\frac{1}{b_1 +}\,\frac{1}{b_2 +}\,\frac{1}{b_3 + \cdots}$$

The continued fraction converges if and only if the series $\sum b_n$ diverges.

Proof. Corollary 2 affords us a means of turning the continued fraction into an alternating series. Indeed, with $c_n = B_n B_{n-1}$ we have

$$\frac{1}{b_1 +}\,\frac{1}{b_2 + \cdots}\,\frac{1}{b_n} = \frac{1}{c_1} - \frac{1}{c_2} + \cdots + (-1)^{n-1}\frac{1}{c_n} \tag{2}$$

We observe now that the numbers B_n are bounded below by $\theta = \min\{1, b_1\}$. This is trivial for $n = 0$ and 1 because $B_0 = 1$ and $B_1 = b_1$. If it is true for indices $\leq n - 1$, then it is true for n since

$$B_n = b_n B_{n-1} + B_{n-2} \geq b_n\theta + \theta > \theta$$

Now, the numbers c_n are monotonically increasing because

$$c_n = B_n B_{n-1} = (b_n B_{n-1} + B_{n-2})B_{n-1} = b_n B_{n-1}^2 + c_{n-1} \geq c_{n-1} + \theta^2 b_n$$

Thus if $\sum b_n = \infty$, then $c_n \uparrow \infty$ since (with $c_0 = 0$)

$$\begin{aligned} c_n &= (c_n - c_{n-1}) + (c_{n-1} - c_{n-2}) + \cdots + (c_1 - c_0) \\ &\geq \theta^2 b_n + \theta^2 b_{n-1} + \cdots + \theta^2 b_1 \end{aligned}$$

Under these circumstances, the right side of (2) converges, by a theorem on alternating series. On the other hand, if $\sum b_n < \infty$, then we can prove that c_n is bounded above. To do so, we prove first that

$$B_n < (1 + b_1)(1 + b_2)\cdots(1 + b_n)$$

For $n = 1$ this inequality is obvious since $B_1 = b_1$. If the inequality is true for indices $\leq n$, then

$$\begin{aligned} B_{n+1} &= b_{n+1}B_n + B_{n-1} \\ &< b_{n+1}(1 + b_1)\cdots(1 + b_n) + (1 + b_1)\cdots(1 + b_{n-1}) \\ &= (1 + b_1)\cdots(1 + b_{n-1})[b_{n+1}(1 + b_n) + 1] \\ &< (1 + b_1)\cdots(1 + b_{n-1})(1 + b_n)(1 + b_{n+1}) \end{aligned}$$

Since $1 + x < e^x$ for $x > 0$, we have

$$B_n < e^{b_1}e^{b_2}\cdots e^{b_n} < \exp \sum_{k=1}^{\infty} b_k \equiv M$$

Hence $c_n = B_n B_{n-1} < M^2$, and the terms of the series in (2) do not converge to zero. ∎

In illustration of this theorem we can easily prove that

$$\sqrt{2} - 1 = \frac{1}{2+} \frac{1}{2+} \cdots$$

Indeed, since $b_n = 2$, $\sum b_n$ diverges, and the continued fraction converges. Let c_n denote its nth convergent. We see that $c_{n+1} = 1/(2 + c_n)$ whence, by taking limits, we obtain $c = (2 + c)^{-1}$. The solution of this quadratic equation is $c = \sqrt{2} - 1$.

It is also of interest to note that the continued-fraction form $\dfrac{1}{b_1 +} \dfrac{1}{b_2 +} \cdots$

is really general, since the following identity is valid,

$$\frac{a_1}{b_1 +} \frac{a_2}{b_2 +} \frac{a_3}{b_3 +} \cdots = \frac{\lambda_1 a_1}{\lambda_1 b_1 +} \frac{\lambda_1 \lambda_2 a_2}{\lambda_2 b_2 +} \frac{\lambda_2 \lambda_3 a_3}{\lambda_3 b_3 +} \cdots$$

where the λ_n are arbitrary nonzero numbers. In particular, if λ_n is chosen appropriately, we may make $\lambda_{n-1}\lambda_n a_n = 1$.

Up to this point, we have not indicated how it is possible to expand functions into continued fractions. One method is the following:

Theorem. $\displaystyle\sum_{k=1}^{n} \frac{1}{D_k} = \frac{1}{D_1 -} \frac{D_1^2}{D_1 + D_2 -} \frac{D_2^2}{D_2 + D_3 -} \cdots \frac{D_{n-1}^2}{D_{n-1} + D_n}$

Proof. If $n = 2$, the assertion to be proved is that

$$\frac{1}{D_1} + \frac{1}{D_2} = \frac{1}{D_1 -} \frac{D_1^2}{D_1 + D_2}$$

This is true since both sides of the equation may be written as

$$\frac{D_1 + D_2}{D_1 D_2}$$

Now assume the validity of the equation for an arbitrary value of n. Then

$$\sum_{k=1}^{n+1} \frac{1}{D_k} = \sum_{k=1}^{n-1} \frac{1}{D_k} + \frac{1}{D_n D_{n+1}/(D_n + D_{n+1})}$$

$$= \frac{1}{D_1 -} \frac{D_1^2}{D_1 + D_2 -} \cdots \frac{D_{n-1}^2}{D_{n-1} + D_n D_{n+1}/(D_n + D_{n+1})}$$

$$= \frac{1}{D_1 -} \frac{D_1^2}{D_1 + D_2 -} \cdots \frac{D_{n-1}^2}{D_{n-1} + D_n -} \frac{D_n^2}{D_n + D_{n+1}}$$ ∎

As an example of this theorem, we obtain the following typical expansion from a Taylor series:

$$e^x = \frac{1}{1} + \frac{1}{x^{-1}} + \frac{1}{2x^{-2}} + \frac{1}{6x^{-3}} + \frac{1}{24x^{-4}} + \cdots$$

$$= \frac{1}{1 -} \frac{1}{1 + x^{-1} -} \frac{x^{-2}}{x^{-1} + 2x^{-2} -} \frac{4x^{-4}}{2x^{-2} + 6x^{-3} -} \frac{36x^{-6}}{6x^{-3} + 24x^{-4} - \cdots}$$

$$= \frac{1}{1 -} \frac{x}{x + 1 -} \frac{x}{x + 2 -} \frac{4x}{2x + 6 -} \frac{36x}{6x + 24 - \cdots}$$

$$= \frac{1}{1 -} \frac{x}{x + 1 -} \frac{x}{x + 2 -} \frac{2x}{x + 3 -} \frac{3x}{x + 4 - \cdots} \frac{kx}{x + k + 1 - \cdots}$$

Let us conclude this section with a remarkable theorem of Stieltjes in which are disclosed several unexpected relations between continued fractions and orthogonal polynomials. Recall (page 107) first that corresponding to any positive continuous weight function w on the interval $[-a,a]$, there exists a sequence of monic, orthogonal polynomials $Q_0, Q_1, \ldots$. These polynomials satisfy a recurrence relation $Q_n = (x - a_n)Q_{n-1} - b_nQ_{n-2}$ with starting values $Q_0 = 1$, $Q_1 = x - a_1$. We also have had occasion (page 112) to employ the same recurrence relation with starting values $\phi_0 = 0$, $\phi_1 = b_1$ to generate a second sequence of polynomials, $\phi_0, \phi_1, \phi_2, \ldots$. In order to describe Stieltjes' result, we require also the *moments* of w. These are the numbers

$$\mu_n = \int_{-a}^{a} x^n w(x)\, dx$$

***Theorem* [Stieltjes].** *For any $x \notin [-a,a]$,*

$$\frac{b_1}{x - a_1 -} \frac{b_2}{x - a_2 - \cdots} = \int_{-a}^{a} \frac{w(t)\, dt}{x - t} = \frac{\mu_0}{x} + \frac{\mu_1}{x^2} + \cdots$$

Furthermore, the nth convergent of the continued fraction is $\phi_n(x)/Q_n(x)$.

Proof. Let us begin with the easier part, viz., the second equality. We have at once

$$\int_{-a}^{a} \frac{w(t)\, dt}{x - t} = \frac{1}{x}\int_{-a}^{a} \frac{w(t)\, dt}{1 - t/x} = \frac{1}{x}\int_{-a}^{a} w(t) \sum_{n=0}^{\infty} \left(\frac{t}{x}\right)^n dt$$

$$= \frac{1}{x}\sum_{n=0}^{\infty} \frac{1}{x^n}\int_{-a}^{a} t^n w(t)\, dt = \sum_{n=0}^{\infty} \frac{\mu_n}{x^{n+1}}$$

For the other half of the equation, let x be fixed outside of $[-a,a]$, say, $x > a$. Let $H_n(t) = 1 + \sum_{k=1}^{n} c_k(x - t)^k$, where the c_k are chosen to render the number

$$\epsilon_n = \int_{-a}^{a} H_n^2(t) \frac{w(t)}{x - t} dt$$

a minimum. It follows that $\partial\epsilon_n/\partial c_k = 0$, or

$$\int_{-a}^{a} H_n(t)(x - t)^k \frac{w(t)}{(x - t)} dt = 0 \qquad (k = 1, \ldots, n)$$

This shows that H_n is orthogonal to $1, x - t, (x - t)^2, \ldots, (x - t)^{n-1}$, and hence orthogonal to $Q_0, Q_1, \ldots, Q_{n-1}$. Thus H_n is a multiple of Q_n, say $H_n = cQ_n$. Putting $t = x$, we discover that $c = 1/Q_n(x)$. The value of ϵ_n may now be computed, with the help of orthogonality relations as

$$\begin{aligned}\epsilon_n &= \frac{1}{Q_n(x)} \int_{-a}^{a} Q_n(t) \frac{H_n(t)}{x - t} w(t)\, dt \\ &= \frac{1}{Q_n(x)} \int_{-a}^{a} Q_n(t) \left[\frac{1}{x - t} + \sum_{k=1}^{n} c_k(x - t)^{k-1}\right] w(t)\, dt \\ &= \frac{1}{Q_n(x)} \int_{-a}^{a} \frac{Q_n(t)}{x - t} w(t)\, dt\end{aligned}$$

Now the theorem of page 112 asserts that

$$\phi_n(x) = \int_{-a}^{a} \frac{Q_n(x) - Q_n(t)}{x - t} w(t)\, dt$$

With the aid of this equation we obtain

$$\begin{aligned}\epsilon_n &= \frac{1}{Q_n(x)} \int_{-a}^{a} \frac{Q_n(x) - [Q_n(x) - Q_n(t)]}{x - t} w(t)\, dt \\ &= \int_{-a}^{a} \frac{w(t)\, dt}{x - t} - \frac{\phi_n(x)}{Q_n(x)}\end{aligned}$$

It remains to be seen that $\epsilon_n \to 0$. Let $d_1, \ldots, d_{n-1}$ be the parameters in $H_{n-1}(t)$ which would make ϵ_{n-1} a minimum. Then for any λ,

$$\epsilon_n \leq \int_{-a}^{a} \{[1 - \lambda(x - t)][1 + \sum_{k=1}^{n-1} d_k(x - t)^k]\}^2 \frac{w(t)\, dt}{x - t}$$

because the expression between the braces competes with $H_n(t)$ in the minimization. Now let $\lambda = (x + a)^{-1}$ and use the inequalities

$$0 \leq 1 - \frac{x - t}{x + a} \leq 1 - \frac{x - a}{x + a} \equiv \theta < 1$$

to obtain $\epsilon_n \leq \theta^2 \epsilon_{n-1}$. This establishes that $\epsilon_n \to 0$ and completes the proof of the asserted equation.

Now if we compute the nth convergent of the continued fraction by means of Theorem 1 on page 182, we find that the quantities A_n and B_n are defined as follows.

$$A_0 = 0 \qquad A_1 = b_1 \qquad A_{n+1} = (x - a_{n+1})A_n - b_{n+1}A_{n-1}$$

$$B_0 = 1 \qquad B_1 = x - a_1 \qquad B_{n+1} = (x - a_{n+1})B_n - b_{n+1}B_{n-1}$$

Since ϕ_n and Q_n are defined in the same way, $A_n/B_n = \phi_n(x)/Q_n(x)$. ∎

Problems

1. For the continued fraction $\frac{1}{2+}\,\frac{1}{2+}\cdots$ compute the numbers A_n and B_n. Prove that $A_n/B_n \to \sqrt{2} - 1$ by direct calculation.

2. If $\lambda > 0$, what can you say about $\frac{1}{\lambda+}\,\frac{1}{\lambda+}\cdots$?

3. Prove that $1 + \frac{1}{2} + \frac{1}{3} + \cdots + \frac{1}{n+1} = \frac{1}{1-}\,\frac{1}{3-}\,\frac{4}{5-}\,\frac{9}{7-}\,\frac{16}{9-}\cdots\frac{n^2}{2n+1}$.

4. From the series $\cot^{-1} x = 1/x - 1/3x^3 + 1/5x^5 - 1/7x^7 + \cdots$ (valid for $x > 1$), deduce the continued fraction

$$\cot^{-1} x = \frac{1}{x+}\,\frac{x}{3x^2 - 1+}\,\frac{9x^2}{5x^2 - 3+}\,\frac{25x^2}{7x^2 - 5+}\,\frac{49x^2}{9x^2 - 7+}\cdots$$

5. Prove that $\frac{1}{e - 1} = \frac{1}{1+}\,\frac{2}{2+}\,\frac{3}{3+}\cdots$. *Hint*: Compute e^{-1} from the continued fraction for e^x given in the text.

6. If $b_n > 0$ for all n and if the continued fraction $\frac{1}{b_1+}\,\frac{1}{b_2+}\cdots$ converges to c, then the *even* convergents converge *upward* to c and the *odd* convergents converge *downward* to c.

7. Assuming $a_n, b_n > 0$, prove the following. The continued fraction $\frac{a_1}{b_1+}\,\frac{a_2}{b_2+}\cdots$ converges if and only if the series

$$\sum_{n=1}^{\infty} (a_1^{-1}a_2a_3^{-1}a_4 \cdots a_{2n-1}^{-1}b_{2n-1} + a_1a_2^{-1}a_3a_4^{-1} \cdots a_{2n}^{-1}b_{2n})$$

diverges.

8. With the help of Prob. 13 of Chap. 4, Sec. 2, and Stieltjes' theorem of this section, prove that

$$\frac{1}{\sqrt{x^2-1}} = \frac{1}{x-}\,\frac{\frac{1}{2}}{x-}\,\frac{\frac{1}{4}}{x-}\,\frac{\frac{1}{4}}{x-}\,\frac{\frac{1}{4}}{x-\cdots}$$

Hint: The integral in Stieltjes' theorem may be evaluated by letting $t = 2z(1+z^2)^{-1}$. A direct proof is also possible by the same technique as was used earlier to prove $\sqrt{2} = 1 + 1/(2+\cdots)$.

9. Obtain the continued fraction

$$-\log(1-x) = \frac{x}{1-}\,\frac{x}{x+2-}\,\frac{4x}{2x+3-}\,\frac{9x}{3x+4-}\,\frac{16x}{4x+5-\cdots}$$

10. Use Corollary 2 to obtain an infinite series for $\tan z$.

11. The $2n$th convergent of $\dfrac{1}{b_1+}\,\dfrac{1}{b_2+\cdots}$ is the nth convergent of

$$\frac{b_2}{1+b_1b_2-}\,\frac{b_4}{b_2+b_2b_3b_4+b_4-}\,\frac{b_6}{b_4+b_4b_5b_6+b_6-}\,\frac{b_8}{b_6+b_6b_7b_8+b_8-}$$

Thus if all $b_i > 0$, the second continued fraction converges, and in fact it converges monotonically upward to its limit. *Hint*: Induction will prove the asserted equality of convergents. Then use Prob. 6.

12. What is the relationship between

$$\frac{1}{b_1+}\,\frac{1}{b_2+\cdots}\,\frac{1}{b_n} \quad \text{and} \quad \frac{1}{b_n+}\,\frac{1}{b_{n-1}+\cdots}\,\frac{1}{b_1}$$

Chapter 6

Some additional topics

1. The Stone Approximation Theorem

If it were necessary to designate one theorem in approximation theory as being of greater significance than any other, that one would probably be the Weierstrass approximation theorem. The influence of this theorem has been felt not only in the obvious way through its use as a tool in analysis but also in the more far-reaching way of enticing mathematicians into generalizing it or providing it with alternative proofs. The names of Bernstein, Landau, de La Vallée Poussin, Fejér, Lebesgue, Müntz, and Stone are all associated with these efforts, many of which have opened new frontiers of investigation. In this section and the next, two quite dissimilar generalizations of Weierstrass' theorem will come under study.

Up to now in this book the space $C[X]$ of continuous functions on the compact metric space X has been treated almost exclusively as a *linear* space. For the present purposes, it is necessary to call attention to the fact that with the natural definition of multiplication

$$(fg)(x) = f(x)g(x)$$

the space $C[X]$ becomes an *algebra*. An algebra is simply a linear space in which a multiplication has been defined satisfying the postulates $f(g + h) = fg + fh$, $(f + g)h = fh + gh$, $f(gh) = (fg)h$, $\alpha(fg) = (\alpha f)g = f(\alpha g)$. In $C[a,b]$, the polynomials form a *subalgebra* since a product of polynomials is another polynomial; the Weierstrass theorem asserts that this subalgebra is dense in $C[a,b]$. The Stone generalization states that the same conclusion may be drawn for *any* subalgebra A in *any* space $C[X]$ if only we have the two properties

(*i*) $1 \in A$
(*ii*) A separates points of X

The meaning of (*ii*) is that for any two distinct points x and y in X there is a function $f \in A$ such that $f(x) \neq f(y)$. Stone's theorem requires only that X be a compact topological space, but since we wish to avoid considerations of general topology, we take X to be a compact metric space.

We shall require a preliminary definition and a lemma. First, a *cover* of a set X is a collection of sets $\{V_\alpha\}$ such that $X \subset \cup V_\alpha$. Thus each point of x lies in at least one of the sets V_α. A *subcover* is simply a subcollection which is also a cover. An *open cover* is a cover in which each set is open. In the next lemma it is convenient to use the notation $S(x,r)$ for the sphere about x with radius r; thus $S(x,r) = \{y: d(x,y) < r\}$.

Lemma. *Every open cover of a compact metric space contains a finite subcover.*

Proof. Let (X,d) be a compact metric space and $\{V_\alpha\}$ an open cover of X. We must prove that a finite number of the sets V_α also cover X. We shall prove first the existence of an $\epsilon > 0$ (termed a Lebesgue number for the cover) such that every sphere $S(x,\epsilon)$ lies in some V_α. If such an ϵ does not exist, then we may find a sequence of spheres, $S(x_n,1/n)$, each of which lies in *no* V_α. By compactness, some subsequence of $\{x_n\}$ converges; say $x_{n_k} \to x^*$. Since $\{V_\alpha\}$ is a cover of X, there is an index β such that $x^* \in V_\beta$. Since V_β is open, there is a number $\delta > 0$ such that $S(x^*,\delta) \subset V_\beta$. Since $x_{n_k} \to x^*$, we may find an index $m > 2/\delta$ such that $d(x_m,x^*) < \delta/2$. Then it follows that $S(x_m,1/m) \subset S(x_m,\delta/2) \subset S(x^*,\delta) \subset V_\beta$, contradicting our assumption about the spheres $S(x_n,1/n)$. Thus for some ϵ, every sphere $S(x,\epsilon)$ lies in some V_α. Now we shall show that X may be covered by a finite number of such spheres $S(x,\epsilon)$. If this is not possible, then, after selecting x_1 arbitrarily, we may select $x_2 \notin S(x_1,\epsilon)$, then $x_3 \notin S(x_1,\epsilon) \cup S(x_2,\epsilon)$, and so forth. The sequence $\{x_n\}$ thus defined possesses no convergent subsequence, because $d(x_i,x_j) \geq \epsilon$ whenever $i \neq j$. This contradicts the compactness of X. Thus it is possible to write $X = \bigcup_{i=1}^{n} S(z_i,\epsilon)$ for appropriate points $z_1, \ldots, z_n$. Finally, since each $S(z_i,\epsilon)$ is contained in a set V_{α_i}, it follows that $X = \bigcup_{i=1}^{n} V_{\alpha_i}$. ∎

The Stone Approximation Theorem. *Let X be a compact metric space and A a subalgebra of $C[X]$ which contains 1 and separates points in X. Then A is dense in $C[X]$.*

Proof. Denote by $\bar{A}$ (the "closure" of A) the set of all f which are limits of uniformly converging sequences in A. Our task is to establish that $\bar{A} = C[X]$.

We begin by proving that

$$f \in \bar{A} \Rightarrow |f| \in \bar{A}$$

By the Weierstrass theorem there exists a sequence of polynomials P_n such that $P_n(\lambda) \to |\lambda|$ uniformly in $[-\|f\|,\|f\|]$. It follows then that $|f| = \lim_n P_n(f)$. Now $\bar{A}$ itself is a subalgebra of $C[X]$, (Prob. 1), so that $P_n(f)$ belongs to $\bar{A}$. Thus each $P_n(f)$ is the limit of a sequence, say f_{nk}, in A. Finally $\| |f| - f_{nk} \| \le \| |f| - P_n(f) \| + \| P_n(f) - f_{nk} \|$, which shows that $|f|$ is the limit of a sequence in A. (See Prob. 2.)

Next we observe that $\bar{A}$ is closed under the two "lattice" operations:

$$(f \vee g)(x) = \max \{ f(x), g(x) \}$$

$$(f \wedge g)(x) = \min \{ f(x), g(x) \}$$

This is an immediate consequence of what we proved above and the two identities

$$f \vee g = \tfrac{1}{2}(f + g + |f - g|)$$

$$f \wedge g = \tfrac{1}{2}(f + g - |f - g|)$$

The final preliminary step in the proof is to show that given two points p and q in X and two numbers λ and μ, there exists an $F \in A$ such that $F(p) = \lambda$ and $F(q) = \mu$. Since A separates points in X, there exists an $f \in A$ such that $f(p) \neq f(q)$. Form a linear combination of f and 1 as follows:

$$F = \frac{\lambda - \mu}{f(p) - f(q)} f + \frac{\mu f(p) - \lambda f(q)}{f(p) - f(q)} 1$$

It is easy to verify that $F(p) = \lambda$, $F(q) = \mu$, and $F \in A$.

Now for the proof proper, if $\bar{A} \neq C[X]$, then we may find an $f \in C[X]$ and an $\epsilon > 0$ such that

$$\text{(1)} \qquad g \in \bar{A} \Rightarrow \|g - f\| > \epsilon$$

For each pair of points p and q, find $f_{pq} \in A$ such that $f_{pq}(p) = f(p)$ and $f_{pq}(q) = f(q)$. Define open sets $V_{pq} = \{x \in X: f_{pq}(x) < f(x) + \epsilon\}$. Fixing q momentarily, we see that $X = \cup \{V_{pq}: p \in X\}$ since $p \in V_{pq}$. By the lemma, a finite number of these sets cover X, say $V_{p_1q}, \ldots, V_{p_nq}$. Put $f_q = \min \{f_{p_1q}, \ldots, f_{p_nq}\}$. For any $x \in X$, there is an index i such that $x \in V_{p_iq}$, whence $f_q(x) \le f_{p_iq}(x) < f(x) + \epsilon$. Thus $f_q < f + \epsilon$. Now define $V_q = \{x: f_q(x) > f(x) - \epsilon\}$. Since $f_{pq}(q) = f(q) > f(q) - \epsilon$ for all p, it follows that $f_q(q) > f(q) - \epsilon$. We conclude that $q \in V_q$ and consequently that the open sets V_q cover X. By the lemma, a finite number of these cover X, say $V_{q_1}, \ldots, V_{q_m}$. Define $g = \max \{f_{q_1}, \ldots, f_{q_m}\}$. Since $f_q < f + \epsilon$ for all q, we have $g < f + \epsilon$. On the other hand, each x lies in some V_{q_i}, so that $g(x) \ge f_{q_i}(x) > f(x) - \epsilon$. We have proved that $f - \epsilon < g < f + \epsilon$. Since $g \in \bar{A}$, this contradicts the implication (1) above. ■

Corollary 1. *The trigonometric polynomials form a dense set in* $C_{2\pi}$.

Corollary 2. *Let X be a compact set in n space. The polynomials in n variables form a dense set in $C[X]$.*

Problems

1. If A is a subalgebra of $C[X]$, then so is $\bar{A}$.
2. If S is a subset of a metric space, denote by $\bar{S}$ the set of all points which are limits of sequences in S. Show that $\bar{\bar{S}} = \bar{S}$, that $\bar{S}$ is closed and that $\bar{S}$ contains each closed set which contains S.
3. The condition that A separate points of X is *necessary* in order that A be dense in $C[X]$. Is the condition $1 \in A$ also necessary?
4. Let f be a monotone element of $C[a,b]$. Then the algebra generated from 1 and f by multiplications, additions, and scalar multiplications is dense in $C[a,b]$. Give two proofs, one using Stone's theorem and the other only Weierstrass'.
5. Prove Corollary 1.
6. Prove Corollary 2.
7. Prove that the set of polynomials $\sum_{k=0}^{n} c_k x^{17k}$ is dense in $C[a,b]$. Generalize.

2. The Müntz Theorem

In order to explain the Müntz theorem, let us first recall what is meant by a fundamental set in a normed linear space. A set S is *fundamental* if the set of linear combinations of elements of S is a dense set. Thus to each element f of the space and to each $\epsilon > 0$ there corresponds a (finite) linear combination $\sum \lambda_i g_i$ of members of S such that $\| f - \sum \lambda_i g_i \| < \epsilon$. With this terminology, the Weierstrass theorem asserts simply that the set $\mathbf{P} = \{1,x,x^2,\ldots\}$ is fundamental in $C[0,1]$. It is natural to ask what other sets are fundamental in $C[0,1]$. In particular, what subsets of $\mathbf{P}$, if any, are fundamental? The theorem of Müntz gives a complete answer to this question. Another theorem of Müntz answers the same question when the linear space $C[0,1]$ is endowed with the least-squares norm $\left[\int_0^1 | f(x) |^2 \, dx\right]^{1/2}$.

The Müntz theorem is all the more interesting because it traces a logical connection between two apparently unrelated facts: the fundamentality of $\{1,x,x^2,\ldots\}$ and the divergence of the series of reciprocal exponents, $1 + \frac{1}{2} + \frac{1}{3} + \cdots$. In fact, if we wish to delete functions from the set while maintaining its fundamentality, this divergence is precisely the property that must be preserved.

There are several lemmas which are of independent interest. Recall first that in an inner-product space the *Gram determinant* of a finite set of vectors

$\{f_1, \ldots, f_n\}$ is defined by the formula

$$G(f_1, \ldots, f_n) = \begin{vmatrix} \langle f_1, f_1 \rangle & \cdots & \langle f_1, f_n \rangle \\ \cdot & \cdots & \cdot \\ \langle f_n, f_1 \rangle & \cdots & \langle f_n, f_n \rangle \end{vmatrix}$$

Its importance in approximation theory stems from the following result.

Gram's Lemma. *Let M denote an n-dimensional subspace of an inner-product space. The distance d of any point g to M is given by the formula*

$$d^2 = \frac{G(f_1, \ldots, f_n, g)}{G(f_1, \ldots, f_n)}$$

where $\{f_1, \ldots, f_n\}$ is any basis for M.

Proof. The existence of an element $f^* \in M$ for which $\| f^* - g \|$ is minimal follows from the theorem on existence of best approximations (Chap. 1, Sec. 6). We show now that $f^* - g \perp M$. Indeed, if u is any element of M, then so is $f^* + \lambda u$, and consequently $0 \le \| f^* + \lambda u - g \|^2 - \| f^* - g \|^2 = 2\lambda \langle f^* - g, u \rangle + \lambda^2 \| u \|^2$. If this is to be true for all real λ, then $\langle f^* - g, u \rangle = 0$, as was to be shown. By taking $\lambda = 1$ we see that f^* is unique; by reversing the argument we see that if $f^* - g \perp M$, then f^* is the point of M closest to g. Thus f^* is uniquely determined by the conditions $f^* - g \perp f_i$ $(i = 1, \ldots, n)$.

Putting $f^* = \sum_{j=1}^{n} \lambda_j f_j$, we obtain a system of linear equations on the coefficients λ_j:

$$\sum_{j=1}^{n} \lambda_j \langle f_j, f_i \rangle = \langle g, f_i \rangle \qquad (i = 1, \ldots, n)$$

The coefficient determinant here is $G(f_1, \ldots, f_n)$ and must be different from zero since the system of equations has a unique solution. From the orthogonality relations we obtain $d^2 = \| g - f^* \|^2 = \langle g - f^*, g - f^* \rangle = \langle g, g \rangle - \langle g, f^* \rangle$. Again putting $f^* = \sum \lambda_j f_j$, we have $\sum \lambda_j \langle f_j, g \rangle + d^2 = \langle g, g \rangle$. This provides a system of $n + 1$ equations in $n + 1$ variables:

$$\begin{aligned} \lambda_1 \langle f_1, f_1 \rangle + \cdots + \lambda_n \langle f_n, f_1 \rangle + 0d^2 &= \langle g, f_1 \rangle \\ \cdots\cdots\cdots & \\ \lambda_1 \langle f_1, f_n \rangle + \cdots + \lambda_n \langle f_n, f_n \rangle + 0d^2 &= \langle g, f_n \rangle \\ \lambda_1 \langle f_1, g \rangle + \cdots + \lambda_n \langle f_n, g \rangle + d^2 &= \langle g, g \rangle \end{aligned}$$

The solution of this system by Cramer's rule gives us

$$d^2 = \frac{G(f_1,\ldots,f_n,g)}{G(f_1,\ldots,f_n)} \qquad \blacksquare$$

We shall presently apply Gram's lemma to the problem of computing the least-squares distance from x^m to the subspace spanned by $\{x^{p_1},x^{p_2},\ldots,x^{p_n}\}$. Since the inner products that will appear in the Gram determinant are of the form

$$\int_0^1 x^p x^q \, dx = \frac{1}{p+q+1}$$

the following result will be useful.

Cauchy's Lemma.

$$\prod_{(i,j)} (a_i + b_j) \begin{vmatrix} \dfrac{1}{a_1+b_1} & \cdots & \dfrac{1}{a_1+b_n} \\ \cdots & \cdots & \cdots \\ \dfrac{1}{a_n+b_1} & \cdots & \dfrac{1}{a_n+b_n} \end{vmatrix} = \prod_{j<i} (a_i - a_j)(b_i - b_j)$$

Proof. On the left side of the equation we have a rational function of $(a_1,\ldots,a_n,b_1,\ldots,b_n)$. But this rational function has no poles because of the presence of the factor outside the determinant. Thus both members of the equation are in fact polynomials. If the right member vanishes, then either $a_i = a_j$ or $b_i = b_j$ for some pair, $i \neq j$. Then the left member vanishes also. Consequently the right member is a factor of the left member. But in each of the variables a_i and b_i both members of the equation are polynomials of the same degree, viz., $n - 1$. Hence the left member is a *constant* multiple of the right. That this constant must be 1 can be proved by taking the limit of both sides of the equation as $b_1 \to -a_1$, $b_2 \to -a_2$, etc., if we first write the left member in the form

$$\prod_{i\neq j} (a_i + b_j) \begin{vmatrix} 1 & \dfrac{a_1+b_1}{a_1+b_2} & \cdots & \dfrac{a_1+b_1}{a_1+b_n} \\ \dfrac{a_2+b_2}{a_2+b_1} & 1 & \cdots & \dfrac{a_2+b_2}{a_2+b_n} \\ \cdots & \cdots & \cdots & \cdots \\ \dfrac{a_n+b_n}{a_n+b_1} & \dfrac{a_n+b_n}{a_n+b_2} & \cdots & 1 \end{vmatrix}$$

The limiting value of the left member is clearly $\prod_{i \neq j} (a_i - a_j)$, and this is the limiting value of the right member as well. ∎

Lemma. *Let $m, p_1, p_2, \ldots, p_n$ be distinct real numbers greater than $-\frac{1}{2}$. The least-squares distance on $[0,1]$ from x^m to the subspace spanned by $\{x^{p_1}, \ldots, x^{p_n}\}$* is

$$d = \frac{1}{\sqrt{2m+1}} \prod_{j=1}^{n} \frac{|m - p_j|}{m + p_j + 1}$$

Proof. By the Gram lemma,

$$d^2 = \frac{G(x^{p_1}, \ldots, x^{p_n}, x^m)}{G(x^{p_1}, \ldots, x^{p_n})}$$

Since $\langle x^p, x^q \rangle = (p + q + 1)^{-1}$, we may use the Cauchy lemma to compute $G(x^{p_1}, \ldots, x^{p_n})$ in the form

$$\begin{vmatrix} \dfrac{1}{p_1 + p_1 + 1} & \cdots & \dfrac{1}{p_1 + p_n + 1} \\ \cdot & \cdot \; \cdot \; \cdot & \cdot \\ \dfrac{1}{p_n + p_1 + 1} & \cdots & \dfrac{1}{p_n + p_n + 1} \end{vmatrix} = \frac{\prod_{j<i} (p_i - p_j)^2}{\prod_{(i,j)} (p_i + p_j + 1)}$$

We may compute $G(x^{p_1}, \ldots, x^{p_n}, x^m)$ in a similar manner, obtaining

$$\frac{\prod_{j<i} (p_i - p_j)^2 \prod_{j=1}^{n} (m - p_j)^2}{\prod_{(i,j)} (p_i + p_j + 1) \prod_{j=1}^{n} (m + p_j + 1)^2 (2m + 1)}$$

The asserted formula for d then follows by taking a quotient and a square root. ∎

Lemma. *Let $0 < a_n \neq 1$ and $a_n \to 0$. Then*

$$\prod_{n=1}^{\infty} (1 - a_n) = 0 \text{ if and only if } \sum_{n=1}^{\infty} a_n = +\infty$$

Proof. Select m such that $a_n < \frac{1}{2}$ whenever $n \geq m$. For such an n we may write

$$\left|\frac{\log(1-a_n)}{a_n} + 1\right| = \left|\frac{-a_n - a_n^2/2 - a_n^3/3 - \cdots}{a_n} + 1\right|$$

$$= \left|\frac{a_n}{2} + \frac{a_n^2}{3} + \frac{a_n^3}{4} + \cdots\right|$$

$$< (\tfrac{1}{2})^2 + (\tfrac{1}{2})^3 + \cdots = \tfrac{1}{2}$$

Thus for $n \geq m$, $-\frac{3}{2} \leq a_n^{-1}\log(1-a_n) < -\frac{1}{2}$, and it follows that the two series $\sum \log(1-a_n)$ and $\sum a_n$ either both converge or both diverge. Therefore, $\sum a_n = +\infty$ iff $\sum \log(1-a_n) = \log \prod (1-a_n) = -\infty$. ∎

Müntz's First Theorem. *Consider the set of functions* $\{x^{p_1}, x^{p_2}, \ldots\}$ *where* $-\frac{1}{2} < p_n \to \infty$. *It is fundamental in the least-squares norm on*

$[0,1]$ *if and only if* $\displaystyle\sum_{p_n \neq 0} \frac{1}{p_n} = +\infty$.

Proof. Since the set of all polynomials is dense, it is enough for the present purpose to find a necessary and sufficient condition that every monomial x^m ($m = 0, 1, 2, \ldots$) be approximable by functions of the form $\sum \lambda_i x^{p_i}$. By a previous lemma, the distance from x^m to the subspace spanned by $\{x^{p_1}, \ldots, x^{p_n}\}$ is given by

$$d_n = \frac{1}{\sqrt{2m+1}} \prod_{j=1}^{n} \left|\frac{m - p_j}{m + p_j + 1}\right|$$

A necessary and sufficient condition that $d_n \downarrow 0$ as $n \uparrow \infty$ is therefore that

$$\prod_{j=1}^{\infty}\left(1 - \frac{2m+1}{m+p_j+1}\right) = 0$$

In order for this to occur (when m is not a p_i) it is necessary and sufficient

(according to the preceding lemma) that $\displaystyle\sum_{j=1}^{\infty} \frac{2m+1}{m+p_j+1} = \infty$. This in turn

is equivalent to the condition $\displaystyle\sum_{p_j \neq 0} \frac{1}{p_j} = +\infty$. (See Prob. 5.) ∎

Müntz's Second Theorem. *The set of functions* $\{1, x^{p_1}, x^{p_2}, \ldots\}$ *where* $1 \leq p_n \to +\infty$ *is fundamental (with the uniform norm) in* $C[0,1]$ *if and only if* $\sum (1/p_n) = \infty$.

Proof. The Cauchy–Schwarz inequality in an inner-product space states that $|\langle f,g\rangle| \leq \|f\| \cdot \|g\|$, the norm being of course defined as $\langle f,f\rangle^{1/2}$. In the linear space $C[0,1]$ with the integral inner product, we may take $g(x) = 1$ and obtain thereby

$$\int_0^1 |f(x)|\, dx \leq \left[\int_0^1 |f(x)|^2\, dx\right]^{1/2}$$

Thus for any $x \in [0,1]$ and for any integer $m \geq 1$ we may write

$$\begin{aligned}
\left| x^m - \sum_{i=1}^n \lambda_i x^{p_i} \right| &= \left| \int_0^x \left(m t^{m-1} - \sum_{i=1}^n \lambda_i p_i t^{p_i-1}\right) dt \right| \\
&\leq \int_0^x \left| m t^{m-1} - \sum_{i=1}^n \lambda_i p_i t^{p_i-1} \right| dt \\
&\leq \int_0^1 \left| m t^{m-1} - \sum_{i=1}^n \lambda_i p_i t^{p_i-1} \right| dt \\
&\leq \left[\int_0^1 \left| m t^{m-1} - \sum_{i=1}^n \lambda_i p_i t^{p_i-1} \right|^2 dt\right]^{1/2}
\end{aligned}$$

Now assume that $\sum (1/p_n) = \infty$. By Müntz's first theorem, the final member of this inequality may be made arbitrarily small by appropriate choice of $n, \lambda_1, \ldots, \lambda_n$. Since x was arbitrary, the norm

$$\max_{0\leq x\leq 1} \left| x^m - \sum_{i=1}^n \lambda_i x^{p_i} \right|$$

also becomes small. The case $m = 0$ is trivial. For the other half of the theorem we observe that if $\sum (1/p_n) < \infty$, then some function f is *not* arbitrarily well approximated in the least-squares sense by the functions $\sum \lambda_i x^{p_i}$. Hence an inequality

$$\int_0^1 \left| f(x) - \sum \lambda_i x^{p_i} \right|^2 dx > \epsilon$$

is true for all choices of λ_i. Since

$$\left[\int_0^1 |g(x)|^2\, dx\right]^{1/2} \leq \max_x |g(x)|$$

it follows that f is also not well approximated in the uniform norm. ∎

Problems

1. Prove, for example by induction, that $G(f_1) > 0$, $G(f_1,f_2) > 0, \ldots$ provided that $\{f_1,f_2,\ldots\}$ is an independent set. How does this generalize the Cauchy-Schwarz inequality?
2. Give another proof of Gram's lemma in the case that $\{f_1,\ldots,f_n\}$ is an orthonormal set.
3. The Cauchy lemma may be proved by induction. First subtract the last row from each preceding row. Then factor a common expression from each row and a common expression from each column. Now subtract the last column from each preceding column and perform further factorizations. Finally use the induction hypothesis.
4. An $n \times n$ matrix H_n having elements $a_{ij} = (i + j - 1)^{-1}$ is called a *Hilbert* matrix. Show that $0 < \det H_n < (2^n n!)^{-1}$.
5. Show that the series $\sum_n \dfrac{2m + 1}{m + p_n + 1}$ and $\sum_n \dfrac{1}{p_n}$ converge or diverge together.
6. If $p_n \geq 1$, if $\sum (1/p_n) < \infty$, and if $m \neq p_n$ for all n, then x^m is not approximable by the functions $1, x^{p_1}, x^{p_2}, \cdots$ in $C[0,1]$.
7. A subspace of $C[0,1]$ generated by $\{1,x^{p_1},x^{p_2},\ldots\}$ where $p_n \geq 1$ and $\sum (1/p_n) < \infty$ is said to be *thin*. The vector sum of any finite number of thin subspaces is still a thin subspace.
8. Which of the following are fundamental in $C[0,1]$:

   (*a*) $\{1,x^2,x^4,x^8,x^{16},\ldots\}$
   (*b*) $\{1,x^{p_1},x^{p_2},\ldots\}$ with p_n being the nth prime number. *Hint*: By the prime-number theorem, $n \log n/p_n \to 1$.
   (*c*) $\{1,x^{p_1},x^{p_2},\ldots\}$ with $p_n = 10^{10} + n$

3. The Converses of the Jackson Theorems

In Chap. 4, Sec. 6, we examined some of the theorems of Jackson which provide upper bounds for the quantity $E_n(f)$ in terms of the "smoothness" of the function f. Recall that $E_n(f)$ is the error in the best uniform approximation of f by a polynomial of degree $\leq n$, either algebraic or trigonometric, depending on the context. Here it will be convenient to use the latter, so that

$$E_n(f) = \inf_{a_k,b_k} \max_{0 \leq x \leq 2\pi} \left| f(x) - \sum_{k=0}^{n} (a_k \cos kx + b_k \sin kx) \right|$$

One of the Jackson theorems states, for example, that if $f \in \text{Lip}\,\alpha$ with $0 < \alpha \leq 1$ then $E_n(f)$ must converge to zero with at least the rapidity of $n^{-\alpha}$. The notation $f \in \text{Lip}\,\alpha$ means that f satisfies a Lipschitz condition of the form

$$|f(x) - f(y)| \leq \lambda |x - y|^\alpha$$

The natural question of whether conversely we may infer anything about the smoothness of f from the behavior of $E_n(f)$ was answered first in a remarkable group of theorems due to Bernstein. In particular, the theorem just cited has the following exact converse in the case $0 < \alpha < 1$: If the sequence $n^\alpha E_n(f)$ is bounded, then $f \in \text{Lip}\ \alpha$. In the case $\alpha = 1$, the smoothness property of f which is equivalent to the boundedness of $n^\alpha E_n(f)$ was discovered by Zygmund.

The theorems of Bernstein and Zygmund are given here for trigonometric approximation, and from these, theorems concerning approximation by algebraic polynomials may be derived.

Bernstein's Theorem I. *If $f \in C_{2\pi}$ and $E_n(f) \le An^{-\alpha}$ with $\alpha \in (0,1)$, then $f \in \text{Lip}\ \alpha$.*

Proof. For each n let P_n denote a trigonometric polynomial of degree $\le n$ such that $\| f - P_n \| \le An^{-\alpha}$. Set $V_0 = P_1$ and $V_n = P_{2^n} - P_{2^{n-1}}$ for $n \ge 1$. Then $f = \sum_{n=0}^{\infty} V_n$, and consequently for any m, with the help of the mean-value theorem and Bernstein's inequality (page 91), we may write

$$(1)\quad |f(x) - f(y)| = \left| \sum_{n=0}^{\infty} [V_n(x) - V_n(y)] \right|$$

$$\le \sum_{n=0}^{m-1} |V_n(x) - V_n(y)| + \sum_{n=m}^{\infty} |V_n(x)| + \sum_{n=m}^{\infty} |V_n(y)|$$

$$\le \sum_{n=0}^{m-1} |V_n'(\xi_n)|\,|x - y| + 2 \sum_{n=m}^{\infty} \| V_n \|$$

$$\le \sum_{n=0}^{m-1} 2^n \| V_n \|\,|x - y| + 2 \sum_{n=m}^{\infty} \| V_n \|$$

In this inequality we next insert the following estimate of $\| V_n \|$:

$$(2)\quad \| V_n \| \le \| P_{2^n} - f \| + \| f - P_{2^{n-1}} \|$$

$$\le A(2^n)^{-\alpha} + A(2^{n-1})^{-\alpha} \le B2^{-n\alpha}$$

Inequality (1) then becomes

$$(3)\quad |f(x) - f(y)| \le B|x - y| \sum_{n=0}^{m-1} 2^{n(1-\alpha)} + 2B \sum_{n=m}^{\infty} 2^{-n\alpha}$$

$$= B|x - y| \frac{2^{m(1-\alpha)} - 1}{2^{1-\alpha} - 1} + 2B2^{-m\alpha} \frac{1}{1 - 2^{-\alpha}}$$

$$\le C[|x - y| 2^{m(1-\alpha)} + 2^{-m\alpha}]$$

Now let $|x - y| = \delta$. In order for our last estimate to be majorized by a term $\lambda\delta^\alpha$, we must establish an inequality of the form

$$\delta 2^{m(1-\alpha)} + 2^{-m\alpha} \leq D\delta^\alpha$$

But this is equivalent to

$$(\delta 2^m)^{1-\alpha} + (\delta 2^m)^{-\alpha} \leq D$$

and this will be achieved if $\delta 2^m$ is bounded above and away from zero. For example, if m is the least integer such that $2^m\delta \geq 1$, then $2^m\delta < 2$. ∎

Bernstein's Theorem II. *If $f \in C_{2\pi}$ and $E_n(f) \leq A/n$, then the modulus of continuity of f satisfies $\omega(\delta) \leq \lambda\delta\,|\log\delta\,|$, for small δ.*

Proof. Inequalities (1) and (2) of the preceding proof are still valid when $\alpha = 1$. In the estimate (3) we now obtain instead

$$|f(x) - f(y)| \leq B\,|x - y|\,m + 4B2^{-m} \leq C[m\delta + 2^{-m}]$$

In order for this to be majorized by a term $\lambda\delta\,|\log\delta\,|$ it suffices again to keep $\delta 2^m$ bounded above and away from zero, as the reader should verify (Prob. 3). ∎

Bernstein's Theorem III. *If $f \in C_{2\pi}$ and $E_n(f) < An^{-p-\alpha}$ where p is a natural number and $\alpha \in (0,1)$, then f possesses continuous derivatives of orders $1, \ldots, p$, and the last one belongs to* Lip α.

Proof. Proceeding as in the first theorem, we have $f = \sum_{n=0}^{\infty} V_n$ and

$$\|V_n\| \leq \|P_{2^n} - f\| + \|f - P_{2^{n-1}}\|$$
$$\leq A(2^n)^{-p-\alpha} + A(2^{n-1})^{-p-\alpha} \leq B2^{-n(p+\alpha)}$$

Thus the Bernstein inequality applied p times gives

$$\|V_n^{(p)}\| \leq (2^n)^p\,\|V_n\| \leq B2^{-n\alpha}$$

The series $\sum_{n=0}^{\infty} V_n^{(p)}$ therefore converges uniformly, by the Weierstrass M test (Prob. 19 of Chap. 4, Sec. 3). Hence $f^{(p)}$ exists, is continuous, and is given by $\sum_{n=0}^{\infty} V_n^{(p)}$. It is obvious that

$$E_{2^m}(f^{(p)}) \leq \|f^{(p)} - \sum_{n=0}^{m} V_n^{(p)}\|$$

since $\sum_{n=0}^{m} V_n^{(p)}$ is a trigonometric polynomial of degree $\leq 2^m$. But

$$\|f^{(p)} - \sum_{n=0}^{m} V_n^{(p)}\| = \|\sum_{n=m+1}^{\infty} V_n^{(p)}\| \leq C2^{-m\alpha}$$

Consequently, $E_{2^m}(f^{(p)}) \leq C(2^m)^{-\alpha}$. Now for any n, we may find an m such that $2^m \leq n < 2^{m+1}$. Then

$$E_n(f^{(p)}) \leq E_{2^m}(f^{(p)}) \leq C(2^m)^{-\alpha} < C\left(\frac{n}{2}\right)^{-\alpha}$$

Applying Bernstein's Theorem I, we conclude that $f^{(p)} \in \text{Lip}\ \alpha$. ∎

To conclude this section we shall examine in more detail the anomaly which is due to the fact that Bernstein's Theorem II and Jackson's Theorem II are not converses of each other. These two theorems state respectively (for $f \in C_{2\pi}$ and ω the modulus of continuity of f)

$$\{nE_n(f)\} \quad \text{is bounded} \Rightarrow \frac{\omega(\delta)}{\delta \log \delta} \quad \text{is bounded}$$

$$\frac{\omega(\delta)}{\delta} \quad \text{is bounded} \Rightarrow \{nE_n(f)\} \quad \text{is bounded}$$

We owe to Zygmund the discovery of the precise structural property of f which is *necessary* and *sufficient* for the sequence $\{nE_n(f)\}$ to be bounded. Let us define as *Zygmund's modulus* the quantity

$$\omega^*(\delta) = \sup_x \sup_{|h| \leq \delta} |f(x+h) - 2f(x) + f(x-h)|$$

It is obvious that $\omega^*(\delta) \leq 2\omega(\delta)$; but for some functions f it happens that $\omega^*(\delta)/\delta$ is bounded while $\omega(\delta)/\delta$ is not. We may now state the desired result.

Zygmund's Theorem. *In order for a function f in the class $C_{2\pi}$ to satisfy an inequality*

$$E_n(f) \leq An^{-1} \qquad (n = 1, 2, \ldots)$$

it is necessary and sufficient that it satisfy an inequality

$$\omega^*(\delta) \leq B\delta \qquad (\delta > 0)$$

Proof. Suppose that $E_n(f) \leq An^{-1}$ for all n. As in Bernstein's Theorem I, we select trigonometric polynomials P_n such that $\| P_n - f \| = E_n(f)$, and then define $V_n = P_{2^n} - P_{2^{n-1}}$. Thus $f = \sum V_n$, and we have for any m,

$$\begin{aligned}
(4) \quad & |f(x+h) - 2f(x) + f(x-h)| \\
&= \left| \sum V_n(x+h) - 2V_n(x) + V_n(x-h) \right| \\
&\leq \sum_{n=0}^{m-1} |V_n(x+h) - 2V_n(x) + V_n(x-h)| + 4 \sum_{n=m}^{\infty} \| V_n \|
\end{aligned}$$

In order to estimate $\| V_n \|$ we write

$$\| V_n \| = \| P_{2^n} - P_{2^{n-1}} \| \le \| P_{2^n} - f \| + \| f - P_{2^{n-1}} \| \tag{5}$$
$$= E_{2^n}(f) + E_{2^{n-1}}(f)$$
$$\le A2^{-n} + A2^{-n+1} = 3A2^{-n}$$

In order to estimate the terms in the first sum of (4), we apply the mean-value theorem twice, then Bernstein's inequality (Chap. 3, Sec. 7), and finally (5) above, getting

$$| V_n(x + h) - 2V_n(x) + V_n(x - h) |$$
$$= | \{V_n(x + h) - V_n(x)\} - \{V_n(x) - V_n(x - h)\} |$$
$$= | h | | V_n'(\xi) - V_n'(\eta) |$$
$$= | h | | \xi - \eta | | V_n''(\zeta) |$$
$$\le 2h^2 \| V_n'' \|$$
$$\le 2h^2 2^{2n} \| V_n \|$$
$$\le 6Ah^2 2^n$$

Returning to (4) with these estimates, we arrive at

$$| f(x + h) - 2f(x) + f(x - h) | \le 6Ah^2 \sum_{n=0}^{m-1} 2^n + 12A \sum_{n=m}^{\infty} 2^{-n}$$
$$\le 6Ah^2 2^m + 24A2^{-m}$$

In order for this last to be majorized by a term $B | h |$, it will be sufficient that m vary with $| h |$ in such a manner that both $2^{-m} | h^{-1} |$ and $2^m | h |$ remain bounded above.

For the second half of the proof, we suppose that $\omega^*(\delta) \le B\delta$ for all $\delta > 0$. As in the proof of Jackson's Theorem II (Chap. 4, Sec. 6), we apply to f the "averaging" operation:

$$\phi(x) \equiv \frac{1}{2\delta} \int_{x-\delta}^{x+\delta} f(t)\, dt \equiv \frac{1}{2\delta} \int_0^{\delta} [f(x + t) + f(x - t)]\, dt$$

Now apply the same operation to ϕ to get the "second average" of f:

$$\psi(x) \equiv \frac{1}{2\delta} \int_0^{\delta} [\phi(x + s) + \phi(x - s)]\, ds$$
$$\equiv \frac{1}{4\delta^2} \int_0^{\delta} \int_0^{\delta} [f(x + s + t) + f(x + s - t)$$
$$+ f(x - s + t) + f(x - s - t)]\, dt\, ds$$

Since f is continuous, ψ has a continuous second derivative. In fact, $\phi'(x) = (1/2\delta)[f(x+\delta) - f(x-\delta)]$ so that $\psi'(x) = (1/2\delta)[\phi(x+\delta) - \phi(x-\delta)]$ and $\psi''(x) = (1/2\delta)[\phi'(x+\delta) - \phi'(x-\delta)] = (1/4\delta^2)[f(x+2\delta) - 2f(x) + f(x-2\delta)]$. It follows at once that $\|\psi''\| \le (1/4\delta^2)\omega^*(2\delta) \le B/2\delta$.

Now we must prove that $\|\psi - f\| \le B\delta$. In fact,

$$\psi(x) - f(x) = \frac{1}{4\delta^2}\int_0^\delta\int_0^\delta [f(x+s+t) + f(x+s-t) + f(x-s+t) + f(x-s-t) - 4f(x)]\,dt\,ds$$

Since the integrand may be written as

$$[f(x+s+t) - 2f(x+s) + f(x+s-t)] + [f(x-s+t) - 2f(x-s) + f(x-s-t)] + [2f(x+s) - 4f(x) + 2f(x-s)]$$

its modulus cannot exceed $\omega^*(t) + \omega^*(t) + 2\omega^*(s) \le 2B(t+s) \le 4B\delta$, whence $|\psi(x) - f(x)| \le B\delta$.

Now to complete the proof, let P be the trigonometric polynomial of degree $\le n$ which best approximates ψ. By Jackson's Theorem IV (Chap. 4, Sec. 6),

$E_n(\psi) \le \frac{\pi}{2}(n+1)^{-2}\|\psi''\|$. Hence

$$E_n(f) \le \|P - f\| \le \|P - \psi\| + \|\psi - f\| \le \frac{\pi}{2}(n+1)^{-2}\frac{B}{2\delta} + B\delta$$

If $\delta = n^{-1}$, then we reach the conclusion that $E_n(f) \le Bn^{-1}(1 + \frac{1}{4}\pi)$. ■

Problems

1. Prove that Lip α is a linear subspace of $C[a,b]$ which is also closed under the following operations.

$$(f \vee g)(x) = \max\{f(x), g(x)\}$$
$$(f \wedge g)(x) = \min\{f(x), g(x)\}$$
$$|f|(x) = |f(x)|$$
$$(fg)(x) = f(x)g(x)$$

2. Try to prove the first Bernstein theorem by writing

$$|f(x) - f(y)| \le |f(x) - P_n(x)| + |P_n(x) - P_n(y)| + |P_n(y) - f(y)| \le 2\|f - P_n\| + |P_n'(\xi)|\,|x - y| \le 2E_n(f) + n\|P_n\|\,|x - y| \quad \text{etc.}$$

3. Complete the proof of the second Bernstein theorem.
4. Zygmund's theorem shows that the class $\Lambda_* = \{f \in C_{2\pi}: \sup_{\delta>0} \delta^{-1}\omega^*(\delta) < \infty\}$ is identical with the class $\{f \in C_{2\pi}: \sup_n nE_n(f) < \infty\}$. But the proof can be made to show considerably more, viz., that for all $f \in \Lambda_*$,

$$\sqrt{\frac{2}{\pi}} \sup_n nE_n(f) \leq \sup_{\delta>0} \delta^{-1}\omega^*(\delta) \leq 36 \sup_n nE_n(f)$$

5. In order that a function f from class $C_{2\pi}$ shall possess derivatives of all orders, it is necessary and sufficient that for each fixed k, $n^k E_n(f) \to 0$ as $n \to \infty$.
6. Bernstein's Theorem I (Chap. 6, Sec. 3) and Jackson's Theorem III (Chap. 4, Sec. 6) tell us that if $0 < \alpha < 1$, then the class $\text{Lip}\,\alpha = \{f \in C_{2\pi}: \sup_{\delta>0} \delta^{-\alpha}\omega(\delta) < \infty\}$ is identical with the class $\{f \in C_{2\pi}: \sup_n n^\alpha E_n(f) < \infty\}$. The proofs actually yield two positive constants, H and K, such that for $f \in \text{Lip}\,\alpha$,

$$H \sup_n n^{-\alpha}E_n(f) \leq \sup_{\delta>0} \delta^{-\alpha}\omega(\delta) \leq K \sup_n n^{-\alpha}E_n(f)$$

7. What structural properties must f have if $E_n(f) \leq A\theta^n$, where $\theta < 1$?
8. Does every continuous function on $[a,b]$ satisfy a Lipschitz condition of the form

$$|f(x) - f(y)| \leq A\,|x - y|^\alpha \qquad (\alpha > 0)$$

4. Polygonal Approximation and Bases in $C[a,b]$

It is an immediate consequence of the Weierstrass theorem that every element f in $C[a,b]$ can be written in the form $f = \sum_{n=0}^{\infty} V_n$, where each V_n is a polynomial. For example, we may put $V_0 = P_0$ and $V_n = P_n - P_{n-1}$ for $n \geq 1$, P_n being selected as the polynomial of degree $\leq n$ which best approximates f. Up to this point, however, we have not discovered how to expand a given continuous function in the form $f = \sum c_n V_n$ with any *fixed* family of polynomials V_n. In the case of the Tchebycheff polynomials, the best that we could prove was that if $\omega(\delta) \log \delta \to 0$, then an expansion $f = \sum c_n T_n$ was valid. It should be emphasized that the convergence in such infinite series is in terms of the *uniform* norm. Thus in the space $C[-1,1]$, we mean by $f = \sum c_n T_n$ that

$$\lim_{n\to\infty} \max_{-1\leq x\leq 1} \left| f(x) - \sum_{k=0}^{n} c_k T_k(x) \right| = 0$$

Of course, we need hardly mention that only for very few continuous functions is it possible to write $f = \sum c_n t^n$. Thus neither the monomials x^n nor the Tchebycheff polynomials have the property that every continuous function may be expanded uniformly in a series of them. What we seek is a *basis*, in the sense of the following definition.

Definition. *A Schauder basis for a normed linear space E is a (finite or infinite) sequence $\{g_1,g_2,\ldots\}$ such that each element $f \in E$ may be written uniquely in the form $f = \sum_{n=1}^{\infty} c_n g_n$.*

We shall prove presently that $C[a,b]$ does in fact possess a basis of polynomials (theorem of Krein, Milman, and Rutman). It may seem strange at first that such a theorem should be devoid of significance in practical problems of approximation. There seems to be a general principle at work here, forcing any linear process of approximation that converges for *all* continuous functions to be necessarily inefficient for *well-behaved* functions. Stated otherwise, there appears to exist a reciprocal relation between generality and efficiency. The Bernstein polynomials, the Tchebycheff series, and the Fejér method all illustrate this principle. In the present case we shall see that the (nonpolynomial) basis explicitly constructed by Schauder is impractical because the base functions are not especially easy to compute and are nondifferentiable. The basis of polynomials which we will construct suffers from the fact that the base polynomials are necessarily of high degree.

In the course of this section we shall have to deal with *polygonal* approximations to a function $f \in C[a,b]$. Such an approximation is linear over subintervals of $[a,b]$ and agrees with f at the ends of these subintervals. The ends of the subintervals are called *nodes*, and we always assume that the points a and b are included among the nodes. Such an approximation has the general appearance shown in the following figure.

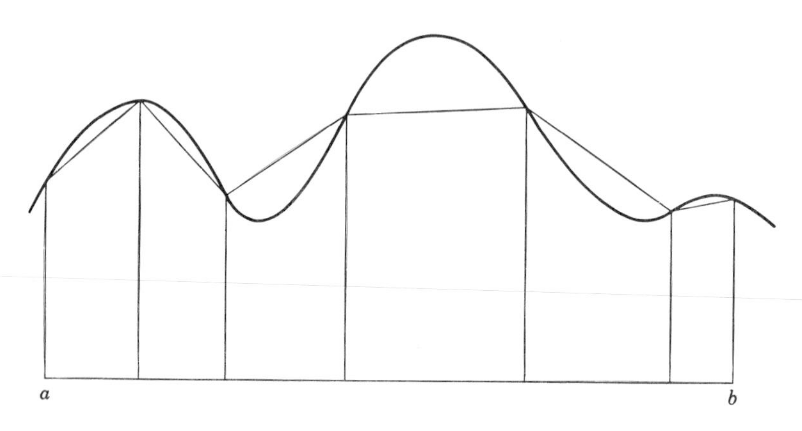

Another bit of terminology: A linear operator L is said to be *bounded* if an inequality $\|Lf\| \leq \lambda \|f\|$ is valid (of course, λ is to be independent of f). The least such λ is called the *bound* of L and is denoted by $\|L\|$. Boundedness implies continuity since $\|Lf - Lg\| = \|L(f-g)\| \leq \lambda \|f-g\|$. The converse is also true but is left to the problems.

Theorem 1. *Given points* $a = x_0 < x_1 < \cdots \leq x_n = b$, *let* Lf *denote the polygonal approximation to* f *with nodes* x_i. *Then* L *is a monotone linear operator of bound* 1.

Proof. Put $\phi = Lf$. Then on $[x_i,x_{i+1}]$, ϕ is given by

$$\phi(x) = \frac{x - x_i}{x_{i+1} - x_i} f(x_{i+1}) + \frac{x_{i+1} - x}{x_{i+1} - x_i} f(x_i)$$

Since f enters this expression linearly, L is linear. It is clear also from this expression that if $f \geq g$, then $Lf \geq Lg$. Finally, the above equation yields immediately the fact that each value of ϕ is a convex linear combination of two values of f, and consequently

$$|\phi(x)| \leq \max\{|f(x_i)|, |f(x_{i+1})|\} \leq \|f\|$$

This proves that $\|Lf\| = \|\phi\| \leq \|f\|$. No constant smaller than $\lambda = 1$ is possible here, as may be seen by taking $f(x) = 1$. ∎

In the next theorem, we have a *dense* sequence $\{x_0,x_1,\ldots\}$ in the interval $[a,b]$. Thus each point of $[a,b]$ is the limit of some subsequence, x_{n_k}. For example, on $[0,1]$ the sequence

$$\{0,1,\tfrac{1}{2},\tfrac{1}{3},\tfrac{2}{3},\tfrac{1}{4},\tfrac{3}{4},\tfrac{1}{5},\tfrac{2}{5},\tfrac{3}{5},\tfrac{4}{5},\tfrac{1}{6},\tfrac{5}{6},\tfrac{1}{7},\ldots\}$$

has the required property. We assume $x_i \neq x_j$ for $i \neq j$.

Theorem 2. *Let* $\{x_0 = a, x_1 = b, x_2, x_3, \ldots\}$ *be a dense sequence in* $[a,b]$. *For each* $n \geq 1$ *let* $L_n f$ *denote the polygonal function which agrees with* f *at nodes* $x_0, \ldots, x_n$. *Then* $L_n f \to f$ *for all* $f \in C[a,b]$.

Proof. The points $x_0, \ldots, x_n$ divide $[a,b]$ into a number of subintervals of which the longest has say length δ_n. Then $\delta_n \downarrow 0$. Given $\epsilon > 0$, select $\delta > 0$ such that $|f(x) - f(y)| < \epsilon$ whenever $|x - y| < \delta$. Then select N so that $\delta_N < \delta$, and let $n \geq N$. Given $x \in [a,b]$, it lies in one of the subintervals created by the nodes $x_0, \ldots, x_n$. Say $x_i \leq x \leq x_j$. Then $|x_i - x_j| \leq \delta_n \leq \delta_N < \delta$, and consequently, using $\phi = L_n f$,

$$\begin{aligned} |f(x) - \phi(x)| &\leq |f(x) - f(x_i)| + |f(x_i) - \phi(x_i)| + |\phi(x_i) - \phi(x)| \\ &\leq \epsilon + 0 + |\phi(x_i) - \phi(x_j)| \\ &\leq \epsilon + |f(x_i) - f(x_j)| < 2\epsilon \end{aligned}$$ ∎

Schauder's Theorem. $C[a,b]$ *possesses a basis.*

Proof. Define L_n as in the preceding theorem, with $(L_0 f)(x) = f(a)$. We are going to define base functions $g_0, g_1, \ldots$ in such a way that the following will be true:

$$f = \lim L_n f = L_0 f + \sum_{n=1}^{\infty} (L_n - L_{n-1})f = \sum_{n=0}^{\infty} c_n g_n \tag{1}$$

Put $g_0(x) = 1$ and $g_1(x) = (x - a)/(b - a)$. For $n \geq 2$, consider the subintervals of $[a,b]$ created by the nodes $x_0, x_1, \ldots, x_{n-1}$. The point x_n lies in one of these. Outside that interval g_n is to be zero. At x_n, g_n takes the value 1,

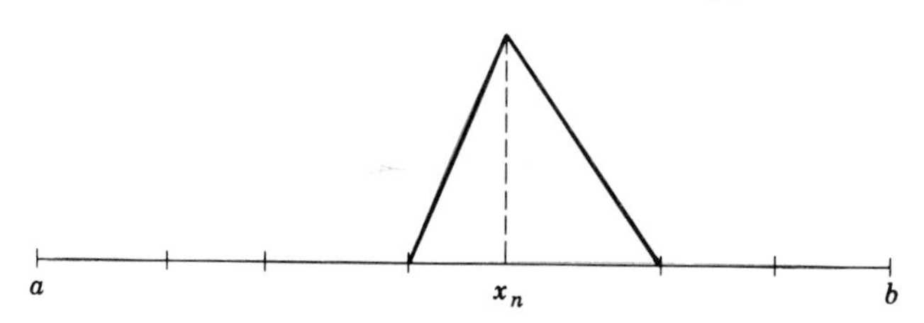

and then varies linearly back to zero as in the figure. In order to prove equation (1), it will be sufficient to prove the following,

$$L_0 f = f(x_0) g_0 \tag{2}$$

$$L_n f = L_{n-1} f + (f - L_{n-1} f)(x_n) g_n \qquad (n \geq 1)$$

for then the coefficients c_n are given by $c_0 = f(x_0)$ and $c_n = (f - L_{n-1}f)(x_n)$. For the proof of (2), define $f_0, f_1, \ldots$ inductively by $f_0 = f(x_0)g_0$ and $f_n = f_{n-1} + (f - f_{n-1})(x_n)g_n$. We will show that $f_n = L_n f$. Now f_n is a polygonal function with corners possible only at the nodes $x_0, \ldots, x_n$. It is therefore sufficient to show that $f_n(x_i) = f(x_i)$ for $i \leq n$. This is obvious for $n = 0$. If it is true for $n - 1$, then it is true for n since

$$\begin{aligned} f_n(x_n) &= f_{n-1}(x_n) + f(x_n)g_n(x_n) - f_{n-1}(x_n)g_n(x_n) \\ &= f_{n-1}(x_n) + f(x_n) - f_{n-1}(x_n) = f(x_n) \\ f_n(x_i) &= f_{n-1}(x_i) + (f - f_{n-1})(x_i)g_n(x_i) \\ &= f_{n-1}(x_i) = f(x_i) \qquad (i < n) \end{aligned}$$

This establishes equation (1). There remains the unicity of the coefficients c_n. If it is possible for a function f to have two expansions $\sum a_n g_n$ and $\sum b_n g_n$, then with $c_n = a_n - b_n$ we have $\sum c_n g_n = 0$. Let c_N be the first nonzero coefficient. Then $\sum_{n=N}^{\infty} c_n g_n = 0$. Remembering that $g_n(x_N) = 0$ for $n > N$, we obtain then the contradiction $0 = \sum_{n=N}^{\infty} c_n g_n(x_N) = c_N g_N(x_N) = c_N$. ∎

In the next theorem we shall need the formulas $c_n = (f - L_{n-1}f)(x_n)$ for $n \geq 1$ and $c_0 = f(x_0)$. Actually, the c_n are functions of f, and render the equation

$$f = \sum_{n=0}^{\infty} c_n(f) g_n$$

true throughout $C[a,b]$. From Theorem 1 and these formulas we see that $| c_n(f) | \leq 2 \| f \|$. It is clear from the formulas that each c_n is a *linear* function: $c_n(\alpha f + \beta h) = \alpha c_n(f) + \beta c_n(h)$. We also require the following lemma, in which I denotes the identity operator.

Lemma. *If L is a linear operator from a Banach space into itself such that $\| L \| < 1$, then $(I - L)^{-1}$ exists and is bounded.*

Proof. If the series $Vf = \sum_{n=0}^{\infty} L^n f$ converges for arbitrary f, then V clearly has the property $LV + I = V$, so that $V = (I - L)^{-1}$. Now

$$\left\| \sum_{n=N}^{M} L^n f \right\| \leq \sum_{n=N}^{\infty} \| L^n f \| \leq \| f \| \sum_{n=N}^{\infty} \| L \|^n = \| f \| \, \| L \|^N \frac{1}{1 - \| L \|}$$

Hence the series for Vf converges, and $\| V \| \leq 1/(1 - \| L \|)$. Certain details in this proof are explored in Probs. 3 to 5. ∎

Theorem of Krein, Milman, and Rutman. *The space $C[a,b]$ possesses a basis of polynomials.*

Proof. Consider the basis $\{g_0, g_1, \ldots\}$ which was defined in Schauder's theorem. By the Weierstrass theorem, we may find polynomials P_0, P_1, ... such that $\sum_{n=0}^{\infty} \| P_n - g_n \| < \frac{1}{2}$. (The subscripts do *not* denote the degrees.) Using the functions c_n introduced above, we define an operator S by putting

$$Sf = \sum_{n=0}^{\infty} c_n(f) P_n$$

If we can show that the operator S has a continuous inverse, then the proof will be completed by observing that each element f has the expansion

$$f = SS^{-1}f = \sum c_n(S^{-1}f) P_n$$

and that this expansion is unique because

$$f = \sum a_n P_n \Rightarrow S^{-1}f = \sum a_n g_n \Rightarrow a_n = c_n(S^{-1}f)$$

In order to show that S^{-1} exists and is continuous, we note first that

$$\| (I - S)f \| \leq \sum_{n=0}^{\infty} | c_n(f) | \, \| g_n - P_n \| \leq 2 \| f \| \sum_{n=0}^{\infty} \| g_n - P_n \| < \| f \|$$

so that $\| I - S \| < 1$, and then apply the lemma. ∎

Problems

1. If $\{g_1,g_2,\ldots\}$ is a fundamental set in $C[a,b]$, then there exists a basis for $C[a,b]$ consisting of linear combinations of the elements $g_1,g_2,\cdots$.
2. A linear operator L which is continuous at 0 must be bounded. *Hint*: If L is not bounded, then there exist vectors f_n satisfying $\| Lf_n \| > n \| f_n \|$. Look at $f_n/(n \| f_n \|)$.
3. A linear operator which is continuous at 0 is continuous everywhere, in fact, uniformly continuous.
4. The bounded linear operators from one normed linear space E_1 into another E_2 form a normed linear space if we define

$$(\alpha_1 L_1 + \alpha_2 L_2)(f) = \alpha_1 L_1 f + \alpha_2 L_2 f \qquad \| L \| = \sup_{\|f\|=1} \| Lf \|$$

5. If L_n are linear operators such that the limit $Lf = \lim\limits_{n\to\infty} L_n f$ exists for all f, then L is also linear.
6. Prove that the norm defined in Prob. 4 has the property $\| L^n \| \leq \| L \|^n$, when $E_1 = E_2$.
7. Prove that the functions c_n which were defined just after Schauder's theorem have the property that $\sum c_n(Lf)g_n = \sum c_n(f)Lg_n$ for any continuous linear operator L.
8. Prove Theorem 2 by use of the monotone-operator theorem (Chap. 3, Sec. 3).
9. The polygonal functions with joints at points $x_1,\ldots,x_n$ form a linear space of dimension n. [Schwerdtfeger, 1960]

5. The Kharshiladze-Lozinski Theorems

We turn now to some recent theorems of "negative character" which assert the impossibility of constructing operators with certain desirable properties. In order to explain what these "desirable" properties are, let us define for each natural number n a nonlinear operator $\mathfrak{Z}_n$ in the space $C[a,b]$ by specifying that $\mathfrak{Z}_n f$ is that polynomial of degree $\leq n$ which best approximates f in the uniform norm. This operator has, of course, the property

(*i*) The range of $\mathfrak{Z}_n$ is the linear span of $\{1,x,\ldots,x^n\}$.

Since $\mathfrak{Z}_n f = f$ when f is already a polynomial of degree $\leq n$, $\mathfrak{Z}_n$ is *idempotent*:

(*ii*) $\mathfrak{Z}_n^2 = \mathfrak{Z}_n$.

The most important property of $\mathfrak{Z}_n$ is given by the Weierstrass theorem:

(*iii*) $\mathfrak{Z}_n f \to f$ for all $f \in C[a,b]$.

Finally, we have seen on page 82 that $\mathfrak{Z}_n$ is continuous:

(*iv*) $\| \mathfrak{Z}_n f - \mathfrak{Z}_n g \| \leq c(n,f) \| f - g \|$.

The property of linearity is conspicuously lacking. For example, on $[0,\pi]$ we have $\mathfrak{Z}_0(\cos x) = 0$ and $\mathfrak{Z}_0(\sin x) = \frac{1}{2}$, while $\mathfrak{Z}_0(\cos x + \sin x) = \frac{1}{2}(\sqrt{2} - 1)$. It is therefore natural to ask whether a *linear* operator can exist having properties (*i*) to (*iv*). Such an operator would presumably be of some practical importance. However, it is now known that no such operator exists.

This theorem is proved below after some preliminaries of great interest in themselves. In the following theorem we use the word *sphere* to mean a set of the form $\{x: d(x,x_0) \leq r\}$, with $r > 0$.

Baire's Theorem. *If a complete metric space is expressed as a countable union of closed sets, then at least one of the closed sets must contain a sphere.*

Proof. If the theorem is false, there exists a complete metric space X which can be expressed in the form $X = \bigcup_{n=1}^{\infty} F_n$, where each F_n is a closed set which contains no sphere. The complements, $\mathcal{O}_n = X \sim F_n$, are open sets which have no point in common but intersect every sphere. We can therefore define a sequence of closed spheres $S_n = \{x: d(x,x_n) \leq \epsilon_n\}$ such that $S_1 \subset \mathcal{O}_1$, $S_2 \subset S_1 \cap \mathcal{O}_2$, $S_3 \subset S_2 \cap \mathcal{O}_3$, etc., and $\epsilon_n \downarrow 0$. The sequence $x_1, x_2, \ldots$ is a Cauchy sequence because if $i, j > n$, then $x_i, x_j \in S_n$, and $d(x_i,x_j) \leq 2\epsilon_n$. Consequently the sequence converges, say to x^*. Since $x_i \in S_n$ when $i > n$, it follows that $x^* \in S_n$ and that $x^* \in \mathcal{O}_n$. This argument is valid for all n, and thus x^* is a point common to all $\mathcal{O}_n$. ∎

Uniform-boundedness Theorem. *Let $\{L_\alpha\}$ be a collection of bounded linear operators from one Banach space into another. If $\sup_\alpha \| L_\alpha f \|$ is finite for each f, then $\sup_\alpha \| L_\alpha \|$ is finite.*

Proof. For $n = 1, 2, \ldots$ let $H_n = \{f \in E: \sup_\alpha \| L_\alpha f \| \leq n\}$. The sets H_n are closed and increasing. By hypothesis every f is in some H_n, and consequently $E = \bigcup_{n=1}^{\infty} H_n$. By the Baire theorem, some H_N contains a sphere $S \equiv \{f: \| f - g \| \leq \epsilon\}$. If $\| f \| \leq 1$, then $g + \epsilon f \in S$, and thus $\| L_\alpha f \| = \| \epsilon^{-1} L_\alpha(g + \epsilon f - g) \| \leq \epsilon^{-1} \| L_\alpha(g + \epsilon f) \| + \epsilon^{-1} \| L_\alpha g \| \leq 2N/\epsilon$. This proves that $\| L_\alpha \| \leq 2N/\epsilon$. ∎

It is convenient to term a linear operator L a *projection* if $L^2 = L$. Such an operator acts like the identity on its range, for if f is in the range of L, then $f = Lg$ for some g, whence $Lf = L^2 g = Lg = f$. We say that L is a *projection onto G* if G is the range of L.

The next theorem states an important extremal property of the Fourier-series projection. This operator, S_n, is such that for any continuous 2π-periodic function f, $S_n f$ is the nth truncation of its Fourier series. Explicit formulas for S_n have been given on page 120, and the estimate $\| S_n \| \leq 3 + \log n$ was derived on page 147.

Theorem. *Let L be any bounded linear projection of the space $C_{2\pi}$ onto the subspace M_n of trigonometric polynomials of degree $\leq n$. Let S_n denote the Fourier-series projection into M_n. Then $\| S_n \| \leq \| L \|$.*

Proof. Define operators T_λ and Φ by the equations

$$(T_\lambda f)(x) = f(x + \lambda)$$

$$(\Phi f)(x) = \frac{1}{2\pi} \int_{-\pi}^{\pi} (T_{-\lambda} L T_\lambda f)(x)\, d\lambda$$

If we can establish that $\Phi = S_n$, then we will be finished because

$$\| S_n f \| = \| \Phi f \| = \max_{0 \leq x \leq 2\pi} \left| \frac{1}{2\pi} \int_{-\pi}^{\pi} (T_{-\lambda} L T_\lambda f)(x)\, d\lambda \right|$$

$$\leq \| T_{-\lambda} L T_\lambda f \| \leq \| L \| \, \| f \|$$

In order to prove that $\Phi = S_n$ it will be enough to prove that $\Phi f_k = S_n f_k$, where f_k is the function $f_k(x) = e^{ikx}$ $(k = 0, \pm 1, \pm 2, \ldots)$, since this family of functions is fundamental in $C_{2\pi}$, while the operators Φ and S_n are linear and continuous. If $|k| \leq n$, then $S_n f_k = f_k$. On the other hand, $T_\lambda f_k \in M_n$, so that $L T_\lambda f_k = T_\lambda f_k$. Thus $T_{-\lambda} L T_\lambda f_k = f_k$ and $\Phi f_k = f_k$. (In the integration, the integrand is independent of λ.) Suppose next that $|k| > n$. Then $S_n f_k = 0$. Since $T_\lambda f_k = e^{ik\lambda} f_k$, it follows that $(T_{-\lambda} L T_\lambda f_k)(x) = e^{ik\lambda}(L f_k)(x - \lambda)$. But $L f_k \in M_n$, and consequently, as functions of λ, $e^{ik\lambda}$ is orthogonal to $(L f_k)(x - \lambda)$. Hence $\Phi f_k = 0$. ∎

The operator S_n was shown on page 120 to have the form

$$(S_n f)(x) = \frac{1}{\pi} \int_{-\pi}^{\pi} f(t + x) D_n(t)\, dt$$

where $D_n(t) = \sin (n + \frac{1}{2})t / 2 \sin \frac{1}{2} t$. We have already observed (page 147) that from the integral form of S_n it follows that

$$\| S_n f \| \leq \| f \| \frac{1}{\pi} \int_{-\pi}^{\pi} | D_n(t) | \, dt \equiv \lambda_n \| f \|$$

(The number λ_n is known as the nth *Lebesgue constant.*) For the function $f(t) = \operatorname{sgn} D_n(t)$ we obtain $\| S_n f \| = \lambda_n \| f \|$, and by taking continuous functions f_ϵ close to f we can get $\| S_n f_\epsilon \| \geq (\lambda_n - \epsilon) \| f_\epsilon \|$. Thus λ_n is the least constant λ such that $\| S_n f \| \leq \lambda \| f \|$, or in short, $\| S_n \| = \lambda_n$. On page 147 we proved that $\lambda_n \leq 3 + \log n$. Now we seek a lower bound on λ_n.

Lemma. *The Lebesgue constants satisfy $\lambda_n > (4/\pi^2) \log n$*

Proof. The change of variable $t \to 2x$ and the inequality $\sin x \leq x$ on $[0,\pi/2]$ give us

$$\lambda_n = \frac{2}{\pi}\int_0^{\pi/2} \frac{|\sin (2n+1)x|}{\sin x}\,dx \geq \frac{2}{\pi}\int_0^{\pi/2} \frac{|\sin (2n+1)x|}{x}\,dx$$

The change of variable $x \to \pi x/(2n+1)$ then yields

$$\lambda_n \geq \frac{2}{\pi}\int_0^{n+1/2} \frac{|\sin \pi x|}{x}\,dx > \frac{2}{\pi}\int_0^{n} \frac{|\sin \pi x|}{x}\,dx$$

Breaking the interval into n subintervals, we obtain

$$\begin{aligned}\lambda_n &\geq \frac{2}{\pi}\left(\int_0^1 + \int_1^2 + \cdots + \int_{n-1}^n\right)\frac{|\sin \pi x|}{x}\,dx \\ &= \frac{2}{\pi}\int_0^1 \left(\frac{1}{x} + \frac{1}{x+1} + \cdots + \frac{1}{x+n-1}\right)\sin \pi x\,dx \\ &\geq \frac{2}{\pi}\int_0^1 \left(\frac{1}{1} + \frac{1}{2} + \cdots + \frac{1}{n}\right)\sin \pi x\,dx \\ &\geq \frac{2}{\pi}\log n \int_0^1 \sin \pi x\,dx = \frac{4}{\pi^2}\log n\end{aligned}$$

That $1 + \frac{1}{2} + \frac{1}{3} + \cdots + 1/n > \log(n+1)$ may be seen from a graph of the function $1/(x+1)$ or by induction. (Cf. Prob. 1.) ∎

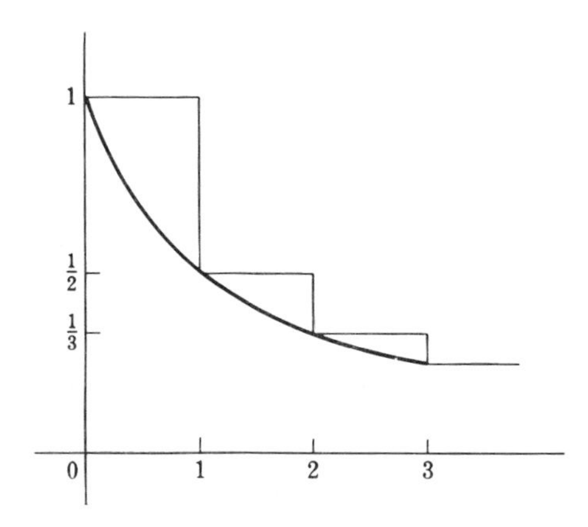

Kharshiladze-Lozinski Theorem I. *For each n let there be given a continuous linear projection L_n of $C_{2\pi}$ onto the space of trigonometric polynomials having degree $\leq n$. Then there exists an $f \in C_{2\pi}$ for which $\| L_n f \|$ is unbounded as $n \to \infty$.*

Proof. By the preceding two theorems, $\| L_n \|$ is unbounded because

$$\| L_n \| \geq \| S_n \| \geq \frac{4}{\pi^2} \log n$$

If $\| L_n f \|$ were bounded for all f, then by the uniform-boundedness theorem, $\| L_n \|$ would be bounded. ∎

Theorem. *Let L be a bounded linear projection of the even part of $C_{2\pi}$ onto the even trigonometric polynomials of degree $\leq n$. Then $\| I - L \| \geq \frac{1}{2}(\lambda_n + 1) > (2/\pi^2) \log n + \frac{1}{2}$.*

Proof. Define the linear operator

$$(\Phi f)(x) = \frac{1}{2\pi} \int_{-\pi}^{\pi} [T_\lambda (I - L)(T_{-\lambda} + T_\lambda) f](x)\, d\lambda$$

in which T_λ denotes the translation operator, $(T_\lambda f)(x) = f(x + \lambda)$. The crux of the proof is in verifying the equation $\Phi = I - S_n$, where S_n is the Fourier-series projection. After that we write

$$\| (I - S_n) f \| = \| \Phi f \| \leq 2 \| I - L \| \, \| f \|$$

whence $2 \| I - L \| \geq \| I - S_n \| = \lambda_n + 1$. (A proof of this last equality is left to Prob. 12.) In order to prove that $\Phi = I - S_n$, it suffices to prove that $\Phi f_k = (I - S_n) f_k$, where $f_k(x) = \cos kx$, since these functions form a fundamental set and the operators in question are continuous. Thus we must show that $\Phi f_k = 0$ when $k \leq n$, and $\Phi f_k = f_k$ for $k > n$. We have

$$(T_\lambda f_k)(x) = \cos k(x + \lambda) = \cos k\lambda \cos kx - \sin k\lambda \sin kx$$

Hence $(T_{-\lambda} + T_\lambda) f_k = 2 \cos k\lambda \cdot f_k$ and $T_\lambda (I - L)(T_{-\lambda} + T_\lambda) f_k = 2 \cos k\lambda \cdot T_\lambda (f_k - L f_k)$. Now if $k \leq n$, $f_k = L f_k$ so that $\Phi f_k = 0$. If $k > n$, then

$$(\Phi f_k)(x) = \frac{1}{\pi} \int_{-\pi}^{\pi} \cos k\lambda [\cos k\lambda \cos kx - \sin k\lambda \sin kx - (L f_k)(x + \lambda)]\, d\lambda$$

Since $(L f_k)(x + \lambda)$ is a trigonometric polynomial of degree $\leq n$ in λ, the integration involving it vanishes, by orthogonality. By the orthonormality relations the remaining integration yields $\cos kx = f_k(x)$. ∎

Kharshiladze-Lozinski Theorem II. *For each n, let L_n denote a bounded linear projection of $C[a,b]$ onto the subspace of algebraic polynomials of degree $\leq n$. Then there exists an $f \in C[a,b]$ for which the sequence $\| f - L_n f \|$ is unbounded.*

Proof. We define a map M from $C[a,b]$ onto the even part of $C_{2\pi}$ by setting

$$(Mf)(\theta) = f\left(\frac{a + b}{2} + \frac{b - a}{2} \cos \theta\right)$$

The map M is an isometric isomorphism. That is, it is one-to-one, linear, and has the property $\| Mf \| = \| f \|$ (Prob. 4). Now the operators $L_n^* = ML_nM^{-1}$ are bounded linear projections of the even part of $C_{2\pi}$ onto the even trigonometric polynomials of degree $\leq n$. By the preceding theorem, then, $\| I - L_n^* \| \to \infty$. Hence $\| I - L_n \| \to \infty$, and by the uniform-boundedness theorem $\| f - L_n f \|$ is unbounded for some f. ∎

***Corollary* [*Nikolaev*].** *For some fixed weight function w, let $L_n f$ denote the polynomial of degree $\leq n$ for which*

$$\int_a^b (f - L_n f)^2 w$$

is a minimum. Then for some continuous f, the (uniform) norms $\| f - L_n f \|$ are unbounded.

Proof. The operators L_n are certainly projections of $C[a,b]$ onto the polynomials of degree $\leq n$. If $Q_0, Q_1, \ldots$ is the sequence of orthonormal polynomials for the inner product $\langle f,g \rangle = \int_a^b f(x)g(x)w(x)$, then L_n is given by the formula (page 14)

$$L_n f = \sum_{k=0}^{n} \langle f,Q_k \rangle Q_k$$

From this the linearity and boundedness of L_n are clear. Thus by the second Kharshiladze-Lozinski theorem $\| f - L_n f \|$ is unbounded for some f. ∎

***Corollary* [*Faber*].** *For each n let a system of $n + 1$ distinct nodes $\xi_0^{(n)}, \ldots, \xi_n^{(n)}$ be specified in the interval $[a,b]$. Let $L_n f$ denote the polynomial of degree $\leq n$ which interpolates to f at the given nodes. Then for some $f \in C[a,b]$, $\| L_n f - f \|$ is unbounded.*

Proof. That $L_n f = f$ when f is a polynomial of degree $\leq n$ follows from the uniqueness part of the interpolation theorem (Chap. 3, Sec. 2). The linearity and boundedness of L_n follow at once from the formulas on page 59:

$$L_n f = \sum_{i=0}^{n} f(\xi_i^{(n)}) l_i^{(n)}$$

Now we apply the Kharshiladze-Lozinski theorem again. ∎

***Corollary* [*Berman*].** *There do not exist bounded linear operators L_n mapping $C[a,b]$ into the polynomials of degree $\leq n$ such that the ratio*

$$\| f - L_n f \| / \| f - \mathfrak{Z}_n f \|$$

remains bounded for all $f \in C[a,b]$.

Proof. (The operator $\mathfrak{I}_n$ was defined on page 210.) If to each f there corresponds a constant c_f such that $\| f - L_n f \| \le c_f \| f - \mathfrak{I}_n f \|$ then L_n must be a projection since $\mathfrak{I}_n f = f$ when f is a polynomial of degree $\le n$. But the same inequality would show that $\| f - L_n f \|$ remains bounded for each f, in contradiction with the Kharshiladze-Lozinski theorem. ∎

In view of these negative results, it is natural to inquire about the sequence L_n (projections of $C[a,b]$ into the polynomials of degree $\le n$), for what functions f is it true that $L_n f \to f$? Certainly this is true for all polynomials. And it is true for those continuous functions which are approached sufficiently rapidly by polynomials.

Lebesgue's Theorem. *Let $G_0 \subset G_1 \subset \cdots$ be a sequence of subspaces in $C[a,b]$, and (for each n) L_n a projection of $C[a,b]$ into G_n. In order that $L_n f \to f$, it is sufficient that* $\| L_n \| \cdot \operatorname{dist}[f, G_n] \to 0$.

Proof. Let g_n be any point of G_n of minimum distance from f. Then

$$\begin{aligned} \| L_n f - f \| &= \| L_n(f - g_n) + (g_n - f) \| \\ &\le \| L_n \| \, \| f - g_n \| + \| g_n - f \| \\ &\le 2 \| L_n \| \, \| f - g_n \| \to 0 \end{aligned}$$

∎

Problems

1. Prove by induction that $1 + \frac{1}{2} + \cdots + 1/n > \log(n + 1)$. *Hint*: The inequality $e^x > 1 + x$ $(x > 0)$ will be useful, and can be obtained from the mean-value theorem.
2. Consider the properties (*i*) to (*iv*) described in the first paragraph of this section. Which of these properties are possessed by the Bernstein operators (page 66)? Answer the same question for the Fejér-Hermite operators (page 69), the Fejér operators (page 122), and the Lagrange interpolating operator (page 59).
3. Prove the minimum property of $\| S_n \|$ by calculations involving $\cos k\theta$ and $\sin k\theta$ rather than $e^{ik\theta}$.
4. For the mapping M defined in the second Kharshiladze-Lozinski theorem, prove that it is linear, one-to-one, and isometric. Prove that $\| L_n^* \| = \| L_n \|$. Give a formula for M^{-1}.

*5. *The Banach theorem.* Consider a sequence of bounded linear operators L_n from one Banach space E into another. A necessary and sufficient condition that $L_n f$ converge for each $f \in E$ is that such convergence occur on a fundamental set in E and that the norms $\| L_n \|$ be bounded. Prove this, using the uniform-boundedness theorem.

6. *Hahn's theorem on interpolation.* For each n let $\xi_0^{(n)}, \ldots, \xi_n^{(n)}$ be a set of $n + 1$ distinct nodes in $[a,b]$. Let the Lagrange interpolation operator for these nodes be written $L_n f = \sum_{i=0}^{n} f(\xi_i^{(n)}) l_i^{(n)}$. A necessary and sufficient condition that $(L_n f)(\xi) \to f(\xi)$ for all $f \in C[a,b]$ is that the numbers $\sum_{i=0}^{n} | l_i^{(n)}(\xi) |$ be bounded. (Here ξ is fixed.) Prove this, using the Banach theorem.

7. *Pólya's theorem on quadrature.* For each n let there be given a numerical integration formula

$$\int_a^b f(x)w(x)\,dx \approx \sum_{i=0}^{n} A_{ni} f(x_{ni}) = L_n f$$

which is *exact* for any polynomial of degree $\le n$. A necessary and sufficient condition that $L_n f \to \int fw$ for all $f \in C[a,b]$ is that the numbers $\sum_{i=0}^{n} |A_{ni}|$ be bounded. Prove this, using the Banach theorem.

8. Berman's corollary may be strengthened as follows: There do not exist linear operators L_n mapping $C[a,b]$ into the polynomials of degree $\le n$ with the property that (for all f and n)

$$\|f - L_n f\| \le (\tfrac{1}{5}\log n)\,\|f - \mathfrak{I}_n f\|$$

9. The Fourier-series projection S_n also minimizes $\|I - S\|$ among all bounded linear projections of $C_{2\pi}$ onto the trigonometric polynomials of degree $\le n$. *Hint*: $(If)(x) = f(x) = \dfrac{1}{2\pi}\displaystyle\int_{-\pi}^{\pi} (T_{-\lambda} I T_\lambda f)(x)\,d\lambda$.

10. The family $\mathfrak{F}$ of all bounded linear projections of $C_{2\pi}$ onto the trigonometric polynomials of degree $\le n$ is a linear manifold. That is, if L and L' are in $\mathfrak{F}$, then $\lambda L + (1 - \lambda)L' \in \mathfrak{F}$ for all λ.

11. The operators T_λ and S_n commute: $T_\lambda S_n = S_n T_\lambda$. Moreover, S_n is the only projection of $C_{2\pi}$ onto the trigonometric polynomials having this property.

12. Prove that $\|I - S_n\| = 1 + \lambda_n$. *Hint*: Consider continuous functions close to the function

$$f_\epsilon(x) = \begin{cases} -1 & |x| \le \epsilon \\ \operatorname{sgn} D_n(x) & |x| > \epsilon \end{cases}$$

13. In $C[0,1]$, consider a least-squares projection onto a one-dimensional subspace, say $Gf = \langle f,g\rangle g$, where $\langle g,g\rangle = 1$. Consider another projection onto the same subspace, $Lf = \langle f,h\rangle g$, where $\langle g,h\rangle = 1$. Show that the necessary and sufficient condition for $\|L\| < \|G\|$ is that $\int |h| < \int |g|$. Thus the projection of minimum norm on an arbitrary subspace is not necessarily the least-squares projection.

14. Given $x_0,\ldots,x_n$ in $[a,b]$, there is a polynomial P of degree $\le n$ such that $\|P\| \ge \frac{1}{2}(\lambda_n - 1)$ and $|P(x_i)| \le 1$. [Fejér, 1930]

15. In the preceding section it is proved that there exists a sequence of polynomials $\{\phi_0,\phi_1,\ldots\}$ such that every $f \in C[a,b]$ is uniquely expressible in a uniformly convergent series, $f = \sum_{n=0}^{\infty} c_n \phi_n$. Show that it is not possible for ϕ_n to be of degree n (for all n).

16. Generalize Lebesgue's theorem.

17. Let $\{g_1,g_2,\ldots\}$ be a sequence in a Banach space E. For each n let G_n denote the subspace spanned by $\{g_1,\ldots,g_n\}$, and let L_n be any linear projection of E onto G_n. Prove that $\sup \|L_n\| < \infty$ if and only if there exists for each $f \in E$ a number $c(f)$ independent of n satisfying $\|L_n f - f\| \le c(f)\,\operatorname{dist}(f,G_n)$.

18. Given $x_k^{(n)} \in [a,b]$ with $n = 1, 2, \ldots$ and $k = 0, \ldots, n+1$, let $L_n f$ denote the polynomial of degree $\le n$ which best approximates f on the set $\{x_0^{(n)},\ldots,x_{n+1}^{(n)}\}$. Show that there exists an $f \in C[a,b]$ for which $\|L_n f\|$ is unbounded. [Curtis, 1962]

19. There exists a continuous 2π-periodic function whose Fourier series does not converge uniformly to it.

20. If f and f_n are elements of $C[a,b]$ such that $\|f_n\| \to \infty$, then $\{f_n\}$ cannot converge uniformly to f, but it may converge pointwise.

6. Approximation in the Mean

In this final section we take up some of the problems associated with approximation in the norm

$$\|f\|_1 = \int_a^b |f(x)|\,dx$$

This is usually termed the L^1 *norm*: L in honor of Lebesgue, and 1 to distinguish this norm from other members of the family

$$\|f\|_p = \left(\int_a^b |f(x)|^p\,dx\right)^{1/p} \qquad (p \geq 1)$$

An approximation with the L^1 norm is also called an *approximation in the mean*, the number $[1/(b-a)]\|f\|_1$ being known as the *mean* or *average value* of $|f|$.

Although the norm $\|f\|_1$ may be defined for functions in a class much wider than $C[a,b]$, many of the interesting questions concern *continuous* functions. We confine ourselves to such questions here, and employ in the theorems and proofs only the Riemann integral. The linear space $C[a,b]$ endowed with the L^1 norm is *incomplete*, and hence not a Banach space (see Prob. 1), but this is no drawback here.

Our first goal is to prove a famous theorem of Jackson regarding the unicity of best L^1 approximations. The normed linear space involved here is not strictly convex, and thus special hypotheses are necessary to ensure unicity of best L^1 approximations. It is remarkable that the Haar condition is a convenient hypothesis in this theory, as it was in the *uniform* approximation problem. Recall that an n-dimensional subspace P of $C[a,b]$ is called a *Haar subspace* if 0 is the only element of P which has n (or more) roots.

For economy, we write $\int f$ in place of $\int_a^b f(x)\,dx$. Also, if f is any real-valued function, sgn f denotes the function whose values are

$$(\operatorname{sgn} f)(x) = \begin{cases} +1 & [f(x) > 0] \\ -1 & [f(x) < 0] \\ 0 & [f(x) = 0] \end{cases}$$

If f has only a finite number of roots, then sgn f has only a finite number of discontinuities and is therefore integrable in the Riemann sense.

Lemma 1. *Let f and h be elements of $C[a,b]$. If f has at most a finite number of roots and if $\int h \operatorname{sgn} f \neq 0$, then for some λ,*

$$\int |f - \lambda h| < \int |f|$$

Proof. Let $x_1, \ldots, x_k$ be the roots of f which lie in the open interval (a,b). For sufficiently small positive ϵ, the set

$$A = [a + \epsilon, x_1 - \epsilon] \cup [x_1 + \epsilon, x_2 - \epsilon] \cup \cdots \cup [x_k + \epsilon, b - \epsilon]$$

consists of $k + 1$ nondegenerate closed intervals. Let B denote the complement of A in $[a,b]$. Without loss of generality we may suppose that $\int h \operatorname{sgn} f > 0$. (In the opposite case we would take λ with different sign.) Select ϵ small enough to ensure that

$$\int_A h \operatorname{sgn} f > \int_B |h|$$

Since A is closed and contains no roots of f, the number

$$\delta = \min \{|f(x)| : x \in A\}$$

is positive. If $0 < \lambda \|h\| < \delta$, then for points in A we have $|\lambda h(x)| < \delta \leq |f(x)|$, and consequently (on A) sgn $(f - \lambda h)$ = sgn f. Thus we have

$$\begin{aligned}
\int |f - \lambda h| &= \int_B |f - \lambda h| + \int_A |f - \lambda h| \\
&= \int_B |f - \lambda h| + \int_A (f - \lambda h) \operatorname{sgn} f \\
&= \int_B |f - \lambda h| + \int_A |f| - \lambda \int_A h \operatorname{sgn} f \\
&= \int_B |f - \lambda h| - \int_B |f| + \int |f| - \lambda \int_A h \operatorname{sgn} f \\
&\leq \lambda \int_B |h| - \lambda \int_A h \operatorname{sgn} f + \int |f| \\
&< \int |f| \qquad \blacksquare
\end{aligned}$$

Jackson's Theorem. *Let P be a Haar subspace of $C[a,b]$. Then each $f \in C[a,b]$ possesses a unique best approximation in the mean from P.*

Proof. Suppose that f has two best approximations from P: p_1 and p_2. Then by the triangle inequality for the L^1 norm, the function $p \equiv \frac{1}{2}(p_1 + p_2)$ is also a best approximation. Consequently

$$\int (|f - p| - \tfrac{1}{2}|f - p_1| - \tfrac{1}{2}|f - p_2|) = 0$$

Since the integrand is continuous and ≤ 0, it must vanish (identically) in $[a,b]$. Now let n be the dimension of P. If $f - p$ has n roots, then $f - p_1$, $f - p_2$, and $p_1 - p_2$ must have the same n roots, and, $p_1 = p_2$ by the Haar condition.

Suppose, therefore, that the function $f_0 = f - p$ has at most $n - 1$ roots. Then there exist points $a = x_0 < x_1 < \cdots < x_n = b$ containing among them all the roots of f_0. By Lemma 1, the expression

$$\int h \operatorname{sgn} f_0 \equiv \sum_{i=1}^{n} \sigma_i \int_{x_{i-1}}^{x_i} h \equiv \sum_{i=1}^{n} \sigma_i \phi_i(h)$$

must vanish for all $h \in P$, for otherwise we can reach the contradiction $\int |f_0 - \lambda h| < \int |f_0|$ by appropriately choosing $h \in P$ and λ. If $\{g_1, \ldots, g_n\}$ is a basis for P, then the matrix $\phi_i(g_j)$ is singular, since $\sum_i \sigma_i \phi_i(g_j) = 0$. Thus we can determine a nonzero n-tuple $[c_1, \ldots, c_n]$ such that $\sum_j c_j \phi_i(g_j) = 0$. This equation asserts that the nonzero function $h = \sum_j c_j g_j$ has the property

$$0 = \phi_i(h) = \int_{x_{i-1}}^{x_i} h \qquad (i = 1, \ldots, n)$$

and this in turn implies that h has a root in each interval (x_{i-1}, x_i), in violation of the Haar condition. ∎

We shall not attempt to give a *general* characterization theorem for best approximations in the mean but content ourselves instead with some important special cases.

Characterization Theorem. *Let P be a subspace and f an element of $C[a,b]$. Let p be an element of P which coincides with f in no more than a finite number of points. This p is a best L^1 approximation to f if and only if* $\operatorname{sgn}(f - p) \perp P$.

Proof. If the condition fails, then for some $h \in P$, $\int h \operatorname{sgn}(f - p) \neq 0$. By the lemma we may find a λ such that

$$\int |f - p - \lambda h| < \int |f - p|$$

On the other hand, if the condition is fulfilled then for any $p_1 \in P$,

$$\int |f - p_1| \geq \int (f - p_1) \operatorname{sgn} (f - p)$$

$$= \int (f - p) \operatorname{sgn} (f - p)$$

$$= \int |f - p| \qquad \blacksquare$$

Now we must recall the definition of a *Markoff* system: It is an ordered set of continuous functions $\{g_0,g_1,\ldots\}$, either finite or infinite in number, such that *each* initial segment $\{g_0,\ldots,g_n\}$ satisfies the condition of Haar. For example, $\{1,x,x^2,\ldots\}$ is a Markoff system on $[-1,1]$ while $\{x,1,x^2,x^3,\ldots\}$ is not. The next lemma provides a substitute for the Gram-Schmidt process in the context of L^1 approximation.

Lemma 2. *Let $\{g_0,g_1,\ldots\}$ be a Markoff system on $[a,b]$. For each n, let h_n be the best L^1 approximation of 0 in the form $h_n = g_n - \sum_{i=0}^{n-1} c_{ni}g_i$. Then $\{h_0,h_1,\ldots\}$ is a Markoff system, and $h_k \perp \operatorname{sgn} h_n$ for $k < n$.*

Proof. From the form of h_n it is clear that $\{h_0,\ldots,h_n\}$ spans the same linear subspace as $\{g_0,\ldots,g_n\}$. Since this linear subspace is a Haar subspace, $\{h_0,\ldots,h_n\}$ satisfies the Haar condition. Since this is true for arbitrary n, $\{h_0,\ldots\}$ is a Markoff system. From the definition of h_n it follows that $\sum_{i=0}^{n-1} c_{ni}g_i$ is a best L^1 approximation of g_n. The preceding theorem then implies that sgn h_n is orthogonal to $g_0,\ldots,g_{n-1}$ and hence orthogonal to $h_0,\ldots,h_{n-1}$. ∎

Markoff's Theorem. *Let $h_0, h_1, \ldots$ be defined as in Lemma 2. If $p \equiv \sum_{i=0}^{n} c_ih_i$ is a generalized polynomial such that $\operatorname{sgn}(f - p) = \operatorname{sgn} h_{n+1}$, then p is the best L^1 approximation to f of the stated form.*

Proof. Since h_{n+1} has finitely many roots, so has $f - p$. Consequently the characterization theorem applies to this situation. Let P denote the linear span of $\{h_0,\ldots,h_n\}$. By Lemma 2, $\operatorname{sgn} h_{n+1} \perp P$. By hypothesis, then, $\operatorname{sgn}(f - p) \perp P$. By the characterization theorem, p is *a* best approximation of f. By the Jackson theorem it is *the* best approximation of f. ∎

If the process described in Lemma 2 is applied to the Markoff system $\{1,x,x^2,\ldots\}$ on the interval $[-1,1]$ the resulting polynomials are multiples of the Tchebycheff polynomials of the second kind. These are defined by the formulas

$$U_n(x) = \frac{\sin (n+1)\theta}{\sin \theta} \qquad (\cos \theta = x)$$

Some properties of U_n have been listed in Probs. 1 and 11 of Chap. 4, Sec. 1. We now establish this important extremal property of the polynomials U_n.

Theorem. *The monic polynomial of degree n which best approximates* 0 *on* $[-1,1]$ *in the* L^1 *norm is* $2^{-n}U_n$.

Proof. From their definition we see that $U_0(x) = 1$ and $U_1(x) = 2x$. The trigonometric identity $\sin (n+1)\theta = 2 \sin n\theta \cos \theta - \sin (n-1)\theta$ yields a recurrence relation for U_n, viz., $U_n(x) = 2xU_{n-1}(x) - U_{n-2}(x)$. Thus by induction, the leading coefficient of U_n is 2^n. Hence $2^{-n}U_n$ is of the required form. In view of the characterization theorem, it will be enough to establish (and here we replace n by $n-1$ for convenience)

$$\int_{-1}^{1} x^k \operatorname{sgn} U_{n-1}(x)\, dx = 0 \qquad (k = 0, \ldots, n-2)$$

By making the change of variable $x = \cos \theta$, and integrating over the interval twice, we find that the following would be sufficient:

$$\int_{-\pi}^{\pi} \sin \theta \cos^k \theta \operatorname{sgn} \sin n\theta\, d\theta = 0 \qquad (k = 0, \ldots, n-2)$$

Since $\sin \theta \cos^k \theta$ is a linear combination of the functions $e^{im\theta}$ with $|m| < n$, it will be enough to prove

$$I \equiv \int_{-\pi}^{\pi} e^{im\theta} \operatorname{sgn} \sin n\theta\, d\theta = 0 \qquad [m = 0, \pm 1, \ldots, \pm(n-1)]$$

But in this integral, if we make the change of variable $\theta = \phi + \pi/n$ and exploit the periodicity of the functions, we obtain

$$I = \int_{-\pi}^{\pi} e^{im(\phi+\pi/n)} \operatorname{sgn} \sin (n\phi + \pi)\, d\phi = -e^{im\pi/n} I$$

Since $-e^{im\pi/n} \neq 1$, $I = 0$. ∎

Problems

1. Show that the linear space $C[a,b]$ equipped with the L^1 norm is incomplete. *Hint*: On $[0,2]$ for example, let $f_n(x) = x^n$ when $0 \leq x \leq 1$ and $f_n(x) = 1$ when $1 < x \leq 2$.
2. The linear space $C[a,b]$ with the L^1 norm is not strictly convex.

***3.** If $f \in C[a,b]$, does it follow that sgn f is Riemann-integrable? In order to answer this, one must know that a function g is Riemann-integrable iff the set of points where g is discontinuous may be covered, for any $\epsilon > 0$, by a sequence of open intervals (α_n,β_n) where $\sum (\beta_n - \alpha_n) < \epsilon$. Prove that the set of discontinuity points of sgn f consists of those roots of f in every neighborhood of which there are nonroots. Show that this set is closed. The following two problems are related to this question.

***4.** Let $0 < \lambda < \frac{1}{3}$, and consider the set S which remains when we perform the following sequence of operations upon the unit interval $[0,1]$. First remove an open interval of width λ from the middle. Then from the middle of each of the two remaining closed intervals remove an open interval of width λ^2. Then from the middle of each of the four remaining closed intervals remove an open interval of width λ^3, and so on. Show that the total length of removed intervals is < 1. Show that S is closed. Show that S contains no interval. Show that if S is covered by intervals (α_n,β_n), then $\sum (\beta_n - \alpha_n) > 1 - 3\lambda$.

***5.** On the unit interval define f by an infinite sequence of steps of which the first three are shown. On each interval of width λ^n the height of the graph is $\frac{1}{2}\lambda^n$. Show that f is

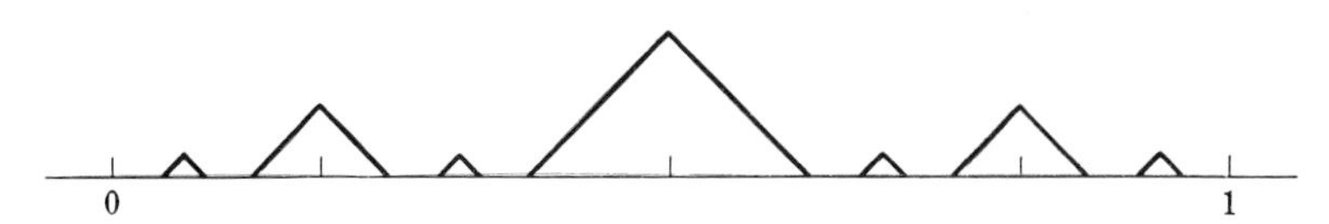

continuous. Indeed, $|f(x) - f(y)| \le |x - y|$. Show that the set of discontinuities of sgn f is precisely the set S of Prob. 4. Show that sgn f is not Riemann-integrable.

6. If the function f has the property $f(\theta + \pi) = -f(\theta)$, then for $|k| < n$ we have $\int_{-\pi}^{\pi} e^{ik\theta} f(n\theta)\, d\theta = 0$. *Hint*: Proceed as in the proof of the last theorem. [Achieser]

7. Prove by means of Prob. 6 that for $k = 0, 1, \ldots, n - 1$, $\int_0^{\pi} \cos k\theta \operatorname{sgn} \cos n\theta\, d\theta = \int_0^{\pi} \sin k\theta \operatorname{sgn} \sin n\theta\, d\theta = 0$.

8. Give an example of a function $f \in C[-1,1]$ which has a nonunique best approximation in the mean by the functions cx. Then look at the same problem on the interval $[0,1]$.

9. Prove that the roots of U_{n+1} are the points $x_k = \cos [k\pi/(n + 2)]$. Prove that if the polynomial p, having degree $\le n$, interpolates to f at the points x_k, then it is the best L^1 approximation to f on $[-1,1]$, provided that $p - f$ *changes sign* at and only at x_k.

10. Prove that if $\{1,x,\ldots,x^n,f\}$ satisfies the Haar condition on $[-1,1]$, then the polynomial of degree $\le n$ which best approximates f in the L^1 norm is the polynomial of *interpolation* for the nodes $x_i = \cos [i\pi/(n + 2)]$.

11. Prove the following theorem, which has a bearing upon Prob. 8: Let P be an n-dimensional subspace of $C[a,b]$ with the property that 0 is the only element of P having n roots in the open interval. Then each $f \in C[a,b]$ has a unique best approximation in the mean from P.

Notes

Chapter 1

The following are general works on approximation theory: Natanson [1949], Achieser [1947], Rice [1964], Timan [1960], Todd [1963], Sard [1963], de La Vallée Poussin [1919], Burkill [1959], Golomb [1960], Davis [1963], Jackson [1930], and Meinardus [1964]. Two survey articles by Buck [1959, 1959a] are especially valuable.

For the necessary background material on linear algebra, the reader may consult any of several standard textbooks, such as Stoll [1952] or Paige and Swift [1961]. There are a number of convenient sources of information about metric spaces and normed linear spaces: for example, Kolmogorov and Fomin [1957], Simmons [1963], and Taylor [1958]. For convexity in general, the books by Eggleston [1958], Fan [1958], and Valentine [1964] are recommended. See also the symposium volume of Klee [1963] and the older book by Bonnesen and Fenchel [1934]. For convex functions, the papers of Green [1954] and Beckenbach [1948], as well as the book by Popoviciu [1944], should be consulted. The theorem of Carathéodory was given in Carathéodory [1911], but is due essentially to Kirchberger [1903]. Helly's theorem was given in Helly [1923]. The proof in Sec. 5 is from Rademacher and Schoenberg [1950]. A recent generalization is to be found in Horn [1949]. The continuity of convex functions was proved in Jensen [1906]. The classical inequalities which are scattered through the problems are to be found in Hardy, Littlewood, and Pólya [1934]. The existence theorem for best approximations (Sec. 6) was enunciated in F. Riesz [1918]. It had been given earlier for polynomials in Kirchberger [1903], often mistakenly ascribed to Tchebycheff. The sequence spaces, from which the example of page 21 is drawn, are treated in a unified manner in Schatten [1950]. Uniform convexity was introduced in Clarkson [1936]. The existence theorem of page 22 was given in Sz.-Nagy [1942]. The observation that strict convexity implies the unicity of best approximations (page 23) has been ascribed to Krein. Recent investigations of the existence and unicity of best approximations in Banach spaces appear in Phelps [1960, 1963, 1964].

Chapter 2

The earliest investigation of systems of linear inequalities occurs in Fourier [1824]. Later important contributions occur in Motzkin [1936], Chernikov [1953], Dines [1936], and Fan [1956]. Algorithms for solving such systems occur, for example, in Motzkin and Schoenberg [1954] and Goldstein and Cheney [1958]. The solution of systems of linear inequalities and of systems of equations in the minimax sense can be attacked by the methods of linear programming. Of the vast literature on linear programming we cite only the treatise by one of the originators of the subject, Dantzig [1963]. The earliest discussion of the minimax problem for systems of linear equations is probably in de La Vallée Poussin [1911]. The algorithm of Pólya was given originally for the approximation of continuous functions by polynomials in Pólya [1913]. Its use in solving systems of equations in the minimax sense is described in Goldstein, Hereshoff, and Levine [1957]. A proof of convergence of the Pólya algorithm when Haar's condition is violated has been given in Descloux [1963]. Extensions of Pólya's ideas have been given in Hoel [1935] and Kripke [1964]. Other algorithms for the finite-dimensional minimax problem have been given in Zuhovickii [1951, 1953, 1961], Remes [1956], Stiefel [1959, 1959a, 1960], Curtis and Frank [1959], and Valentine and Van Dine [1963]. What we have termed the *Haar condition* in Sec. 6 played a crucial role in the paper of Haar [1918], although it had been employed earlier in de La Vallée Poussin [1911]. The exchange theorem is from Stiefel [1959]. The algorithm for convex programming is from Cheney and Goldstein [1959]. See also Beale [1955]. Recent work on the degenerate cases of discrete Tchebycheff approximation occurs in Descloux [1961, 1961a] and in Bittner [1961].

Chapter 3

The topic of interpolation is a vast one, and is the subject of several treatises, such as Steffensen [1927], Thiele [1909], Markoff [1896], and Nörlund [1924]. The first systematic work was by Newton. For modern extensions of the classical theory, especially to several variables, infinite intervals, and for specialized numerical purposes, see the extensive work of Salzer, as, for example, Salzer [1946, 1951a, 1956, 1957, 1959a,b,c, 1960a,b,c,d]. For interpolation in several variables consult also Thatcher [1960].

The extremal properties of the Tchebycheff polynomials were first investigated in Tchebycheff [1859, 1881]. The properties of these polynomials and their numerical values are the subject of the book by Lanczos [1952]. See also Dickinson [1939]. Tchebycheff polynomials may be defined for any domain in the complex plane. There is an extensive theory. See, for example, Ullman [1959] and the citations given there.

The general case of Hermite interpolation (in which at various points different numbers of derivatives are specified) appears first in Hermite [1878.] Recent investigations occur in Sharma [1964] and P. Szász [1959].

The divergence phenomenon for interpolating polynomials of successively higher degree was first observed in Méray [1884, 1896]. One of his examples was $f(x) = 1/(1 - x^2)$. In Runge [1901] the example $f(x) = 1/(1 + x^2)$ was thoroughly investigated. See Steffensen [1927, p. 35] or Montel [1910, p. 51] for the analysis of this example. Faber's theorem, that for nodes prescribed in advance there always exists a continuous function exhibiting the divergence phenomenon, is in Faber [1914]. A simplification of the proof was given in Fejér [1930]. Bernstein's investigation of the interpolants for $| x |$ occurs in Bernstein [1912b]. Further investigations of convergence problems are in Grünwald [1936, 1938, 1941, 1943], Feldheim [1939], Erdös [1950], Marcinkiewicz [1937a], and Runck [1961].

The original proof of the Weierstrass approximation theorem is in Weierstrass [1885]. The two papers of Runge [1885, 1885a] together give an independent and nearly simultaneous proof. Weierstrass' memoir is translated into French in Weierstrass [1886]; see also his collected works [1894]. Many other proofs of the theorem followed shortly thereafter: Picard [1891], Volterra [1897], Lebesgue [1898], Mittag-Leffler [1900], Fejér [1900], Lerch [1903], Landau [1908], de La Vallée Poussin [1912], and Bernstein [1912]. Fejér's theorem of 1900 is amplified in Fejér [1904]. Several of the proofs of the Weierstrass theorem depend upon singular integrals, the general theory of which is extensive. See, for example, Butzer [1960]. The Weierstrass theorem cannot be extended to regions in the complex plane without assuming analyticity of the function being approximated. The theorem here is due to Walsh [1926]. The first extension of the Weierstrass theorem to several variables occurs in Picard [1891]. The Fejér-Hermite operator is studied in Fejér [1916]. Fejér's proof of the Weierstrass theorem via the Hermite interpolation formula is in Fejér [1930]. The proof given here is from Korovkin [1959].

The first proof that every continuous function on $[a,b]$, where $0 < a < b < 1$, is uniformly approximable by polynomials with integer coefficients occurs in Pál [1914]. His proof is outlined in Prob. 19 of Chap. 3, Sec. 3. See also Kakeya [1914], Okada [1923], Bernstein [1930], Hewitt and Zuckermann [1959], Chlodovsky [1925], Kantorovich [1931a], Gelfond [1955], and Trigub [1961].

A method of approximating continuous functions by polynomials of the form $\sum f(k/n)\phi_{nk}(x)$ was first given in Borel [1905, p. 80]. Bernstein's proof of the Weierstrass theorem in Bernstein [1912] followed Borel's ideas and had the merit of employing the simple functions $\phi_{nk} = \binom{n}{k} x^k(1 - x)^{n-k}$. The general theory of such operators is dealt with in Favard [1944] (written in captivity).

The theorem on monotone operators is essentially proved in Bohman [1952], where the special case of operators $(L_n f)(x) = \sum_k f(x_{nk})\phi_{nk}(x)$ is investigated. The special case of integral operators is due to Korovkin [1953]. The book by Korovkin [1959] is a convenient reference. Recent studies of monotone operators occur in Strang [1962], Baskakov [1957], Hsu [1961], and Cheney and Sharma [1964, 1965]. The closely related *variation diminishing* operators are studied in Schoenberg [1959] and Pólya and Schoenberg [1958].

The Bernstein polynomials have been the object of much study, of which the book by Lorentz [1953] gives a general account. Modifications of these polynomials for purposes of approximation on infinite intervals have been given in Chlodovsky [1937], O. Szász [1950], and Meyer-König and Zeller [1960]. Convergence of the polynomials in regions of the complex plane has been discussed in Wright [1930] and Kantorovich [1931]. The rapidity of convergence of $B_n f$ has been investigated in Popoviciu [1935], Voronovskaya [1932], and Bajsanski and Bojanič [1964]. The approximation of discontinuous functions by modified Bernstein polynomials has been discussed in Chlodovsky [1929], Kantorovich [1930], Lorentz [1937], Butzer [1953a], and Herzog and Hill [1946]. Two-dimensional Bernstein polynomials are studied in Butzer [1953].

The problem of finding best uniform approximations to a function was, according to Tchebycheff, first considered by Poncelet. It was taken up in Tchebycheff [1854]. Uniform approximation by rather general n-parameter families of functions was first considered in Tchebycheff [1859]. The next important works are Kirchberger [1903] and J. W. Young [1907]. A characterization theorem similar to that in Sec. 4 was first proved in Kirchberger [1903]. Related results have been given in Remes [1935], Schnirelmann [1938], Zuhovickiĭ [1956], Lawson [1961], Bram [1958], Barbuti [1960], Cheney and Goldstein [1959, 1962, 1965], and Rivlin and Shapiro [1961]. The complex case was treated in Kolmogorov [1948]. The alternation theorem for polynomial approximation was first proved in Borel [1905], often incorrectly ascribed to Tchebycheff [1859]. The alternation theorem for arbitrary linear families satisfying the Haar condition was proved in J. W. Young [1907], often incorrectly attributed to Bernstein [1926]. The theorem of de La Vallée Poussin occurs in de La Vallée Poussin [1910]. The theorem on page 76 arose from conversations with P. C. Curtis, Jr. General investigations of systems of functions which satisfy the Haar condition occur in Remes [1957], Bernstein [1926, 1937], Laasonen [1949], and Krein [1951].

Much of the theory of Sec. 3 has been extended to nonlinear approximating families. Tchebycheff considered several special such cases, and Young gave the first existence theorem. Subsequent work occurs in Motzkin [1949], Tornheim [1950, 1956], Curtis [1959], and Rice [1960a,b, 1961a, 1962a, etc.].

The theory of approximation by polynomials over domains in the complex plane is quite different from the real theory. Much of our knowledge here is due to Walsh [1935a] and Sewell [1942].

A theory also exists for various "constrained" linear approximations: for example, Jurkat and Lorentz [1961], Rice [1963], and Burov [1961]. The theory of "one-sided" approximation has been begun in Freud [1955]. The theory of approximation by generalized polynomials which are required to interpolate to the function at one or more prescribed points has been termed *approximation with nodes* and developed in a series of papers by Paszkowski [1955, 1956, 1957, 1962].

The unicity theorem for best approximation by polynomials was proved first in Borel [1905, p. 85]. Unicity in the complex case is due to Tonelli [1908]. For generalized polynomials, with Haar condition, the unicity is due to J. W. Young [1907]. For linear functions of two variables it is due to Collatz [1956]. The strong unicity theorem is due to Newman and Shapiro [1962]. The proof here is from Cheney [1962]. Haar's theorem was given in Haar [1918]. His proof that without the Haar condition unicity may fail is the one used here. Other proofs of Haar's theorem are in Singer [1960] and Pták [1958]. The complex case is in Kolmogorov [1948]. Other cases are discussed in Rivlin and Shapiro [1960]. The conditions necessary and sufficient that the set of best approximations have dimension $\leq k$ have been given in Rubinstein [1955]. The continuity of the best-approximation operator $\mathfrak{T}$ was first proved in Borel [1905, p. 89]. The stronger form, giving a Lipschitz condition, occurs in Freud [1958], independently proved in Maehly and Witzgall [1960]. The first result to the effect that only certain compact sets X are suitable environments for systems of functions satisfying the Haar condition is found in Haar [1918], where it is shown that X cannot contain any sphere of dimension greater than 1. In Mairhuber [1956] it is proved that if X is in n space, it must lie on an arc. Progress in removing the restriction that X lie in n space has been registered in Curtis [1959], Sieklucki [1958] and Schoenberg [1960]. A generalization of Mairhuber's theorem to the complex case has been made in Schoenberg and Yang [1961]. Unicity problems for infinite-dimensional subspaces have been dealt with in Singer [1957], Phelps [1960, 1963], and Garkavi [1964].

The results of Sec. 6 are taken from Cheney [1962]. The word *fundamental*, introduced in Banach [1932], is used in preference to the more usual but overworked *closed*. The first proof of convergence of best approximations on discrete sets to the best approximation on a continuum is in Motzkin and Walsh [1956].

Markoff's inequality appeared first in A. A. Markoff [1889]. Bernstein's inequality for purely cosine polynomials occurs in Bernstein [1912b]. The general case was observed in Fejér [1914] and in a letter from Landau to Bernstein mentioned in Bernstein [1926]. Another proof was given in M. Riesz [1914, 1914a]. The proof in the text is from Pólya and Szegö [1925, vol. II, p. 90]. In Duffin and Schaeffer [1941] a stronger form of Markoff's inequality is proved: $\max | P'(x) | \leq n^2 \max | P(x_i) |$ where the first max is over $[-1,1]$ and the second is for $x_i =$ extrema of T_n. The inequality of Prob. 3 is far from best possible: in V. A. Markoff [1892] it is proved that

$$\| P^{(k)} \| \leq \frac{n^2(n^2-1)\cdots[n^2-(k-1)^2]}{1\cdot 3\cdot 5\cdots(2k-1)} \| P \|$$

The simplest proof of the latter is probably that of Rogosinski [1955]. Other papers directly concerned with generalizations of the inequalities of Markoff or Bernstein are by Soble [1957], Rogosinski [1954], Lax [1944], Bary [1954], Erdös [1940], and Achieser and Levin [1957]. Lemma 3 and Theorem 1 of Sec. 7 are taken from Cheney [1962]. Theorem 2 is from Curtis [1962]. The device used in the proof part (*iii*) of Lemma 3 is from Bernstein [1932].

The algorithms of Remes are given in Remes [1935]. His second algorithm occurs first in Remes [1934]. See also the treatise by Remes [1957]. The explicit equations for the error in approximating by a polynomial of degree $\leq n$ on $n+2$ points were first given in de La Vallée Poussin [1910]. The extension of the second algorithm to nonlinear approximating families is due to Novodvorskii and Pinsker [1951]. A proof of quadratic convergence in the second Remes algorithm has been given in Veidinger [1960]. The analogue for polynomials of the descent algorithm of Chap. 2 has been given in Zuhovickii [1958]. An important algorithm also capable of quadratic convergence is that of Murnaghan and Wrench [1959, 1960]. Other algorithms have been proposed in Rice [1959, 1961b] and Akilov and Rubinov [1964]. Papers concerned with various computation aspects of these algorithms are by Shenitzer [1957], Bittner [1961], Wenzl [1954], and Briones [1962, 1964]. The problem of approximating a function and its derivative has been treated in Moursund [1964] and Moursund and Stroud [1965]. The difficult problem of discovering classes of functions whose best approximations can be obtained algebraically has been attacked in Rivlin [1962] and Talbot [1962].

For an extensive tabulation of best approximations to elementary functions, with a variety of intervals, accuracies, and forms, see the handbook by Hart [1965]. The handbook by Abramowitz and Stegun [1964] should also be consulted. Other collections of approximations for digital computation are found in Hastings [1955], Langdon [1955], and Duijvestijn and Dekkers [1961]. Approximations for physical data have been tabulated in Rice [1962b].

Chapter 4

Orthogonal polynomials are the subject of the treatise by Szegö [1939] and of the shorter work by Shohat [1934]. A bibliography on this topic is Shohat, Hille, and Walsh [1940]. Orthogonal functions generally are the subject of the treatises by Kaczmarz and Steinhaus [1935], Sansone [1951], and Tricomi [1955]. For trigonometric series see the treatises of Zygmund [1935, 1959] and Bary [1961]. Part II of Natanson [1949] is especially recommended for its exposition of the theory of orthogonal polynomials. The theorems about the three-term recurrence relation and the roots of the orthogonal polynomials are from Stieltjes [1884]. The result (page 111) that best approximations in the least-squares norm must also interpolate is due to Shohat [1933].

The topic of numerical integration is the subject of the bibliography by Stroud [1961]. A modern treatise covering this and many related topics is Sard [1963]. The Gaussian quadrature formula (with constant weight function) goes back to Gauss [1814], and its generalization to arbitrary weights to Christoffel [1858]. The theorem that orthogonal polynomials must have their roots on (a,b) has been generalized to orthogonal systems which are also Markoff systems; see Videnskii [1963]. For the use of orthogonal polynomials in numerical linear algebra, see the lectures of Stiefel [1958]. The theorem of Stieltjes concerning Gaussian quadrature occurs in Stieltjes [1884].

The history of the Fourier series and its role in the development of modern mathematics may be read in Hobson [1926] or Jeffrey [1956]. The theorem by Dirichlet on the convergence of Fourier series is found in Dirichlet [1829]. The first continuous function having a divergent Fourier series was constructed in du Bois-Reymond [1876]. Simpler examples have been given in Fejér [1909, 1909a]. An example of an $f \in C_{2\pi}$ such that every *subsequence* of partial sums of the Fourier series diverges is given in Menchoff [1944]. In Kolmogorov [1926] there is constructed a Lebesgue-integrable function whose Fourier series diverges everywhere. The theorem of Fejér on Césarò summability of Fourier series is in Fejér [1900, 1904]. Korovkin's theorem on monotone operators in $C_{2\pi}$ is given in Korovkin [1959]. For the convergence of orthogonal expansions generally see the book by Alexits [1960] or the article by Shohat [1935].

The expansion of common functions in series of Tchebycheff polynomials is a problem of numerical analysis which is studied in Clenshaw [1962]. This work gives tables of coefficients to aid in the construction of computer programs. See also Baraket [1961], Cody [1965], Thacher [1965], and Wimp [1961, 1962]. The use of Tchebycheff expansions in numerical integration has been investigated in Clenshaw and Curtis [1960]. For the conversion of power series into Tchebycheff series see Minnick [1957], de Vogelaere [1959], Thacher [1965], and Elliott [1964]. The possibility of making slight corrections to the coefficients in a Tchebycheff expansion in order to obtain best approximations in the uniform norm was explored in a series of papers by Hornecker [1958, 1959, 1959a, 1960, 1961]. The advantages of making a linear fractional transformation of the variable before expanding are discussed in Thacher [1965a].

The so-called "lethargy" theorem was proved in the more exact form, $E_n(f) = \epsilon_n$, in Bernstein [1938], and his proof is the one generalized in Timan [1960, p. 40] and Golomb [1960; 1962, p. 94]. Functions all of whose best uniform approximations are simply truncations of their Tchebycheff expansions first appeared in Bernstein [1912a]. See also Bernstein [1913]. In Bernstein [1936] it is proved that if the function $f = \sum a_k T_{n_k}$ with $a_k > 0$ has this property, then n_{k+1}/n_k is always an odd integer. Hardy's investigation of the Weierstrass nondifferentiable function is in Hardy [1916]. For a graph of one of these functions and a discussion of its numerical properties see Salzer and Levine [1961].

The earliest results relating $E_n(f)$ to the coefficients in the expression $f \sim \sum a_k T_k$ are to be found in Bernstein [1912b]. Parts (*ii*) and (*iii*) of Theorem 5, Sec. 4, are found there explicitly. Part (*i*) of that theorem is given in Bernstein [1952, p. 169]. See also Blum and Curtis [1961], Rivlin [1962a], and Clenshaw [1964].

The numerical problem of "fitting" data by polynomials in the least-squares sense is a very old one, and its history can be read in Eisenhart [1963]. Polynomials which are orthogonal with respect to a discrete inner product were investigated in Tchebycheff [1859a]. More recent treatments are given in Lanczos [1938] and Forsythe [1957]. See also Ascher and Forsythe [1958], Dent and Newhouse [1959], Herzberger [1949], Spitzbart and Shell [1958], Lewis [1947], and the treatise by Guest [1961].

The theorem of Erdös and Turán was proved in Erdös and Turán [1937]. The special case $w(x) = 1$ had been given in Fejér [1932], and the special case $w(x) = (1 - x^2)^{-1/2}$ in Erdös and Feldheim [1936]. An analogue of the Erdös-Feldheim result in the complex plane has been given in Walsh and Sharma [1964].

Theorems in which the rate of decrease of $E_n(f)$ is established for a family of functions (usually fulfilling some differentiability conditions) are generally called theorems of *Jackson type*, in recognition of the pioneering investigation of Jackson [1911]. (This work was a prize-winning dissertation written under Landau's direction.) See also Jackson [1912, 1930]. Conversely, theorems in which the smoothness of f is deduced from the sequence $E_n(f)$ are called theorems of *Bernstein type*, in honor of the work of Bernstein [1912b]. This memoir won a prize (800 Fr.) which had been established a few years earlier by the Belgian Academy. In it Bernstein independently arrived at one of Jackson's theorems, viz., if $f \in \text{Lip}\,\alpha$, with $\alpha < 1$, then $n^\alpha E_n(f)$ is bounded. These early investigations were stimulated by the concrete problems posed in de La Vallée Poussin [1908] and Lebesgue [1908]. The problem of determining the exact constant in the first Jackson theorem was solved simultaneously but by dissimilar methods in Favard [1937] and Achieser and Krein [1937]. The determination of the best form for Jackson's third theorem has been accomplished in Korneicuk [1962]. Theorems of Jackson type in which the derivatives are replaced by difference quotients have been given in Whitney [1957, 1959]. Theorems of Jackson type in which the function satisfies a condition $f^{(k)}(x) \geq 0$ have been given in Shisha [1965].

The study of sets of functions by means of their approximability by polynomials (or other classes) has been carried on vigorously in the recent past. The interested reader should consult the paper by Lorentz [1962] for an introduction to this subject.

Chapter 5

Approximation by rational functions is the subject of the treatise by Walsh [1935], but the point of view there is different from the one here. The earliest discussion of rational approximation occurs in the memoir of Tchebycheff [1859]. Here is found a theorem of the form, "the error of the best approximation must reach its maximum magnitude in at least k points." (The alternation in the signs is not proved.) The existence theorem (Sec. 2) for best rational approximations was first proved in Walsh [1931]. Extensions of the theorem to more general families of functions have been given in Boehm [1964, 1965], Newman and Shapiro [1965], Collatz [1960], and Goldstein [1963]. The existence theorem for rational trigonometric forms is from Cheney and Loeb [1964]. See also Rivlin [1964].

The arrangement of a rational function in continued-fraction form illustrates the fact that the "obvious" manner of evaluating an algebraic expression is not necessarily the most efficient. The first hint that a polynomial of degree n may possibly be evaluated using *fewer* than n multiplications was given in Ostrowski [1954]. The results there obtained for $n = 4$ were extended in Motzkin [1955] and Pan [1959]. See also Hart et al. [1965].

The corollary in Sec. 3, which characterizes best approximations by ordinary rational functions, was first proved in Achieser [1930]. The unicity (corollary, page 164) is due to Achieser [1947]. Generalizations of both these theorems to other nonlinear families have been given in Rice [1961]. The continuity theorem for ordinary rational approximation occurs in Maehly and Witzgall [1960]. The remaining theorems of Secs. 3 and 4 are taken from Cheney and Loeb [1964] and Cheney [1965a]. The generalizations of de La Vallée Poussin's theorem (Probs. 8 and 9, Sec. 3) are probably new. The alternation theorem has been put to work in Boehm [1964a] to investigate functions whose best approximations in $\mathbf{R}_m^n[a,b]$ are polynomials or reciprocals of polynomials. For the family $\mathbf{R}_m^n[a,b]$, the points of $C[a,b]$ at which the best-approximation operator $\mathfrak{T}$ is continuous have been characterized in Werner [1965]. The convergence question for sequences of best rational approximations in the complex domain has been dealt with in many papers by Walsh. See, for example, Walsh [1929, 1931a, 1955]. Results in the real case have been obtained in Boehm [1965a].

The earliest discussions of numerical procedures for computing best rational approximations appear to be those of Wenzl [1954], Loeb [1957, 1959, 1960], and Maehly [1960]. The first two algorithms of Sec. 5 are due to Loeb. The "differential correction algorithm" is from Cheney and Loeb [1962]. The known algorithms up to 1963 are summarized in Cheney and Southard [1963]. Much work has been done on the Remes second algorithm because of its quadratic convergence. See Werner [1962, 1962a, 1963], Wetterling [1963, 1963a], Fraser and Hart [1962], and Stiefel [1964]. The possibility of subdividing the interval in an optimum way and approximating a function in pieces has been dealt with in Lawson [1963, 1964].

For the subject of interpolation with rational functions see Stoer [1961] and the references cited there.

The Padé approximations take their name from the memoir of Padé [1892], although they had been studied earlier in Frobenius [1881], and Frobenius ascribes their invention to Cauchy. For a modern treatment see Wall [1948]. The idea of passing *directly* from a differential equation to various rational approximations of its solutions seems to have originated in Laguerre [1885] and has been developed more recently in Lanczos [1956] and Luke [1955, 1957, 1958]. The Padé approximation can be obtained also as a limit of Tchebycheff approximations when the interval shrinks to a point. See Walsh [1964]. The problem of devising algorithms with which to obtain various types of rational approximations from power series has been studied in Spielberg [1961] and Wynn [1960]. The "quotient-difference algorithm" is applicable to this problem also. See Rutishauser [1956], Henrici [1958], and Bauer [1959].

For tables of rational approximations to specific functions, the handbook by Hart [1965] should be consulted. See also Miller and Hurst [1958], Macon [1955], Fröberg [1961], Spielberg [1961, 1962] Luke [1957, 1957a, 1960], Hershey [1962], Clendenin [1961], Werner and Collinge [1961], Rice [1964a], Teichroew [1952], and Kogbetliantz [1957, 1960].

The standard works on continued fractions are Perron [1929] and Wall [1948]. Other (more limited) textbooks are Khovanski [1956] and Khintchine [1956]. The convergence theorem of page 184 goes back to Seidel [1846]. The theorem of Stieltjes was given in Stieltjes [1895].

Chapter 6

The Stone approximation theorem appears in embryonic form in Stone [1937]. The later expositions of Stone [1948, 1962], which take advantage of intervening work such as Kakutani [1941], are recommended. Progress in lifting the restrictions on X and on the approximating family is continually being made. See, for example, Arens [1949], Buck [1958], Hewitt [1947], Isbell [1958], Anderson [1962], and Banaschewski [1957].

The theorem of Gram was given first in Gram [1879], and translated in Gram [1883]. The lemma of Cauchy is from Cauchy [1841]. Müntz's theorems were given in Müntz [1914]. Bernstein came close to discovering them in Bernstein [1912b], and in fact proved that if $0 < \alpha_1 < \alpha_2 < \cdots < K$, then $\{1, x^{\alpha_1}, x^{\alpha_2}, \ldots\}$ is fundamental in $C[0,1]$. The name of Otto Szász is sometimes associated with these theorems because of the nearly simultaneous investigation by Szász [1916]. If $\{p_n\}$ is an increasing sequence of natural numbers, with $\sum p_n^{-1} < \infty$, then the polynomials $\sum \lambda_n x^{p_n}$ do not fill out $C[0,1]$; the problem of describing the functions which can be approximated by these special polynomials has been completely solved in Clarkson and Erdös [1943]. Their theorem asserts that such functions must be analytic on $[0,1)$. See also Carleman [1922]. Similar problems about the family $\{e^{\lambda_n x}\}$ have been investigated in Hirschmann [1949]. The determination of general criteria for fundamentality (or "closure," as it is sometimes unfortunately called) is studied in Fichtenholz [1926]. See Szász [1953], Boas [1946], Kosloff [1950], and Hsieh-chang [1964] for related material.

The Bernstein theorems in Sec. 3 are from the prize-winning work of Bernstein [1912b]. The proofs of Sec. 3 indicate why Bernstein was at the same time interested in establishing the inequality (page 91) which now bears his name. Zygmund's theorem is from Zygmund [1945]. For related theorems pertaining to ordinary polynomial approximation see Freud [1959] and the references given there.

The bases that are now distinguished by Schauder's name were first considered in Schauder [1927]. In this paper bases were constructed for a number of important spaces, including $C[a,b]$. The Krein-Milman-Rutman theorem is from Krein, Milman, and Rutman [1940]. For further information about bases in general see Banach [1932], Karlin [1948], James [1950], Gelbaum [1950], Singer [1961], and Semadeni [1963]. For other bases in $C[a,b]$ see Ellis and Kuehner [1960] and Cielsielski [1963]. Polygonal approximation is studied in Stone [1961] and Schwerdtfeger [1960, 1963].

The Baire theorem and the uniform-boundedness theorem are standard results which can be found in books on functional analysis. The former was proved for $X = [a,b]$ in Baire [1899]. The latter is sometimes termed the Banach-Steinhaus theorem and was given in Banach and Steinhaus [1927]. The theorem that the Fourier projection operator is of minimal norm is due to Lozinski [1948]. The formula on which the proof is based, $S_n f = (2\pi)^{-1} \int T_{-\lambda} L T_\lambda f \, d\lambda$, goes back essentially to Marcinkiewicz [1937]. Another related formula occurred in Lozinski [1944]. The term *Lebesgue constant* is in recognition of the memoir by Lebesgue [1909], in which the divergence $\lambda_n \to \infty$ was first established. The Kharshiladze-Lozinski theorem first appeared in Lozinski [1948]. Several special cases are much older. For the Fourier projection it is in Lebesgue [1909] and for the Lagrange interpolation operator (page 215), it is given in Faber [1914]. A generalization to arbitrary Markoff systems was given in Banach [1940]. The case of weighted least-squares projection (page 215) was given in Nicolaev [1948]. The Berman corollary of Sec. 5 is given in Berman [1959]. The Hahn theorem of Prob. 6, Sec. 5, is from Hahn [1918]. The Pólya theorem of Prob. 7, Sec. 5, is from Pólya [1933]. The result of Faber has been generalized and made more precise in Curtis [1962]. See also Curtis [1963]. The state of the subject up to 1939 is summarized in Feldheim [1939]. Various stronger divergence phenomena have been established. In Marcinkiewicz [1937] and Grünwald [1936] it is shown that there exists an $f \in C[-1,1]$ such that the Lagrange interpolants, with nodes at the zeros of T_n, diverge everywhere. In Bernstein [1931] it is proved that for any disposition of nodes there exists a continuous function and a point ξ such that the Lagrange interpolants are unbounded at ξ.

Approximation in the mean apparently was first considered in Tchebycheff [1859]. Markoff's theorem occurred in A. A. Markoff [1898]. The analogue of the Weierstrass theorem was provided in Weyl [1916], where it is shown that if f is bounded and Riemann-integrable, then polynomials P_1 and P_2 exist satisfying $P_1 \leq f \leq P_2$ and $\int_a^b (P_2 - P_1) < \epsilon$. Jackson's theorem was given first in Jackson [1921]. Other proofs have been given in Walsh and Motzkin [1959], Pták [1958a], Kripke and Rivlin [1965], and Krein [1938]. Jackson's proof is essentially reproduced in Achieser [1956, p. 77], and Krein's proof in Timan [1963, p. 38]. The proof given here is from Cheney [1965]. It borrows all its ideas from other sources, particularly Walsh and Motzkin, Jackson, and Pták. The question of characterizing subspaces which always provide unique L^1 approximations has been considered in Phelps [1960] and [1966] and Kripke and Rivlin [1965]. The characterization theorem given here is an elementary form of a much more general statement involving measure theory and integrable functions. The orthogonality condition was given first in Nikolski [1940]. Extensive treatments of the entire subject are to be found in Achieser and Krein [1938], Nikolskii [1940], Havinson [1958], Kripke and Rivlin [1965], and Krein [1951]. The minimization of $\int_{-1}^{1} | x^n + a_{n-1} x^{n-1} + \cdots |$ was posed as a problem in Korkin and Solotareff [1873], and was solved in Stieltjes [1914, pp. 11–20]. See also Visser [1945] and Geronimus [1935]. The theorem that the polynomials of best pth-power integral approximation converge to the polynomial of best L^1 approximation as $p \to 1$ was proved in Hoel [1935]. This is the analogue of Pólya's result for $p \to \infty$. The L^1 approximation on discrete sets is discussed in Motzkin and Walsh [1955]. Nonlinear L^1 approximation has been investigated in Rice [1964c,d].

References

Abbreviations

ACM Association for Computing Machinery
ACMC ACM Communications
ACMJ ACM Journal
ACTA Acta Mathematica
AECT Atomic Energy Commission Translations (available from the Office of Technical Services, Department of Commerce, Washington)
AFST Annales de la Faculté des Sciences de l'Université de Toulouse
AJM American Journal of Mathematics
AM Annals of Mathematics
AMAF Arkiv för Matematik, Astronomi och Fysik
AMASH Acta Mathematica Academiae Scientiarum Hungaricae
AMM American Mathematical Monthly
AMPA Annali di Matematica Pura ed Applicata
AMS American Mathematical Society
AMSC AMS Colloquium Publications
AMST AMS Translation
ARK Arkiv för Matematik
ARMA Archive for Rational Mechanics and Analysis
ASENS Annales Scientifiques de l'École Normale Supérieure
ASM Acta Scientiarum Mathematicarum (Szeged)
ASNP Annali della Scuola Normale Superiore di Pisa
BAMS AMS Bulletin
BAP Bulletin de l'Académie Polonaise des Sciences
BSM Bulletin des Sciences Mathématique
CJM Canadian Journal of Mathematics
CKMS Communications of the Kharkov Mathematical Society
CMB Canadian Mathematical Bulletin
CR Comptes Rendus Hebdomadaries, Séances de l'Académie des Sciences, Paris
DAN Doklady Akademii Nauk SSSR
DMV Deutsche Mathematiker-Vereinigung Jahresbericht
DUKE Duke Mathematical Journal
FM Fundamenta Mathematicae
GM Symposium on the Approximation of Functions, sponsored by General Motors Laboratory, Sept., 1964; proceedings published by Elsevier Publishing Co., Amsterdam
IAN Izvestia Akademii Nauk SSSR
IBM International Business Machines Corp.
JLMS Journal of the London Mathematical Society
JMM Journal of Mathematics and Mechanics
JMP Journal of Mathematics and Physics
JMPA Journal de Mathématiques Pures et Appliquées
JRAM Journal für die Reine und Angewandte Mathematik
MA Mathematische Annalen
MSB Mathematiceskii Sbornik
MSM Mémorial des Sciences Mathématiques
MTAC Mathematical Tables and Other Aids to Computation = Mathematics of Computation
MZ Mathematische Zeitschrift
NAMS AMS Notices
NBS National Bureau of Standards
NBSAM NBS Applied Mathematics Series
NBSJ NBS Journal of Research

NM Numerische Mathematik
NYAS New York Academy of Science Annals
ONA *On Numerical Approximation*, University of Wisconsin Press, Madison, 1959
PAMS AMS Proceedings
PJ Pacific Journal of Mathematics
RCMP Rendiconti del Circolo Matematico di Palermo
SIAM Society for Industrial and Applied Mathematics
SIAMB SIAM Journal Series B Numerical Analysis
SIAMJ SIAM Journal
SIAMR SIAM Review
SM Studia Mathematica
SOVM Soviet Mathematics Doklady
TAMS AMS Transactions
TMJ Tôhoku Mathematical Journal
UMN Uspehi Mathematiceskii Nauk
ZAMM Zeitschrift für Angewandte Mathematik und Mechanik

References

ABRAMOWITZ, M., and I. A. STEGUN (eds.). [1964] *Handbook of Mathematical Functions with Formulas, Graphs, and Mathematical Tables*, NBSAM **55.**

ACHIESER, N. I. [1930] *On extremal properties of certain rational functions*, DAN, 495–499 (Russian). [1947] *Lectures on the Theory of Approximation*, Gostekhizdat, Moscow (Russian). = [1955] *Vorlesungen über Approximationstheorie*, Akademie-Verlag, Berlin. = [1956] *Theory of Approximation*, Ungar, New York.

ACHIESER, N. I., and M. G. KREIN. [1937] *Best approximation of differentiable periodic functions by means of trigonometric sums*, DAN **15,** 107–112 (Russian). [1938] *Some Questions in the Theory of Moments*, Kharkov (Russian). = [1962] Translation, AMS.

ACHIESER, N. I., and B. YA. LEVIN. [1957] *Inequalities for derivatives analogous to Bernstein's inequality*, DAN **117,** 735–738 (Russian).

AKILOV, G. P., and A. M. RUBINOV. [1964] *The method of successive approximations for determining the polynomial of best approximation*, DAN **157,** 503–505. = SOVM **5.**

ALEXITS, G. [1960] *Konvergenzprobleme der Orthogonalreihen*, Budapest.

ANDERSON, F. W. [1962] *Approximation in systems of real-valued continuous functions*, TAMS **103,** 249–271.

ARENS, R. [1949] *Approximation in, and representation of, certain Banach algebras*, AJM **71,** 763–790.

ARUFFO, G. [1952] *Un' osservatione sull' approssimazione di una funzione continua per mezzo di una successione di funzione razionali*, Unione Matematica Italiana, Bolletino **7,** 44–47.

ASCHER, M., and G. E. FORSYTHE. [1958] *SWAC experiments on the use of orthogonal polynomials for data-fitting*, ACMJ **5,** 9–21.

BAIRE, R. [1899] *Sur les fonctions de variables réelles*, AMPA **3,** 1–123.

BAJŠANSKI, B., and R. BOJANIĆ. [1964] *A note on approximation by Bernstein polynomials*, BAMS **70,** 675–677.

BANACH, S. [1922] *An example of an orthogonal development whose sum is everywhere different from the developed function*, Proceedings of the London Mathematical Society **21,** 95–97. [1932] *Théorie des Opérations Linéaires*, Warsaw. Reprinted by Hafner, New York, 1950; Chelsea Publ. Co., New York, 1955. [1940] *Sur la divergence des interpolations*, SM **9,** 156–163.

BANACH, S., and H. STEINHAUS. [1927] *Sur le principe de la condensation de singularités*, FM **9,** 50–61.

BANASCHEWSKI, B. [1957] *On the Weierstrass-Stone approximation theorem*, FM **44,** 249–252.

BARAKET, R. [1961] *Evaluation of the incomplete gamma function of imaginary argument by Chebyshev polynomials*, MTAC **15,** 7–11.

BARBUTI, U. [1960] *Sulla teoria della migliore approssimazione nel senso di Tchebychev*, Rendiconti del Seminario Matematico dell' Universita di Padova **30,** 82–96, 302–308.

BARY, N. K. [1954] *Generalization of inequalities of S. N. Bernstein and M. M. Markov*, IAN **18,** 159–176 (Russian). [1961] *A Treatise on Trigonometric Series*, Moscow (Russian). = Translation, Macmillan, New York, 1964.

BASKAKOV, V. A. [1957] *An instance of a sequence of linear positive operators in the space of continuous functions*, DAN **113,** 249–251 (Russian).

BAUER, F. L. [1959] *The quotient-difference and epsilon algorithms*, ONA, 361–370.

BEALE, E. M. L. [1955] *On minimizing a convex function subject to linear inequalities*, Journal of the Royal Statistical Society **B17**, 173–177.

BECKENBACH, E. F. [1948] *Convex functions*, BAMS **54**, 439–460.

BERMAN, D. L. [1959] *On the impossibility of constructing a linear polynomial operator, giving an approximation of best order*, UMN **14**, 141–142 (Russian).

BERNSTEIN, S. N. [1912] *Démonstration du théorème de Weierstrass fondée sur le calcul de probabilités*, CKMS **13**, 1–2. [1912a] *Sur la valeur asymptotique de la meilleure approximation des fonctions analytiques*, CR **155**, 1062–1065. [1912b] *Sur l'ordre de la meilleure approximation des fonctions continues par des polynomes*, Académie Royale de Belgique, Classe des Sciences, Mémoires Collection in 4°, ser. II, **4**(1922). = Russian translation in CKMS **13**(1912), 49–194. Edited Russian version in Bernstein [1952]. = English translation of latter in Bernstein [1958]. [1913] *On asymptotic values of the best approximation of analytic functions*, CKMS **13**, 263–273 (Russian). = Bernstein [1958], 99–108. [1926] *Leçons sur les Propriétés Extrémales et la Meilleure Approximation des Fonctions Analytiques d'une Variable Réelle*, Gauthier-Villars, Paris. [1930] *Several remarks on polynomials of least deviation with integer coefficients*, DAN, 411–418 (Russian). = Bernstein [1958], 140–144. [1931] *Sur la limitation des valeurs d'un polynome $P_n(x)$ de degré n sur tout un segment par ses valeurs en $(n + 1)$ points du segment*, IAN, 1025–1050. [1932] *Sur une modification de la formule d'interpolation de Lagrange*, CKMS **5**, 49–57. [1936] *On periodic functions for which the best converging series is a Fourier series*, Trudy Leningrad Industr. In-ta, Phys.-Math. Science Section, no. 3, 1–8 (Russian). [1937] *Extremal Properties of Polynomials and the Best Approximation of Continuous Functions of a Single Real Variable, part I*, Leningrad-Moscow (Russian). [1938] *Sur le problème inverse de la théorie de la meilleure approximation des fonctions continues*, CR **206**, 1520–1523. [1952] *Collected Works*, Moscow (Russian). [1958] translation of vol. I of *Collected Works*, AECT no. 3460.

BITTNER, L. [1961] *Das Austauschverfahren der linearen Tschebyscheff-Approximation bei nicht erfüllter Haarscher Bedingung*, ZAMM **41**, 238–256.

BLUM, E. K., and P. C. CURTIS. [1961] *Asymptotic behavior of the best polynomial approximation*, ACMJ **8**, 645–647.

BOAS, R. P. [1946] *Fundamental sets of entire functions*, AM **47**, 21–32.

BOAS, R. P., and R. C. BUCK. [1958] *Polynomial Expansions of Analytic Functions*, Academic, New York.

BOEHM, B. W. [1964] *Existence, Characterization, and Convergence of Best Rational Tchebycheff Approximations*, thesis, University of California, Los Angeles. = RAND Corp. Report R-427-PR. [1964a] *Functions whose best rational Chebyshev approximations are polynomials*, NM **6**, 235–242. [1965] *Existence of best rational Tchebycheff approximations*, PJ, **15**, 19–28. [1965a] *Convergence of best rational Tchebycheff approximations*, TAMS, to appear.

BOHMAN, H. [1952] *On approximation of continuous and of analytic functions*, ARK **2**, 43–56.

BOJANIĆ, R. See Bajšanski, B.

BONNESEN, T., and W. FENCHEL. [1934] *Theorie der Konvexen Körper*, Springer, Berlin. Reprinted by Chelsea Publ. Co., New York, 1948.

BOREL, E. [1905] *Leçons sur les Fonctions de Variables Réelles*, Gauthier-Villars, Paris.

BRAM, J. [1958] *Chebychev approximation in locally compact spaces*, PAMS **9,** 133–136.

BRIONES, F. [1962] *Une nouvelle méthode pour calculer des alternantes au sens de Tchebichef*, CR **254,** 4417–4419. [1964] *On the alternants appearing in Chebyshev's best approximation problem*, NM **6,** 211–223.

BUCK, R. C. [1958] *Bounded continuous functions on a locally compact space*, Michigan Mathematical Journal **5,** 95–104. [1959] *Survey of recent Russian literature on approximation*, ONA, 341–359. [1959a] *Linear spaces and approximation theory*, ONA, 11–23. [1962] (Ed.) *Studies in Modern Analysis*, Prentice-Hall, Englewood Cliffs, N.J.

BUCK, R. C. See Boas, R. P.

BURKILL, J. C. [1959] *Lectures on Approximation by Polynomials*, Bombay.

BUROV, V. N. [1961] *The approximation of functions by polynomials satisfying non-linear relations*, DAN **138,** 515–517 (Russian). = SOVM **2,** 640–642.

BUTZER, P. L. [1953] *On two-dimensional Bernstein polynomials*, CJM **5,** 107–113. [1953a] *Linear combinations of Bernstein polynomials*, CJM **5,** 559–567. [1960] *Representation and approximation of functions by general singular integrals*, Nederlandse Akademie van Wetenschappen Indagationes Mathematicae **22,** 1–24.

CARATHÉODORY, C. [1907] *Über den Variabilitätsbereich der Koeffizienten von Potenzreihen*, MA **64,** 95–115. [1911] *Über den Variabilitätsbereich der Fourier'schen Konstanten von positiven harmonischen Funktionen*, RCMP **32,** 193–217.

CARLEMAN, T. [1922] *Über die Approximation analytischer Funktionen durch lineare Aggregate von vorgegebenen Potenzen*, AMAF **17.**

CARLSSON, S. O. [1962] *Orthogonality in normed linear spaces*, ARK **4,** 297–318.

CAUCHY, A. [1841] *Mémoire sur les fonctions alternées et sur les sommes alternées*, Exercices d'Analyse et de Phys. Math. **II,** 151–159, Bachelier, Paris. Reproduced in Oeuvres Complètes, 2d ser., **XII,** 173–182, Gauthier-Villars, Paris, 1916.

CHENEY, E. W. [1962] *Some relationships between the Tchebycheff approximations on an interval and on a discrete subset of that interval*, Mathematical Note no. 262, Boeing Scientific Research Laboratories, Seattle, Wash. [1965] *An elementary proof of Jackson's theorem on mean-approximation*, Mathematics Magazine **38,** 189–191. [1965a] *Approximation by generalized rational functions*, GM, 101–110.

CHENEY, E. W., and A. A. GOLDSTEIN. [1958a] *Note on a paper by Zuhovickii concerning the Tchebycheff problem for linear equations*, SIAMJ **6,** 233–239. [1959] *Newton's method for convex programming and Tchebycheff approximation*, NM **1,** 253–268. [1962] *Tchebycheff approximation in locally convex spaces*, BAMS **68,** 449–450. [1965] *Tchebycheff approximation and related extremal problems*, JMM, **14,** 87–98.

CHENEY, E. W., and H. L. LOEB. [1962] *On rational Chebyshev approximation*, NM **4,** 124–127. [1964] *Generalized rational approximation*, SIAMB **1,** 11–25.

CHENEY, E. W., and A. SHARMA. [1964] *Bernstein power series*, CJM **16,** 241–252. [1965] *On a generalization of Bernstein polynomials*, Rivista di Matematica Parma, (2) **5,** 77–84.

CHENEY, E. W., and T. H. SOUTHARD. [1963] *A survey of methods for rational approximation, with particular reference to a new method based on a formula of Darboux*, SIAMR **5,** 219–231.

CHENEY, E. W. See Goldstein, A. A.

CHERNIKOV, S. N. [1953] *Systems of linear inequalities*, UMN **8,** 7–73 (Russian).

CHLODOVSKY, I. [1925] *Une rèmarque sur la représentation des fonctions continues par des polynomes à coefficients entiers*, MSB **32,** 472–475. [1929] *Sur la représentation des fonctions discontinues par des polynomes de M. S. Bernstein*, FM **13,** 62–72. [1937] *Sur le développement des fonctions définies dans un intervalle infini en séries de polynomes de M. S. Bernstein*, Compositio Math. **4,** 380–393.

CHRISTOFFEL, E. B. [1858] *Über die Gaussische Quadratur und eine Verallgemeinerung derselben*, JRAM **55,** 61–82.

CHURCHILL, R. V. [1941] *Fourier Series and Boundary Value Problems*, McGraw-Hill, New York.

CIELSIELSKI, A. [1963] *Properties of the orthonormal Franklin system*, SM **23,** 141–157.

CLARKSON, J. A. [1936] *Uniformly convex spaces*, TAMS **40,** 396–414.

CLARKSON, J. A., and P. ERDÖS. [1943] *Approximation by polynomials*, DUKE **10,** 5–11.

CLENDENIN, W. W. [1961] *Notes on the construction of rational approximations for the error function and for similar functions*, ACMC **4,** 354–355.

CLENSHAW, C. W. [1955] *A note on the summation of Chebyshev series*, MTAC **9,** 118–120. [1962] *Chebyshev Series for Mathematical Functions*, Her Majesty's Stationery Office, London. [1964] *A comparison of "best" polynomial approximations with truncated Chebyshev series expansions*, SIAMB **1,** 26–37.

CLENSHAW, C. W., and A. R. CURTIS. [1960] *A method for numerical integration on an automatic computer*, NM **2,** 197–205.

CODY, W. J. [1965] *Chebyshev polynomial expansions of complete elliptic integrals*, MTAC **19,** 249–259.

COLLATZ, L. [1956] *Approximation von Funktionen bei einer und bei mehreren unabhängigen Veränderlichen*, ZAMM **36,** 198–211. [1960] *Tschebyscheffsche Annäherung mit rationalen Funktionen*, Abhandlungen aus dem Mathematischen Seminar der Universität Hamburg **24,** 70–78.

COLLINGE, R. See Werner, H.

CURTIS, A. R. See Clenshaw, C. W.

CURTIS, P. C. [1959] *N-parameter families and best approximation*, PJ **9,** 1013–1027. [1962] *Convergence of approximating polynomials*, PAMS **13,** 385–387. [1963] *Divergence of approximating polynomials*, PAMS **14,** 713–717.

CURTIS, P. C., and W. L. FRANK. [1959] *An algorithm for the determination of the polynomial of best minimax approximation to a function defined on a finite point set*, ACMJ **6,** 395–404.

CURTIS, P. C. See Blum, E. K.

DANTZIG, G. B. [1963] *Linear Programming and Extensions*, Princeton, Princeton, N.J.

DAVIS, P. J. [1963] *Interpolation and Approximation*, Blaisdell, New York.

DE LA VALLÉE POUSSIN, C. J. [1908] *Note sur l'approximation par un polynome d'une fonction dont la derivée est à variation bornée*, Bulletin Académie Belgique. [1910] *Sur les polynomes d'approximation et la représentation approchée d'un angle*, Académie Royale de Belgique, Bulletins de la Classe des Sciences, **12.** [1911] *Sur la methode de l'approximation minimum*, Societé Scientifique de Bruxelles, Annales, seconde partie, Mémoires **35,** 1–16. [1912] *Cours d'Analyse Infinitésimale*, 2d ed., Gauthier-Villars, Paris. [1918] *Sur la meilleure approximation des fonctions* *d'une variable réelle par des expressions d'ordre donné*, CR **166,** 799–802. [1919] *Leçons sur l'Approximation des Fonctions d'une Variable Réelle*, Gauthier-Villars, Paris. Reprinted in 1952.

DE VOGELAERE, R. [1959] *Remarks on the paper "Tchebysheff approximations for power series,"* ACMJ **6,** 111–114.

DEKKERS, A. J. See Duijvestijn, A. J. W.

DENT, B. A., and A. NEWHOUSE. [1959] *Polynomials orthogonal over discrete domains*, SIAMR **1,** 55–59.

DESCLOUX, J. [1961] *Contribution au Cacul des Approximations de Tschebicheff*, thesis, Eidgenössische Technische Hochschule, Zurich. [1961a] *Dégénérescence dans les approximations de Tschebyscheff linéaires et discrètes*, NM **3,** 180–187. [1963] *Approximations in L^p and Chebyshev approximations*, SIAMJ **11,** 1017–1026.

DICKINSON, D. R. [1939] *On Tschebyscheff polynomials*, Quarterly Journal of Mathematics, Oxford ser. **10,** 227–282; continued in **12**(1940), 184–191, and in JLMS **17**(1942), 211–217.

DINES, L. L. [1936] *Convex extension and linear inequalities*, BAMS **42,** 353–365.

DIRICHLET, G. L. [1829] *Sur la convergence des séries trigonométriques qui servent à représenter une fonction arbitraire entre des limites données*, JRAM **4,** 157–169.

DU BOIS-REYMOND, P. [1876] *Untersuchungen über die Convergenz und Divergenz der Fourierschen Darstellungsformeln* (mit drei lithographirten tafeln), Abhandlungen der Mathematisch-Physicalischen Classe der K. Bayerische Akademie der Wissenshaften **12,** 1–103.

DUFFIN, R. J., and A. C. SCHAEFFER. [1941] *A refinement of an inequality of the brothers Markoff*, TAMS **50,** 517–528.

DUIJVESTIJN, A. J. W., and A. J. DEKKERS. [1961] *Chebyshev approximations of some transcendental functions for use in digital computing*, Philips Research Reports **16,** 145–174.

EGGLESTON, H. G. [1958] *Convexity*, Cambridge, London.

EISENHART, C. [1963] *The background and evolution of the method of least squares*, International Statistical Institute, Ottawa, Canada.

ELLIOTT, D. [1964] *The evaluation and estimation of the coefficients in the Chebyshev series expansion of a function*, MTAC **18,** 274–284.

ELLIS, H. W., and D. G. KUEHNER. [1960] *On Schauder bases for spaces of continuous functions*, CMB **3,** 173–184.

ERDÖS, P. [1940] *Extremal properties of derivatives of polynomials*, AM **41,** 310–313. [1950] *Some theorems and remarks on interpolation*, ASM **12,** 11–17.

ERDÖS, P., and E. FELDHEIM. [1936] *Sur le mode de convergence pour l'interpolation de Lagrange*, CR **203,** 913–915.

ERDÖS, P., and P. TURÁN. [1937] *On interpolation*, AM **38,** 142–155; **39**(1938), 703–724; **41**(1940), 510–553. [1955] *On the role of the Lebesgue functions in the theory of Lagrange interpolation*, AMASH **6,** 47–60.

ERDÖS, P. See Clarkson, J. A.

FABER, G. [1914] *Über die interpolatorische Darstellung stetiger Funktionen*, DMV **23,** 192–210.

FAN, K. [1956] *On systems of linear inequalities*, pp. 99–156 in *Linear Inequalities and Related Systems*, Princeton, Princeton, N.J. [1958] *Convex Sets and Their Applications*, Argonne National Laboratory, Lemont, Ill.

FAVARD, J. [1937] *Sur les meilleurs procédés d'approximation de certaines classes de fonctions par des polynomes trigonométriques*, BSM **61,** 209–224, 243–256. [1944] *Sur les multiplicateurs d'interpolation*, JMPA **23,** 219–247.

FEJÉR, L. [1900] *Sur les fonctions bornées et intégrables*, CR **131,** 984–987. [1904] *Untersuchungen über Fouriersche Reihen*, MA **58,** 51–69. [1909] *Beispiele stetiger Funktionen mit divergenter Fourierreihen*, JRAM **137,** 1–5. [1909a] *Eine stetige Funktion, deren Fouriersche Reihe divergiert*, RCMP **28,** 402–404. [1910] *Lebesguesche Konstanten und divergente Fourierreihen*, JRAM **138,** 22–53. [1914] *Über konjugierte trigonometrische Reihen*, JRAM **144,** 48–56. [1916] *Über Interpolation*, Nachrichten der Akademie der Wissenschaften in Göttingen, 66–91. [1930] *Über Weierstrassche Approximation besonders durch Hermitesche Interpolation*, MA **102,** 707–725. [1930a] *Die Abschätzung eines Polynoms in einem Intervalle, wenn shranken für seine Werte und ersten Ableitungswerte in . . .*, MZ **32,** 426–457. [1932] *Bestimmung derjenigen Abszissen eines Intervalles . . .*, ASNP **1,** 263–276. [1934] *On the characterization of some remarkable systems of points of interpolation by means of conjugate points*, AMM **41,** 1–14.

FELDHEIM, E. [1939] *Théorie de la Convergence des Procédés d'Interpolation et de Quadrature Mécanique*, MSM **95.**

FELDHEIM, E. See Erdös, P.

FENCHEL, W. See Bonnesen, T.

FICHTENHOLZ, G. M. [1926] *Sur la notion de fermeture des systèmes de fonctions*, RCMP **50,** 385–398.

FOMIN, S. V. See Kolmogorov, A. N.

FORSYTHE, G. E. [1957] *Generation and use of orthogonal polynomials for data-fitting with a digital computer*, SIAMJ **5,** 74–88.

FORSYTHE, G. E. See Ascher, M.

FOURIER, J. B. J. [1824] *Solution d'une question particulière du calcul des inégalités, second extrait*, Histoire de l'Académie des Sciences, 48–51. Also in Oeuvres de Fourier **II,** 325–328, Paris, 1890.

FRANK, W. L. See Curtis, P. C.

FRASER, W., and J. F. HART. [1962] *On the computation of rational approximations to continuous functions*, ACMC **5,** 401–403, 414.

FREUD, G. [1955] *Über einseitige Approximation durch Polynome, I*, ASM **16,** 12–28. [1958] *Eine Ungleichung für Tschebyscheffsche Approximationspolynome*, ASM **19,** 162–164. [1959] *Über die Approximation reelter stetigen Funktionen durch gewöhnliche Polynome*, MA **137,** 17–25.

FROBENIUS, G. [1881] *Ueber Relationen zwischen den Näherungsbrüchen von Potenzreihen*, JRAM **90,** 1–17.

FRÖBERG, C.-E. [1961] *Rational Chebyshev approximations of elementary functions*, Nordisk Tidskrift för Informations Behandling **1,** 256–262.

GARKAVI, A. L. [1964] *Approximative properties of subspaces with finite defect in the space of continuous functions*, DAN **155,** 513–516 (Russian). = SOVM **5,** 440–443.

GAUSS, C. F. [1814] *Methodus nova integralium valores per approximationem inveniendi*, Collected Works, vol. 3, 163–196.

GELBAUM, B. R. [1950] *Expansions in Banach spaces*, DUKE **17,** 187–196.

GELFOND, A. O. [1955] *On uniform approximations by polynomials with integral coefficients*, UMN **10,** 11–65 (Russian).

GERONIMUS, J. [1935] *Sur quelques propriétés extrémales de polynomes dont les coefficients premiers sont donnés*, CKMS **12,** 49–59. [1960] *Polynomials Orthogonal on a Circle and Interval*, Pergamon Press, New York.

GOLDSTEIN, A. A. [1963] *On the stability of rational approximation*, NM **5,** 431–438. [1965] *Rational approximations on finite point sets*, GM. [1966] *Constructive Real Analysis*, Harper & Row, New York.

GOLDSTEIN, A. A., and E. W. CHENEY. [1958] *A finite algorithm for the solution of consistent linear equations and inequalities and for the Tchebycheff approximation of inconsistent linear equations*, PJ **8,** 415–427.

GOLDSTEIN, A. A., N. LEVINE, and J. B. HERESCHOFF. [1957] *On the "best" and "least qth" approximation of an overdetermined system of linear equations*, ACMJ **4,** 341–347.

GOLDSTEIN, A. A. See Cheney, E. W.

GOLOMB, M. [1960] *Approximation Theory*, Argonne National Laboratory, Applied Mathematics Division. Revised version, *Lectures on Theory of Approximation*, 1962.

GONCHAROV, V. L. [1934] *Theory of Interpolation and Approximation of Functions*, Moscow (Russian). New edition, Moscow, 1954.

GRAM, J. P. [1879] *Om Rackkendvilklinger bestemte ved Hjaelp af de mindste Kvadraters Methode*, Copenhagen. = [1883] *Über die Entwicklung reeller Funktionen in Reihen mittels der Methode der kleinsten Quadrate*, JRAM **94,** 41–73.

GREEN, J. W. [1954] *Recent applications of convex functions*, AMM **61,** 449–454.

GRÜNWALD, G. [1936] *Über Divergenzerscheinungen der Lagrangeschen Interpolationspolynome stetiger Funktionen*, AM **37,** 908–918. [1941] *Note on interpolation*, BAMS **47,** 257–260. [1943] *On the theory of interpolation*, ACTA **75,** 219–245.

GRÜNWALD, G., and P. TURÁN [1938] *Über Interpolation*, ASNP **7,** 137–146.

GUEST, P. G. [1961] *Numerical Methods of Curve Fitting*, Cambridge, London.

HAAR, A. [1918] *Die Minkowskische Geometrie und die Annäherung an stetige Funktionen*, MA **78,** 294–311.

HAHN, H. [1918] *Über das Interpolationsproblem*, MZ **1,** 115–142.

HARDY, G. H. [1916] *Weierstrass' non-differentiable function*, TAMS **17,** 301–325. [1942] *Note on Lebesgue's constants . . .*, JLMS **17,** 4–13.

HARDY, G. H., J. E. LITTLEWOOD, and G. PÓLYA. [1934] *Inequalities*, Cambridge, London.

HART, J. F., et al. [1968] *Handbook of Computer Approximations*, Wiley, New York.

HART, J. F. See Fraser, W.

HASTINGS, C., et al. [1955] *Approximations for Digital Computers*, Princeton, Princeton, N.J.

HAVINSON, S. YA. [1958] *On uniqueness of functions of best approximation in the metric of the space L^1*, IAN **22,** 243–270.

HELLY, E. [1923] *Über Mengen könvexer Körper mit gemeinschaftlichen Punkten*, DMV **32,** 175–176.

HENRICI, P. [1958] *The quotient-difference algorithm*, NBSAM **46,** 23–46. [1963] *Some applications of the quotient-difference algorithm*, AMS Proceedings of Symposia in Applied Mathematics **15,** 159–183.

HERESCHOFF, J. B. See Goldstein, A. A., N. Levine, and J. B. Hereschoff.

HERMITE, C. [1878] *Sur la formule d'interpolation de Lagrange*, JRAM **84,** 70–79.

HERSHEY, A. V. [1962] *Computing Programs for the Complex Fresnel Integral*, U. S. Naval Weapons Laboratory, Dahlgren, Va., Report 7670.

HERZBERGER, M. [1949] *The normal equations of the method of least squares and their solution*, Quarterly of Applied Mathematics **7**, 217–223.

HERZOG, F., and J. D. HILL. [1946] *The Bernstein polynomials for discontinuous functions*, AJM **68**, 109–124.

HEWITT, E. [1947] *Certain generalizations of the Weierstrass approximation theorem*, DUKE **14**, 419–427.

HEWITT, E., and H. S. ZUCKERMAN. [1959] *Approximation by polynomials with integral coefficients, a reformulation of the Stone-Weierstrass theorem*, DUKE **26**, 305–324.

HILL, J. D. See Herzog, F.

HILLE, E. See Shohat, J. A.

HIRSCHMAN, I. J. [1949] *Approximation by non-dense sets of functions*, AM **50**, 666–675.

HOBSON, E. W. [1926] *The Theory of Functions of a Real Variable and the Theory of Fourier's Series*, Cambridge, London. = Dover, New York, 1957.

HOEL, P. G. [1935] *Certain problems in the theory of closest approximation*, AJM **57**, 891–901.

HORN, A. [1949] *Some generalizations of Helly's theorem on convex sets*, BAMS **55**, 923–929.

HORNECKER, G. [1958] *Évaluation approchée de la meilleure approximation polynomiale d'ordre n de f(x) sur un segment fini [a,b]*, Chiffres **1**, 157–169. [1959] *Approximations rationnelles voisines de la meilleure approximation au sens de Tchebycheff*, CR **249**, 939–941. [1959a] *Determination des meilleures approximations rationnelles (au sens de Tchebichef) de fonctions réelles d'une variable sur un segment fini et des bornes d'erreur correspondantes*, CR **249**, 2265–2267. [1960] *Méthodes pratiques pour la détermination approchée de la meilleure approximation polynomiale ou rationnelle*, Chiffres **3**, 193–228. [1961] *Nouvelle évaluation d'une borne supérieure de E_n, meilleure approximation polynomiale d'ordre n*, Chiffres **4**, 37–40.

HSIEH-CHANG, S. [1964] *Completeness of the sequence of functions $f(\lambda_n z)$*, Acta Math. Sinica **14**, 103–118 (Chinese). = Chinese Math. **5**, 112–128.

HSU, L. C. [1961] *Approximation of non-bounded continuous functions by certain sequences of linear positive operators or polynomials*, SM **21**, 37–43.

HURST, R. P. See Miller, J.

ISBELL, J. R. [1958] *Algebras of uniformly continuous functions*, AM **68**, 96–125.

JACKSON, D. [1911] *Über die Genauigkeit der Annäherung Stetiger Funktionen Durch Ganze Rationale Funktionen Gegebenen Grades und Trigonometrische Summen Gegebener Ordnung*, dissertation, Göttingen. [1912] *On approximation by trigonometric sums and polynomials*, TAMS **13**, 491–515. [1921] *Note on a class of polynomials of approximation*, TAMS **22**, 320–326. [1930] *The theory of approximation*, AMSC **XI**.

JAMES, R. C. [1950] *Bases and reflexitivity of Banach spaces*, AM **52**, 518–527.

JEFFREY, R. L. [1956] *Trigonometric Series*, Canadian Math. Congress, Toronto.

JENSEN, J. L. W. V. [1906] *Sur les fonctions convexes et les inégalités entre les valeurs moyennes*, ACTA **30**, 175–193.

JURKAT, W. B., and G. G. LORENTZ. [1961] *Uniform approximation by polynomials with positive coefficients*, DUKE **28**, 463–474.

KACZMARZ, S., and H. STEINHAUS. [1935] *Theorie der Orthogonalreihen*, 1st ed., Warsaw; 2d ed., Chelsea Pub. Co., New York, 1951.

KAKEYA, S. [1914] *On approximate polynomials*, TMJ **6,** 182–186.

KAKUTANI, S. [1941] *Concrete representation of abstract* (M)*-spaces*, AM **42,** 994–1024.

KANTOROVICH, L. V. [1930] *Sur certains développements suivant les polynomes de la forme de S. Bernstein, I, II*, DAN, 563–568, 595–600. [1931] *Sur la convergence de la suite des polynomes de S. Bernstein en dehors de l'interval fundamental*, IAN, 1103–1115. [1931a] *Quelques observations sur l'approximation de fonctions au moyen de polynomes à coefficients entiers*, IAN, 1163–1168.

KARLIN, S. [1948] *Bases in Banach spaces*, DUKE **15,** 971–985.

KHINTCHINE, A. [1956] *Kettenbrüche*, Teubner, Leipzig.

KHOVANSKI, A. N. [1956] *The Application of Continued Fractions and Their Generalizations to Problems in Approximation Theory*, Moscow (Russian). = English translation, Noordhoff, Groningen, Netherlands, 1963.

KIRCHBERGER, P. [1903] *Über Tchebychefsche Annäherungsmethoden*, dissertation, Göttingen, 1902. = MA **57,** 509–540.

KLEE, V. [1963] (Ed.) *A Symposium on Convexity*, Proceedings of Symposia in Pure Mathematics **7,** AMS.

KOGBETLIANTZ, E. G. [1957] Four papers on the computation of elementary functions, IBM Journal of Research and Development **1**(1957), 110–115; **2**(1958), 43–53, 218–222; **3**(1959), 147–152. [1960] *Generation of elementary functions*, pp. 7–35 in Ralston and Wilf [1960].

KOLMOGOROV, A. N. [1926] *Une série de Fourier-Lebesgue divergente partout*, CR **183,** 1327–1328. [1948] *A remark on the polynomials of P. L. Čebyšev deviating the least from a given function*, UMN **3,** 216–221 (Russian).

KOLMOGOROV, A. N., and S. V. FOMIN. [1957] *Elements of the Theory of Functions and Functional Analysis*, Graylock Press, Rochester, N.Y.

KORKIN, A. N., and E. I. SOLOTAREFF. [1873] *Sur une certain minimum*, Nouvelles Annales de Mathématiques **2.**

KORNEICUK, N. P. [1962] *The exact constant in D. Jackson's theorem on best uniform approximation of continuous periodic functions*, DAN **145,** 514–515. = SOVM **3,** 1040–1041.

KOROVKIN, P. P. [1953] *On convergence of linear positive operators in the space of continuous functions*, DAN **90,** 961–964 (Russian). [1959] *Linear Operators and Approximation Theory*, Fizmatgiz, Moscow (Russian). = translation, Hindustan Publ. Corp., Delhi, 1960.

KOSLOFF, V. YA. [1950] *On completeness of function systems*, DAN **73,** 441–444 (Russian).

KREIN, M. G. [1938] *The L-problem in an abstract linear normed space*, pp. 175–204 in Achieser and Krein [1938]. [1951] *The ideas of P. L. Čebyšev and A. A. Markov in the theory of limiting values of integrals and their further development*, UMN **6,** 3–120 (Russian). = AMST (II)**12,** 1–122.

KREIN, M. G. See Achieser, N. I.

KREIN, M., D. MILMAN, and M. RUTMAN. [1940] *A note on basis in Banach space*, Comm. Inst. Sci. Math. Mec. Univ. Kharkoff [Zapiski Inst. Mat. Mech.], ser. 4, **16,** 106–110 (Russian and English).

KRIPKE, B. [1964] *Best approximation with respect to nearby norms*, NM **6,** 103–105.

KRIPKE, B. R., and T. J. RIVLIN. [1965] *Approximation in the metric of* $L^1(X, \mu)$, TAMS **119,** 101–122.

KUEHNER, D. G. See Ellis, H. W.

KUHN, H., and A. W. TUCKER. [1956] *Linear Inequalities and Related Systems*, Princeton, Princeton, N.J.

LAASONEN, P. [1949] *Einige Sätze über Tschebyscheffsche Funktionensysteme*, Annales Academiae Scientiarum Fennicae, ser. AI, **52,** 3–24.

LAGUERRE, E. [1885] *Sur la réduction en fractions continues d'une fonction qui satisfait à une équation différentielle linéaire du premier ordre dont les coefficients sont rationnels*, JMPA **1,** 135–165.

LANCE, G. N. [1960] *Numerical Methods for High Speed Computers*, Iliffe and Sons Ltd., London.

LANCZOS, C. [1938] *Trigonometric interpolation of empirical and analytical functions*, JMP **17,** 123–199. [1952] *Tables of Chebyshev polynomials* $S_n(x)$ *and* $C_n(x)$, NBSAM **9.** [1956] *Applied Analysis*, Prentice-Hall, Englewood Cliffs, N.J.

LANDAU, E. [1908] *Über die Approximation einer stetigen Funktionen durch eine ganze rationale Funktion*, RCMP **25,** 337–345.

LANGDON, L. R. [1955] *Approximating functions for digital computers*, Industrial Mathematics **6,** 79–100.

LAWSON, C. L. [1961] *Contributions to the Theory of Linear Least Maximum Approximation*, dissertation, University of California, Los Angeles. Reprinted by Jet Propulsion Laboratory, Pasadena, Calif. [1963] *Segmented Rational Minmax Approximation, Characteristic Properties and Computational Methods*, Jet Propulsion Laboratory, Pasadena, Calif., Report 32–579. [1964] *Characteristic properties of the segmented rational minmax approximation problem*, NM **6,** 293–301.

LAX, P. D. [1944] *Proof of a conjecture of P. Erdös on the derivative of a polynomial*, BAMS **50,** 509–513.

LEBESGUE, H. [1898] *Sur l'approximation des fonctions*, BSM **22,** 278–287. [1908] *Sur la représentation approchée des fonctions*, RCMP **26,** 325–328. [1909] *Sur les intégrales singulières*, AFST **1,** 25–117.

LERCH, M. [1903] *Sur un point de la théorie des fonctions génératrices d'Abel*, ACTA **27,** 339–352.

LEVIN, B. YA. See Achieser, N. I.

LEVINE, N. See Goldstein, A. A.; see Salzer, H. E.

LEWIS, D. C. [1947] *Polynomial least square approximations*, AJM **69,** 273–278.

LITTLEWOOD, J. E. See Hardy, G. H.

LOEB, H. L. [1957] *On rational fraction approximations at discrete points*, Convair Astronautics Applied Mathematics, ser. 9. [1960] *Algorithms for Chebyshev approximations using the ratio of linear forms*, SIAMJ **8,** 458–465.

LOEB, H. L. See Cheney, E. W.

LORENTZ, G. G. [1937] *Zur theorie der polynome von S. Bernstein*, MSB **2,** 543–556. [1951] *Deferred Bernstein polynomials*, PAMS **1,** 72–76. [1953] *Bernstein Polynomials*, University of Toronto Press, Toronto, Canada. [1960] *Approximation of smooth functions*, BAMS **66,** 124–125. [1960a] *Lower bounds for the degree of approximation*, TAMS **97,** 25–34. [1962] *Metric entropy, widths, and superpositions of functions*, AMM **69,** 469–485.

LORENTZ, G. G. See Jurkat, W. B.

LOZINSKI, S. [1944] *On convergence and summability of Fourier series and interpolation processes*, MSB **56**, 175–263 (English). [1948] *On a class of linear operators*, DAN **61**, 193–196 (Russian).

LUKE, Y. L. [1955] *Remarks on the τ-method for the solution of linear differential equations with rational coefficients*, SIAMJ **3**, 179–191. [1957] *Rational approximations to the exponential function*, ACMJ **4**, 24–29. [1957a] *On the computation of log Z and arctan Z*, MTAC **11**, 16–18. [1958] *The Padé and the τ-method*, JMP **37**, 110–127. [1960] *On economic representations of transcendental functions*, JMP **38**, 279–294.

MACON, N. [1955] *On the computation of exponential and hyperbolic functions using continued fractions*, ACMJ **2**, 262–266.

MAEHLY, H. J. [1960] *Methods for fitting rational approximations*, ACMJ **7**, 150–162; **10**(1963), 257–277.

MAEHLY, H., and C. WITZGALL. [1960] *Tschebyscheff-Approximationen in kleinen Intervallen*, NM **2**, 142–150, 293–307.

MAIRHUBER, J. C. [1956] *On Haar's theorem concerning Chebychev approximation problems having unique solutions*, PAMS **7**, 609–615.

MARCINKIEWICZ, J. [1937] *Quelques remarques sur l'interpolation*, Acta Litterarum ac Scientiarum, Szeged **8**, 127–130. [1937a] *Sur la divergence des polynomes d'interpolation*, ibid., 131–135. [1964] *Collected Papers*, Krakowskie Przedmiescie, Warsaw.

MARKOFF, A. A. [1889] *Sur une question posée par Mendeleieff*, IAN **62**, 1–24. [1896] *Differenzenrechnung*, Teubner, Leipzig. [1898] *On the asymptotic values of integrals in relation to interpolation*, Papers of the Academy of Science, St. Petersburg, ser. 8, **6**.

MARKOFF, V. A. [1892] *On functions deviating the least from zero on a given interval*, St. Petersburg (Russian). = *Über Polynome, die in einem gegebenen Intervalle möglichst wenig von Null abweichen*, MA **77**(1916), 213–258.

MEINARDUS, G. [1961] *Über Approximationen analytischer Funktionen in einem reellen Intervall*, ARMA **7**, 143–159. [1962] *Über Tschebyscheffsche Approximationen*, ARMA **9**, 329–351. [1964] *Approximation von Funktionen und Ihre Numerische Behandlung*, Springer, Berlin.

MEINARDUS, G., and D. SCHWEDT. [1964] *Nicht-lineare Approximation*, ARMA **17**, 297–326.

MENCHOFF, D. [1944] *Sur les sommes partielles des séries de Fourier des fonctions continues*, MSB **57**, 385–430.

MÉRAY, C. [1884] *Observations sur la légitimité de l'interpolation*, ASENS **1**, 165–176. [1896] *Nouveaux exemples d'interpolations illusoires*, BSM **20**, 266–270.

MEYER-KÖNIG, W., and K. ZELLER. [1960] *Bernsteinsche Potenzreihen*, SM **19**, 89–94.

MILLER, J., and R. P. HURST. [1958] *Simplified calculation of the exponential integral*, MTAC **12**, 187–193.

MILMAN, D. See Krein, M.

MILNE, W. E. See Thacher, H. C.

MINNICK, R. C. [1957] *Tchebysheff approximations for power series*, ACMJ **4**, 487–504.

MITTAG-LEFFLER, G. [1900] *Sur la représentation analytique des fonctions d'une variable réelle*, RCMP **14**, 217–224.

MONTEL, P. [1910] *Leçons sur les Séries de Polynomes à une Variable Complexe*, Gauthier-Villars, Paris.

MORONEY, R. M. [1961] *The Haar problem in L_1*, PAMS **12,** 793–795.

MOTZKIN, T. S. [1936] *Beiträge zur Theorie der Linearen Ungleichungen*, dissertation, Basel. = translation, RAND Corp., Santa Monica, Calif., 1958. [1949] *Approximation by curves of a unisolvent family*, BAMS **55,** 789–793. [1955] *Evaluation of polynomials*, BAMS **61,** 163, abstract no. 315B.

MOTZKIN, T. S., and I. J. SCHOENBERG. [1954] *The relaxation method for linear inequalities*, CJM **6,** 393–404.

MOTZKIN, T. S., and J. L. WALSH. [1955] *Least pth power polynomials on a real finite point set*, TAMS **78,** 67–81. [1956] *Least pth power polynomials on a finite point set*, TAMS **83,** 371–396.

MOTZKIN, T. S. See Walsh, J. L.

MOURSUND, D. G. [1964] *Chebyshev approximations of a function and its derivatives*, MTAC **18,** 382–389. [1965] *Some computational aspects of the uniform approximation of a function and its derivative*, SIAMJ, to appear.

MOURSUND, D. G., and A. H. STROUD. [1965] *The best Čebyšhev approximation to a function and its derivative on $n + 2$ points*, SIAMB, **2,** 15–23.

MÜNTZ, C. [1914] *Über den Approximationssatz von Weierstrass*, H. A. Schwarz Festschrift, Mathematische Abhandlungen, Berlin, 303–312.

MURNAGHAN, F. D., and J. W. WRENCH. [1959] *The determination of the Chebyshev approximating polynomial for a differentiable function*, MTAC **13,** 185–193. [1960] *The Approximation of Differentiable Functions*, David Taylor Model Basin, Washington, D.C., Report 1175.

NATANSON, I. P. [1949] *Constructive Theory of Functions*, Gostekhizdat, Moscow. = [1955] *Konstructive Funktionentheorie*, Akademie-Verlag, Berlin. = [1961] English translation, AECT no. 4503.

NEWHOUSE, A. See Dent, B. A.

NEWMAN, D. J., and H. S. SHAPIRO. [1962] *Some theorems on Čebyšev approximation*, DUKE **30**(1963), 673–682; abstract, NAMS **9**(1962), 143. [1965] *Approximation by generalized rational functions*, to appear.

NIKOLAEV, V. F. [1948] *On the question of approximation of continuous functions by means of polynomials*, DAN **61,** 201–204 (Russian).

NIKOLSKI, S. M. [1940] *Mean approximation of functions by trigonometric polynomials*, IAN **4,** 207–256 (Russian).

NITSCHE, J. C. C. [1962] *Über die Abhängigkeit der Tschebyscheffschen Approximierenden einer differenzierbaren Funktion vom Intervall*, NM **4,** 262–276.

NÖRLUND, N. E. [1924] *Vorlesungen über Differenzenrechnung*, Springer, Berlin.

NOVODVORSKII, E. N., and I. SH. PINSKER. [1951] *On a process of equalization of maxima*, UMN **6,** 174–181 (Russian).

OKADA, Y. [1923] *On approximate polynomials with integral coefficients only*, TMJ **23,** 26–35.

OLDS, C. D. [1950] *The best polynomial approximation of functions*, AMM **57,** 617–621.

OSTROWSKI, A. M. [1954] *On two problems in abstract algebra connected with Horner's rule*, pp. 40–43 in Studies in Mathematics and Mechanics Presented to R. von Mises, Academic, New York.

PADÉ, H. [1892] *Sur la représentation approchée d'une fonction par des fractions rationnelles*, ASENS **9** supplement, 1–93.

PAIGE, L. J., and J. D. SWIFT. [1961] *Elements of Linear Algebra*, Ginn, Boston.

PAINLEVÉ, P. [1898] *Sur le développement des fonctions analytiques pour les valeurs réelles des variables*, CR **126,** 385–388.

PÁL, J. [1914] *Zwei kleine Bemerkungen*, TMJ **6,** 42–43.

PAN, V. YA. [1959] *Schemes for the computation of polynomials with real coefficients*, DAN **127,** 266–269 (Russian).

PASZKOWSKI, S. [1955] *Sur l'approximation uniforme avec des noeuds*, Annales Polonici Mathematica **2,** 129–146. [1956] *On the Weierstrass approximation theorem*, Colloquium Math. **4,** 206–210. [1957] *On approximation with nodes*, Rozprawy Matematyczne **14,** Warsaw. [1962] *The theory of uniform approximation, I: Non-asymptotic theoretical problems*, Rozprawy Matematyczne **26,** Warsaw.

PERRON, O. [1929] *Die Lehre von Kettenbrüchen*, Teubner, Leipzig. [1941] *Über die Approximation stetiger Funktionen durch trigonometrische Polynome*, MZ **47,** 57–65.

PHELPS, R. R. [1960] *Uniqueness of Hahn-Banach extensions and unique best approximation*, TAMS **95,** 238–255. [1963] *Čebyšev subspaces of finite codimension in $C(X)$*, PJ **13,** 647–655. [1966] *Čebyšev subspaces of finite dimension in L_1*, PAMS **17,** 646–652.

PICARD, E. [1891] *Sur la représentation approchée des fonctions*, CR **112,** 183–186.

PINSKER, I. S. See Novodvorskii, E. N.

PÓLYA, G. [1913] *Sur un algorithme toujours convergent pour obtenir les polynomes de meilleure approximation de Tchebycheff pour une fonction continue quelconque*, CR **157,** 840–843. [1933] *Über die Konvergenz von Quadraturverfahren*, MZ **37,** 264–286.

PÓLYA, G., and I. J. SCHOENBERG. [1958] *Remarks on de La Vallée Poussin means and convex conformal maps of the circle*, PJ **8,** 295–334.

PÓLYA, G., and G. SZEGÖ. [1925] *Aufgaben und Lehrsätze aus der Analysis*, Springer, Berlin.

PÓLYA, G. See Hardy, G. H.

POPOVICIU, T. [1935] *Sur l'approximation des fonctions convexes d'ordre supérieur*, Mathematica (Cluj) **10,** 49–54. [1944] *Les Fonctions Convexes*, Hermann & Cie., Paris.

PTÁK, V. [1958] *A remark on approximation of continuous functions*, Czechoslovak Math. J. **8,** 251–256. [1958a] *On approximation of continuous functions in the metric* $\int_a^b |x(t)|\,dt$, ibid., 267–273.

RADEMACHER, H., and I. J. SCHOENBERG. [1950] *Helly's theorem on convex domains and Tchebycheff's approximation problem*, CJM **2,** 245–256.

RALSTON, A., and H. S. WILF. [1960] (Eds.) *Mathematical Methods for Digital Computers*, Wiley, New York.

REMES, E. YA. [1934] *Sur le calcul effectif des polynomes d'approximation de Tchebichef*, CR **199,** 337–340. [1935] *On the Best Approximation of Functions in the Tchebycheff Sense*, Kiev (Ukrainian). [1956] *On effective solution of a system of incompatible linear equations according to the Tchebycheff principle of optimum uniform approximation*, Dopovidi Akademiia Nauk URSS, 315–320 (Ukrainian). [1957] *General Computational Methods of Tchebycheff Approximation*, Kiev (Russian). AECT no. 4491.

RICE, J. R. [1959] *On the convergence of an algorithm for best Tchebycheff approxi-*

mations, SIAMJ **7,** 133–142. [1960] *The characterization of best nonlinear Tchebycheff approximations*, TAMS **96,** 322–340. [1960a] *Criteria for the existence and equioscillation of best Tchebycheff approximations*, NBSJ **64,** 91–93. [1960b] *Chebyshev approximation by $ab^x + c$*, SIAMJ **8,** 691–702. [1961] *Tchebycheff approximations by functions unisolvent of variable degree*, TAMS **99,** 298–302. [1961a] *Best approximations and interpolating functions*, TAMS **101,** 477–498. [1961b] *Algorithms for Chebyshev approximation . . .*, SIAMJ **9,** 571–583. [1962] *Chebyshev approximation by exponentials*, SIAMJ **10,** 149–161. [1962a] *Tchebycheff approximation in a compact metric space*, BAMS **68,** 405–410. [1962b] *Computer Approximations for Physical Tables*, General Motors Research Laboratories, Warren, Mich., Report 387. [1963] *Approximation with convex constraints*, SIAMJ **11,** 15–32. [1963a] *Tchebycheff approximation in several variables*, TAMS **109,** 444–466. [1964] *The Approximation of Functions, vol. I: Linear Theory*, Addison-Wesley, Reading, Mass. [1964a] *On the existence and characterization of best nonlinear Tchebycheff approximations*, TAMS **110,** 88–97. [1964b] *On the L_∞ Walsh arrays for $\Gamma(x)$ and* Erfc (x), MTAC **18,** 617–626. [1964c] *On nonlinear L_1 approximation*, ARMA **17,** 61–66. [1964d] *On the computation of L_1 approximations by exponentials, rationals, and other functions*, MTAC **18,** 390–396.

RIESZ, F. [1918] *Über lineare Funktionalgleichungen*, ACTA **41,** 71–98.

RIESZ, M. [1914] *Eine trigonometrische Interpolationsformel und einige Ungleichungen für Polynome*, DMV **23,** 354–368. [1941a] *Formule d'interpolation pour la dérivée d'un polynome trigonométrique*, CR **158,** 1152–1154.

RIVLIN, T. J. [1962] *Polynomials of best uniform approximation to certain rational functions*, NM **4,** 345–349. [1962a] *Čebyšev expansions and best uniform approximation*, IBM Research Report RZ-93, Zurich. [1964] *A property of the ratio of trigonometric polynomials*, SIAMB **1,** 131–132.

RIVLIN, T. J., and H. S. SHAPIRO. [1960] *Some uniqueness problems in approximation theory*, Communications on Pure and Applied Mathematics **13,** 35–47. [1961] *A unified approach to certain problems of approximation and minimization*, SIAMJ **9,** 670–699.

RIVLIN, T. J. See Kripke, B. R.

ROGOSINSKI, W. W. [1954] *Extremum problems for polynomials and trigonometrical polynomials*, JLMS **29,** 259–275. [1955] *Some elementary inequalities for polynomials*, Mathematical Gazette **39,** 7–12.

RUBINOV, A. M. See Akilov, G. P.

RUBINSTEIN, G. S. [1955] *A method of studying convex sets*, DAN **102,** 451–454 (Russian).

RUNCK, P. O. [1961] *Über Konvergenzfragen bei Polynominterpolation mit äquidistanten Knoten*, JMPA **208,** 51–69; **210,** 175–204.

RUNGE, C. [1885] *Zur Theorie der eindeutigen analytischen Funktionen*, ACTA **6,** 229–245. [1885a] *Über die Darstellung willkürlicher Funktionen*, ACTA **7,** 387–392. [1901] *Über empirische Funktionen und die Interpolation zwischen äquidistanten Ordinaten*, Zeitschrift für Mathematik und Physik **46,** 224–243.

RUTISHAUSER, H. [1956] *Der Quotienten-Differenzen-Algorithmus*, Birkhauser Verlag, Basel.

RUTMAN, M. See Krein, M.

SALZER, H. E. [1944] *A new formula for inverse interpolation*, BAMS **50,** 513–516. [1951] *Checking and interpolation of functions tabulated at certain irregular logarithmic intervals*, NBSJ **46,** 74–77. [1951a] *Formulas for finding the argument for which a function has a given derivative*, MTAC **5,** 213–215. [1956] *Osculatory* *extrapolation and a new method for numerical integration of differential equations*, Journal of the Franklin Institute **262,** 111–120. [1957] *Numerical integration of $y'' = \phi(x,y,y')$ using osculatory interpolation*, ibid. **263,** 401–409. [1959] *Formulae for hyperosculatory interpolation, direct and inverse*, Quarterly Journal of Mechanics and Applied Mathematics **12,** 100–110. [1959a] *Tables of Osculatory Interpolation Coefficients*, NBSAM **56.** [1959b] *Best approximation of mixed type*, SIAMJ **7,** 345–360. [1959c] *Some new divided difference algorithms for two variables*, ONA, 61–98. [1960] *New formulas for trigonometric interpolation*, JMP **39,** 83–96. [1960a] *Alternative formulas for osculatory and hyperosculatory inverse interpolation*, MTAC **14,** 257–261. [1960b] *Formulae for complex Cartesian hyperosculatory interpolaiion*, JMP **39,** 300–307. [1960c] *Hermite's general osculatory interpolation formula and a finite difference analogue*, SIAMJ **8,** 18–27. [1960d] *Optimal points for numerical differentiation*, NM **2,** 214–227.

SALZER, H. E., and N. LEVINE. [1961] *Table of a Weierstrass continuous non-differentiable function*, MTAC **15,** 120–130.

SANSONE, G. [1951] *Orthogonal Functions*, translation, Interscience, New York, 1959.

SARD, A. [1948] *Integral representation of remainders*, DUKE **15,** 333–345. [1949] *Best approximate integration formulae; best approximation formulae*, AJM **71,** 80–91. [1963] *Linear Approximation*, AMS Math. Surveys **9.**

SCHAEFFER, A. C. See Duffin, R. J.

SCHATTEN, R. [1950] *"Closing-up" of sequence spaces*, AMM **57,** 603–616.

SCHÅUDER, J. [1927] *Zur Theorie stetiger Abbildungen in Funktionalräumen*, MZ **26,** 47–65.

SCHNIRELMANN, L. G. [1938] *On uniform approximation*, IAN **1,** 53–60 (Russian).

SCHOENBERG, I. J. [1959] *On variation diminishing approximation methods*, ONA, 249–274. [1960] *On the question of unicity in the theory of best approximation*, NYAS **86,** 682–692.

SCHOENBERG, I. J., and C. T. YANG. [1961] *On the unicity of solutions of problems of best approximation*, AMPA **54,** 1–12.

SCHOENBERG, I. J. See Motzkin, T. S.

SCHOENBERG, I. J. See Pólya, G.

SCHOENBERG, I. J. See Rademacher, H.

SCHWEDT, D. See Meinardus, G.

SCHWERDTFEGER, H. [1960] *Notes on numerical analysis II, interpolation and curve fitting by sectionally linear functions*, CMB **3,** 41–57. Also CMB **4**(1961), 53–55. [1963] *Fonctions polygonales et relations récurrentes*, CR **256,** 4350–4353.

SEIDEL, L. [1846] *Untersuchung über die Convergenz und Divergenz Kettenbrüche*, dissertation, Munich.

SEMADENI, Z. [1963] *Product Schauder bases and approximation with nodes in spaces of continuous functions*, BAP **11,** 387–391.

SEWELL, W. E. [1942] *Degree of Approximation by Polynomials in the Complex Domain*, Princeton, Princeton, N.J.

SHAPIRO, H. S. See Newman, D. J.

SHAPIRO, H. S. See Rivlin, T. J.

SHARMA, A. [1964] *Remarks on Quasi-Hermite-Fejér Interpolation*, CMB **7,** 101–119.

SHARMA, A. See Cheney, E. W.

SHARMA, A. See Walsh, J. L.

SHELL, D. L. See Spitzbart, A.

SHENITZER, A. [1957] *Chebyshev approximation of a continuous function by a class of functions*, ACMJ **4,** 30–35.

SHISHA, O. [1965] *Monotone approximation*, PJ, **15,** 667–671.

SHOHAT, J. [1933] *A simple proof of a formula of Tchebycheff*, TMJ **36,** 230–235. [1934] *Théorie Générale des Polynomes Orthogonaux de Tchebichef*, MSM **66.** [1935] *On the development of functions in series of orthogonal polynomials*, BAMS **41,** 49–82. [1941] *The best polynomial approximation of functions possessing derivatives*, DUKE **8,** 376–385.

SHOHAT, J. A., E. HILLE, and J. L. WALSH. [1940] *A Bibliography on Orthogonal Polynomials*, Bulletins of the National Research Council **103.**

SIEKLUCKI, K. [1958] *Topological properties of sets admitting the Tschebycheff systems*, BAP **6,** 603–606.

SIMMONS, G. F. [1963] *Introduction to Topology and Modern Analysis*, McGraw-Hill, New York.

SINGER, I. [1957] *On uniqueness of the element of best approximation in arbitrary Banach spaces*, Academia Republicii Populare Romine. Institutul de Matematica Studii si Cercetari Matematicae **8,** 235–244 (Roumanian). [1960] *On best approximation of continuous functions*, MA **140,** 165–168. [1961] *Weak* bases in conjugate Banach spaces*, SM **21,** 75–81.

SMIRNOV, V. I. [1961] *Investigations on Contemporary Problems in the Constructive Theory of Functions*, Moscow (Russian).

SOBLE, A. B. [1957] *Majorants of polynomial derivatives*, AMM **64,** 639–643.

SOLOTAREFF, E. I. See Korkin, A. N.

SOUTHARD, T. H. [1957] *Approximation and table of the Weierstrass $\wp$ function in the equianharmonic case for real argument*, MTAC **11,** 99–100.

SOUTHARD, T. H. See Cheney, E. W.

SPIELBERG, K. [1961] *Efficient continued fraction approximations to elementary functions*, MTAC **15,** 409–417. [1961a] *Representation of power series in terms of polynomials, rational approximation and continued fractions*, ACMJ **8,** 613–627. [1962] *Polynomial and continued-fraction approximations for logarithmic functions*, MTAC **16,** 205–217.

SPITZBART, A., and D. L. SHELL. [1958] *A Chebycheff fitting criterion*, ACMJ **5,** 22–31.

STEFFENSEN, J. F. [1927] *Interpolation*, Chelsea Publ. Co., New York. Reprinted in 1950.

STEGUN, I. See Abramowitz, M.

STEINHAUS, H. See Banach, S.

STEINHAUS, H. See Kaczmarz, S.

STESIN, I. M. [1957] *Conversion of orthogonal expansions into a sequence of convergents*, Akademiya Nauk SSSR Vycislitelnaya Matematika **1,** 116–119 (Russian).

STIEFEL, E. L. [1958] *Kernel polynomials in linear algebra and their numerical applications*, NBSAM **49,** 1–22. [1959] *Numerical methods of Tchebycheff approximation*, ONA, 217–232. [1959a] *Über diskrete und lineare Tschebyscheff-Approximationen*, NM **1,** 1–28. [1960] *Note on Jordan elimination, linear programming and Tchebycheff approximation*, NM **2,** 1–17. [1964] *Methods—old and new—for solving the Tchebycheff approximation problem*, SIAMB **1,** 164–176.

STIELTJES, T. J. [1884] *Quelques recherches sur la théorie des quadratures dites mécaniques*, ASENS **1,** 409–426. [1894] *Recherches sur les fractions continues*, AFST **8,** 1–122; **9,**(1895), 1–47. = Oeuvres **II,** 402–559. [1914] Oeuvres, Noordhoff, Groningen, Netherlands.

STOER, J. [1961] *Über zwei Algorithmen zur Interpolation mit rationalen Funktionen,* NM **3,** 285–304. [1964] *A direct method for Chebyshev approximation by rational functions,* ACMJ **11,** 59–69.

STOLL, R. R. [1952] *Linear Algebra and Matrix Theory,* McGraw-Hill, New York.

STONE, H. [1961] *Approximation of curves by line segments,* MTAC **15,** 40–47.

STONE, M. H. [1937] *Applications of the theory of Boolean rings to general topology,* TAMS **41,** 375–481. [1948] *The generalized Weierstrass approximation theorem,* Mathematics Magazine **21,** 167–183, 237–254.

STRANG, G. [1962] *Polynomial approximation of Bernstein type,* TAMS **105,** 525–535.

STROUD, A. H. [1961] *A bibliography on approximate integration,* MTAC **15,** 52–80.

STROUD, A. H. See Moursund, D. G.

SWIFT, J. D. See Paige, L. J.

SZ.-NAGY, B. [1942] *Spektraldarstellung Linearer Transformationen des Hilbertschen Raumes,* Springer, Berlin.

SZÁSZ, O. [1916] *Über die Approximation stetiger Funktionen durch lineare Aggregate von Potenzen,* MA **77,** 482–496. [1950] *Generalization of S. Bernstein's polynomials to the infinite interval,* NBSJ **45,** 239–245. [1953] *On closed sets of rational functions,* AMPA **34,** 195–218. [1955] *Collected Works,* University of Cincinnati Press, Cincinnati, Ohio.

SZÁSZ, P. [1959] *On quasi-Hermite-Fejér interpolation,* AMASH **10,** 413–439.

SZEGÖ, G. [1939] *Orthogonal Polynomials,* AMSC **23.**

SZEGÖ, G. See Pólya, G.

TALBOT, A. [1962] *On a class of Tchebysheffian approximation problems solvable algebraically,* Proceedings of the Cambridge Philosophical Society **58** (part 2), 244–267.

TAYLOR, A. E. [1958] *Introduction to Functional Analysis,* Wiley, New York.

TCHEBYCHEFF, P. L. [1854] *Théorie des mécanismes connus sous le nom de parallélogrammes,* Oeuvres **I,** 111–143. [1859] *Sur les questions de minima qui se rattachent a la représentation approximative des fonctions,* Oeuvres **I,** 273–378. [1859a] *Sur l'interpolation par la méthode des moindres carrés,* Oeuvres **I,** 473–498. [1881] *Sur les fonctions qui s'écartent peu de zéro pour certaines valeurs de la variable,* Oeuvres **II,** 335–356. [1899] Oeuvres, St. Petersburg. = [1962] reprint, Chelsea Publ. Co., New York.

TEICHROEW, D. [1952] *Use of continued fractions in high speed computing,* MTAC **6,** 127–133.

THACHER, H. C. [1960] *Derivation of interpolation formulas in several independent variables,* NYAS **86,** 758–775. [1965] *Conversion of a power (series) to a series of Chebyshev polynomials,* ACMC **7,** 181–182. [1965a] *Independent variable transformations in approximation,* to appear. [1965b] *Chebyshev series for the natural logarithm,* to appear.

THACHER, H. C., and W. E. MILNE. [1960] *Interpolation in several variables,* SIAMJ **8,** 33–42.

THIELE, T. N. [1909] *Interpolationsrechnung,* Teubner, Leipzig.

TIMAN, A. F. [1960] *Theory of Approximation of Functions of a Real Variable,* Moscow. = [1963] translation, Macmillan, New York.

TITCHMARSH, E. C. [1932] *The Theory of Functions,* Oxford, London.

TODD, J. [1963] *Introduction to the Constructive Theory of Functions,* Birkhäuser Verlag, Basel.

TONELLI, L. [1908] *I polinomi d'approssimazione di Tchebychev,* AMPA **15,** 47–119.

TORNHEIM, L. [1950] *On n-parameter families of functions and associated convex functions*, TAMS **69**, 457–467. [1956] *Approximation by families of functions*, PAMS **7**, 641–643.

TRICOMI, F. G. [1955] *Vorlesungen über Orthogonalreihen*, Springer, Berlin.

TRIGUB, R. M. [1961] *Approximation of functions by polynomials with integral coefficients*, DAN **140**, 773–775 (Russian). = SOVM **2**, 1278–1280.

TUCKER, A. W. See Kuhn, H.

TURÁN, P. See Erdös, P.

TURÁN, P. See Grünwald, G.

ULLMAN, J. L. [1959] *On Tchebycheff polynomials*, PJ **9**, 913–923.

VALENTINE, C. W., and C. P. VAN DINE. [1963] *An algorithm for minimax polynomial curve-fitting of discrete data*, ACMJ **10**, 283–290.

VALENTINE, F. A. [1964] *Convex Sets*, McGraw-Hill, New York.

VAN DINE, C. P. See Valentine, C. W.

VEIDINGER, L. [1960] *On the numerical determination of the best approximations in the Chebyshev sense*, NM **2**, 99–105.

VIDENSKIĬ, V. S. [1963] *On zeros of orthogonal polynomials*, DAN **152**, 1038–1041 (Russian). = SOVM **4**, 1479–1482.

VISSER, C. [1945] *A simple proof of certain inequalities for polynomials*, Nederlandse Akademie van Wetenschappen Proceedings **A48**, 276–285.

VOLTERRA, V. [1897] *Sul principio di Dirichlet*, RCMP **11**, 83–86.

VORONOVSKAYA, E. V. [1932] *Détermination de la forme asymptotique d'approximation des fonctions par les polynomes de M. Bernstein*, DAN, 79–85.

WALL, H. S. [1948] *Analytic Theory of Continued Fractions*, Van Nostrand, Princeton, N.J.

WALSH, J. L. [1926] *Über die Entwicklung einer analytischen Funktion nach Polynomen*, MA **96**, 430–436. [1929] *On approximation by rational functions to an arbitrary function of a complex variable*, TAMS **31**, 477–502. [1931] *The existence of rational functions of best approximation*, TAMS **33**, 668–689. [1931a] *On the overconvergence of certain sequences of rational functions of best approximation*, ACTA **57**, 411–435. [1932] *On the overconvergence of sequences of rational functions*, AJM **54**, 559–570. [1935] *Interpolation and Approximation by Rational Functions in the Complex Domain*, AMSC **20**. [1935a] *Approximation by Polynomials in the Complex Domain*, MSM **73**. [1955] *Sur l'approximation par fonctions rationnelles et par fonctions holomorphes bornées*, AMPA **39**, 267–277. [1964] *Padé approximants as limits of rational functions of best approximation*, JMM **13**, 305–312.

WALSH, J. L., and T. S. MOTZKIN. [1959] *Polynomials of best approximation on an interval*, Proceedings of the National Academy of Science, U.S.A. **45**, 1523–1528.

WALSH, J. L., and A. SHARMA. [1964] *Least squares and interpolation in roots of unity*, PJ **14**, 727–730.

WALSH, J. L. See Motzkin, T. S.; see Shohat, J. A.

WEIERSTRASS, K. [1885] *Über die analytische Darstellbarkeit sogenannter willkürlicher Funktionen einer reelen Veränderlichen*, Sitzungsberichte der Akademie zu Berlin, 633–639, 789–805. = [1886] *Sur la possibilité d'une représentation analytique des fonctions dites arbitraires d'une variable réelle*, JMPA **2**, 105–138. [1915] *Mathematische Werke*, Berlin.

WENZL, F. [1954] *Über Gleichungssysteme der Tschebyscheffschen Approximation*, ZAMM **34**, 385–391.

WERNER, H. [1962] *Tschebyscheff-Approximation im Bereich der rationalen Funktionen bei Vorliegen einer guten Ausgangsnäherung,* ARMA **10,** 205–219. [1962a] *Die konstruktive Ermittlung der Tschebyscheff-Approximierenden im Bereich der rationalen Funktionen,* ARMA **11,** 368–384. [1963] *Rationale Tschebyscheff-Approximation, Eigenwert-theorie und Differenzenrechnung,* ARMA **13,** 330–347. [1964] *On the rational Tschebyscheff-operator,* MZ **86,** 317–326.

WERNER, H., and R. COLLINGE. [1961] *Tchebycheff approximation to* $\Gamma(x)$, MTAC **15,** 195–197.

WETTERLING, W. [1963] *Ein Interpolationsverfahren zur Lösung der linearen Gleichungssysteme, die bei der rationalen Tschebyscheff-Approximation auftreten,* ARMA **12,** 403–408. [1963a] *Anwendung des Newtonschen Iterationsverfahrens bei der Tschebyscheff-Approximation, insbesondere mit nichtlinear auftretenden Parametern,* Zeitschrift Moderne Rechentechnik und Automation **10,** 61–63, 112–115.

WEYL, H. [1916] *Über die Gleichverteilung der Zahlen mod. Eins,* MA **77,** 313–352.

WHITNEY, H. [1957] *On functions with bounded nth differences,* JMPA **36,** 67–95. [1959] *On bounded functions with bounded nth differences,* PAMS **10,** 480–481.

WHITTAKER, J. M. [1949] *Sur les Séries de Base de Polynomes Quelconques,* Gauthier-Villars, Paris.

WIGERT, S. [1930] *Sur l'approximation par polynomes des fonctions continues,* AMAF **22B.**

WILF, H. S. See Ralston, A.

WIMP, J. [1961] *Polynomial approximations to integral transforms,* MTAC **15,** 174–178. [1962] *Polynomial expansions of Bessel functions and some associated functions,* MTAC **16,** 446–458.

WITZGALL, C. See Maehly, H.

WRENCH, J. W. See Murnaghan, F. D.

WRIGHT, E. M. [1930] *The Bernstein approximation polynomials in the complex plane,* JLMS **5,** 265–269.

WYNN, P. [1960] *The rational approximation of functions which are formally defined by a power series expansion,* MTAC **14,** 147–186.

YANG, C. T. See Schoenberg, I. J.

YOUNG, J. W. [1907] *General theory of approximation by functions involving a given number of arbitrary parameters,* TAMS **8,** 331–344.

ZELLER, K. See Meyer-König, W.

ZUCKERMANN, H. S. See Hewitt, E.

ZUHOVICKII, S. I. [1951] *An algorithm for the solution of the Čebyšev approximation problem in the case of a finite system of incompatible linear equations,* DAN **79,** 561–564 (Russian). [1953] *On the best approximation in the sense of P. L. Tchebycheff of a finite system of incompatible linear equations,* MSB **33,** 327–342 (Russian). [1956] *On the approximation of real functions in the sense of P. L. Tchebycheff,* UMN **11,** 125–159 (Russian). = AMST (II)**19,** 221–252. [1958] *An algorithm for constructing the Tchebycheff polynomial approximation to a continuous function,* DAN **120,** 693–699 (Russian). [1961] *On a new numerical scheme of the algorithm for the Tchebycheff approximation of an incompatible system of linear equations and a system of linear inequalities,* DAN **139,** 534–537 (Russian). = SOVM **2,** 959–962.

ZYGMUND, A. [1945] *Smooth functions,* DUKE **12,** 47–76. [1945a] *On the degree of approximation of functions by their Fejér means,* BAMS **51,** 274–278. [1959] *Trigonometric Series,* 2d ed., Cambridge, London.

Index

Index